Molybdenum Chemistry
of Biological Significance

Molybdenum Chemistry of Biological Significance

Edited by

William E. Newton

Charles F. Kettering Research Laboratory
Yellow Springs, Ohio

and

Sei Otsuka

Osaka University
Toyonaka, Osaka, Japan

Plenum Press • New York and London

Library of Congress Cataloging in Publication Data

International Symposium on Molybdenum Chemistry of Biological Significance,
    Kyoto, 1979.
    Molybdenum chemistry of biological significance.

    "Held at Lake Biwa, Japan, April 10—13, 1979."
    Includes index.
    1. Molybdenum — Physiological effect — Congresses. 2. Molybdenum — Metab-
olism—Congresses.   3.   Enzymes — Congresses.   4.   Micro-organisms,   Nitrogen-
fixing — Congresses. I. Newton, William Edward, 1938-   II. Otsuka, Sei.
III. Title.
QP535.M7I57 1979   574.1'92   79-6374
ISBN 0-306-40352-8

Proceedings of the International Symposium on Molybdenum Chemistry of
Biological Significance, held at Lake Biwa, Japan, April 10—13, 1979.

© 1980 Plenum Press, New York
A Division of Plenum Publishing Corporation
227 West 17th Street, New York, N.Y. 10011

Printed in the United States of America

PREFACE

In retrospect, it was obvious that we were both, quite independently, contemplating a conference on the role of molybdenum in
biology and related chemistry. At the time though, the meeting of
minds on this matter was quite surprising. Although this subject
has been treated in previous meetings within the overall context
of, say, magnetic resonance or nitrogen fixation, it was apparent
to us both that research in molybdenum-containing enzymes and molybdenum chemistry had progressed rapidly in the last several years.
Jointly, we decided to organize the first meeting on Molybdenum
Chemistry of Biological Significance which was held at the Hotel
Lake Biwa, Shiga, Japan, on April 10-13, 1979. This volume constitutes the Proceedings of that international conference and
covers the broad spectrum of interests from enzymes to coordination
chemistry. It should serve not only as a source of new information
on the latest research results in this area and as a useful reference tool, but should also allow a newcomer or other peripherally
interested researcher to become conversant very rapidly with the
"state-of-the-art" in this specialized and important area of
research.

The conference was sponsored by the Japan Society for the
Promotion of Science, the Japan World Exposition Commemerative Fund
the Yamada Science Foundation, the Nissan Science Foundation, the
Chemical Society of Japan (Kinki Regional Office) and the Agri-
cultural Chemical Society of Japan (Kansai Branch). We thank
these organizations sincerely for their interest and generosity.
We also acknowledge a donation from the Climax Molybdenum Company
of Michigan. We also wish to thank our Symposium President, Shiro
Akabori, Emeritus Professor of Osaka University, and our colleagues
on the Organizing Committee, S. Hino (Hiroshima), Y. Maruyama
(Tokyo), A. Nakamura (Osaka), M. Nakamura (Tokyo), K. Saito
(Sendai), Y. Sasaki (Tokyo), T. Tanaka (Osaka) and F. Egami (Tokyo)
for their very worthy assistance.

The current interest in molybdenum chemistry stems from its
importance in biology as an essential element in metabolism, particularly with respect to nitrogen. Our hope is that this volume
will engender increased research endeavor on this vitally important
subject.

William E. Newton   Yellow Springs, Ohio
Sei Otsuka   Osaka, Japan

v

CONTENTS

MOLYBDOENZYMES

        IRON-MOLYBDENUM COFACTOR AND MO-FE-S CUBANE CLUSTERS

        REACTIONS AND PROPERTIES OF MOLYBDENUM COMPLEXES

MOLYBDOENZYMES

ELECTROCHEMICAL AND KINETIC STUDIES OF NITROGENASE: BRIEF REVIEW

AND RECENT DEVELOPMENTS[*]

G.D. Watt

Charles F. Kettering Research Laboratory
Yellow Springs, Ohio   USA

INTRODUCTION

Nitrogenase is a two-component redox enzyme that catalyzes the reduction of nitrogen to ammonia in aqueous solution according to reaction (1). The two redox component proteins that comprise the

$$N_2 + 6e- + 6H^+ = 2NH_3 \qquad\qquad (1)$$

nitrogen-reducing enzyme are the MoFe protein containing 32 iron and 2 molybdenum atoms (MW = 230,000), and the $MgATP^{2-}$-binding iron protein, containing 4 iron atoms (MW = 65,000). Both proteins can be independently purified and when recombined in the presence of a source of electrons and the energy-yielding substrate $MgATP^{2-}$ catalyze reaction (1). The electron transfer sequence from the low potential electron source through the component proteins to the reducible substrate N_2 constitutes a major area of research in attempting to understand the mechanism of nitrogenase catalysis. Such questions as which protein component first accepts the low potential electrons, how MgATP and electrons are coupled during catalysis, what types of electron transfer centers are present in the proteins and what are their redox properties are all fundamental questions that are being actively investigated.

The kinetic aspects of nitrogenase catalysis constitute yet another area of nitrogenase research actively undergoing rapid development. It has been recognized by several workers[1-4] that the numerous reducible substrates of nitrogenase are all reduced at

*Contribution No. 677 from Charles F. Kettering Research Laboratory

approximately the same rate even though their diversity of chemical
properties suggest this would be unlikely. Thus, a number of questions arise. What are the kinetically discernible events during
nitrogenase catalysis, which of these are rate limiting, how many
substrate-reducing sites are there and what is their nature?
Attempts to answer these and related queries have produced a rather
extensive base of kinetic information about nitrogenase catalysis.[1]

Historically, studies of both the redox properties and kinetic
properties of nitrogenase catalysis have developed more or less
separately, but it is clear that both aspects must, at some point,
be considered together. However, at the present level of understanding of nitrogenase catalysis, the simplification that results
from separate consideration of redox reactions and kinetic properties seems to justify the approximation of separability. This
article considers each area separately, first in brief review followed by recent results from our laboratory impinging on these areas.

KINETIC STUDIES OF NITROGENASE

Both steady state[1,3] and presteady state[5,6] (stopped-flow)
kinetic studies of nitrogenase have provided useful insights into
its mechanism of catalysis. The stopped-flow measurements of
Thorneley and his colleagues[5,6] have provided interesting results
for the rapid reduction of the Fe protein by reductant in the presence of the substrate MgATP and the inhibitor MgADP. They find that
reduction of the Fe protein by SO_2^- is unaffected by MgATP but is
significantly retarded by $MgADP^{2-}$, a known inhibitor of nitrogenase.
Although such studies have been extended[7] to include the entire
nitrogen-fixing system (i.e., $S_2O_4^{2-}$, MgATP, Fe Protein, MoFe protein), sufficient complexity has arisen as to make interpretation
less than direct. However, these authors do conclude that a tightly
bound complex forms between the Fe protein and the MoFe protein
during catalysis and that internal electron transfer occurs from the
Fe protein to the MoFe protein within this complex with a rate constant of 2×10^2 sec^{-1}.

Numerous kinetic studies of substrate reduction by nitrogenase
have been reported using crude nitrogenase preparations[8], nitrogenase complex[3,4,9] and purified protein components[10-13] from various
bacterial sources. These studies monitored the rate of product
formation as either the various substrates and inhibitors or the
conditions (temp, pH etc.) of catalysis were varied. Also, studies
of the competition among reducible substrates (or reducible substrate
against inhibitor) for available electrons in the enzyme have been
made in an attempt to define the nature of the reducible substrate
binding site(s).[9,12,13] In spite of the large number of these
studies, the nature and number of binding sites still remains obscure

and the explanation for substrate reduction selectivity rests with two quite dissimilar models.

Kinetic Models

One model[13] postulates a number of sites, perhaps partially related, where selective substrate reduction occurs. An extensive body of information produced by Hardy et al.[10] and by Hwang et al[13] was interpreted in terms of this model. Considering the wide range of chemical reactivity, electronic properties and structures of the various nitrogenase substrates and recognizing the large size and chemical composition of the MoFe and Fe proteins and thus, the potential for numerous and varied binding sites, it is not unreasonable to propose such a model. However, this model is not only intuitively unsatisfying but also newer experimental results are not easily accommodated within its scope. Consequently, a second model was proposed that suggests that only one site is required to explain reduction of the various substrates of nitrogenase. According to this view, the enzyme becomes a reservoir for electrons sufficiently energetic to be transferred to the reducible substrates in an enzyme-controlled reduction step. The electron level in the enzyme during turnover is a balance between the outflow of activated electrons to the reducible substrates and the influx of activated electrons presumably resulting from MgATP-driven electron transfer from the low potential reductant. The specificity of substrate reduction results from the number of activated electrons present in the enzyme at any given moment. When the electron level is low, substrates requiring only two electrons for reduction are preferred ($H-C{\equiv}C-H$, $2H^+$, N_2O etc.) to substrates requiring four ($HC{\equiv}N \rightarrow CH_3NH_2$) or six electrons ($N_2$, CH_2CN etc.) for reduction This explanation implies that an "all or nothing" binding-reduction interaction of the reducible substrates occurs with the enzyme, especially for those substrates requiring four or six electrons for reduction.

An interesting possibility arises which this latter model has not dealt with as yet. Suppose that a substrate requiring six electrons for complete reduction binds to an enzyme form possessing fewer electrons. It is not likely that the substrate would accept whatever electrons are present and "wait" until further activated electrons are made available to complete the reduction sequence? This aspect, which allows for intermediate reduction states of the six electron substrates, needs further development, particularly in view of the trapping of an enzyme-bound intermediate during nitrogen reduction[14] and the interpretation of HD production[15,16] by a nitrogen-fixing system reported by the Kettering group. Both types of experiments are clearly indicative of enzyme-bound intermediates (at the diimide or hydrazine levels) occurring during the six-electron nitrogen reduction process.

Activated Electrons

As mentioned above, MgATP is though to be involved in the production of activated electrons required for substrate reduction.
Silverstein and Bulen,[9] using a nitrogenase complex from A.
vinelandii with a fixed Fe protein to MoFe protein ratio of 2:1,
demonstrated that, at low levels of MgATP, the two-electron reduction of protons was favored over the six-electron reduction of
nitrogen. This result clearly establishes a role for MgATP in the
production of activated electrons. In an extension of these studies,
Davis et al[12] determined the product distribution of various reducible substrates as a function of component ratio. These experiments
confirmed the conclusions of Silverstein and Bulen and further
demonstrated the involvement of the Fe protein in the formation of
activated electrons. The results show that, at high ratios of Fe
protein/MoFe protein, six-electron substrates compete favorably
with two-electron substrates while at low ratios, two-electron
substrates are selectively reduced and little if any reduction of
six-electron substrates occurs.

The concentration of MgATP and the ratio of Fe protein/MoFe
protein are both involved in determining the electron partitioning
to reducible substrates. Perhaps other unrecognized or uninvesti-
gated variables are also involved. For example, temperature is
known to vary the ATP/2e ratio[17] and may also influence substrate
reduction selectivity. Similarly, effects of varying pH and ionic
strength might also be interesting. Another possible variable is
the potential of the reductant used as an electron source. Watt
and Bulen[18] and Albrecht and Evans[19] have reported the voltage
threshold value to be −380mV and −440mV for operation of A.
vinelandii and chromatium nitrogenases respectively. The rate of
nitrogenase activity at potentials more negative than −380mV for the
Azotobacter system[18] follows a Nernstian-type curve with an operational $E_{1/2}$ of −450mV and n = 2. Conceivably, the nitrogenase system
(composed of a high Fe protein/MoFe protein ratio and excess MgATP)
could be poised at suboptimum potentials, where turnover is limited
by the availability and energy of the electrons, to evaluate how
these constraints affect the distribution of electrons to products.
Results similar for limiting MgATP and low Fe protein/MoFe protein
ratios might be expected.

Kinetics of $S_2O_4^{2-}$ utilization

The above experimental approach determined the amount of product formed from the reducible substrates, the rate at which product
forms, and for the competition experiments, the distribution of
electrons among the various reducible substrates. These results
give only one view of nitrogenase catalysis, from the product formation side only. Another perspective is from the substrate or input

side which might also provide useful insights into the catalytic mechanism. Using electrochemical techniques for monitoring $S_2O_4^{2-}$ concentration and the nitrogenase complex as catalyst, Watt and Burns[3] determined the rate law (equation 2) for $S_2O_4^{2-}$ utilization under a variety of conditions. This rate law indicates that SO_2^- is

$$\frac{-d(S_2O_4^{2-})}{dt} = \frac{kE_T \, (S_2O_4^{2-})^{1/2} \, (ATP)^2}{(ATP)^2 + K_1(ATP) + K_2} \tag{2}$$

the active electron transfer agent to the nitrogenase proteins (a result also obtained from stopped-flow studies[5]) and that two MgATP's are involved in the catalytic mechanism. The result[3] that the enzyme never becomes saturated with respect to SO_2^- is interpreted to indicate that MgATP is first bound to the Fe protein during ntirogenase catalysis followed by SO_2^- reduction.

The nonsaturation[3] of nitrogenase by SO_2^- has been questioned by Hageman and Burris[20] (a problem to be discussed more fully in a later section) who studied the kinetics of $S_2O_4^{2-}$ disappearance by optical means. Using various ratios of purified nitrogenase components, they find that the rate of $S_2O_4^{2-}$ consumption by the enzyme-catalyzed reaction is dependent on the ratio of component proteins and becomes saturated at high levels of $S_2O_4^{2-}$. From a rather involved analysis of their data, they conclude that the ATP-binding Fe protein and the MoFe protein undergo rapid association and dissociation with fleeting formation of a transient complex. During complex formation, the Fe protein transfers one activated electron to the MoFe protein which, when sufficient of these have accumulated, becomes the substrate-reducing site. Consistent with this view, Hageman and Burris[21] find, under conditions of low Fe protein to MoFe proten ratios, that hydrogen is not immediately evolved when the reaction is initiated, which suggests that electrons are stored in the MoFe protein until sufficient numbers are made available to evolve H_2. Independent studies,[22,23] using heterologous mixtures of Fe protein and MoFe proteins from different organisms which form strongly associating complexes that have low activities, are also consistent with the view that the Fe protein and MgATP act together to transfer activated electrons to the MoFe protein where substrate reduction occurs.

This brief sketch rationalizes a broad and extensive body of experimental evidence into a unified and reasonable model of nitrogenase catalysis. However, lest we become too optimistic, it should be kept in mind that many of these results have not been independently confirmed and some experiments need to be extended to include other sets of conditions. However, these views do form a useful working hypothesis from which more detailed studies of nitrogenase can be made.

The observation of Watt and Burns[3] that the nitrogenase complex from A. vinelandii becomes saturated with MgATP but not with $S_2O_4^{2-}$ is interpreted to indicate that MgATP is first bound to the Fe protein followed by SO_2^- reduction of the Fe protein. Hageman and Burris,[20] using purified component proteins from A. vinelandii and an optical method for following $S_2O_4^{2-}$ concentration, have observed saturation with $S_2O_4^{2-}$. The disagreement may be due to differences in properties of the nitrogenase complex[3] versus the purified components[20] or to the different techniques and methods of data analysis used for following $S_2O_4^{2-}$ utilization. Because the early sequence of kinetic events involving the reductant, MgATP and the Fe protein is of such fundamental importance in understanding how activated electrons are produced and ultimately appear in the reduced product, we have remeasured the kinetics for $S_2O_4^{2-}$ consumption. These newer experiments reported below were measured at various ratios of purified recombined component proteins using three independent methods for following $S_2O_4^{2-}$ utilization. These are: 1) the original polarographic method[3]; 2) the optical method used by Hageman and Burris,[20] and 3) a calorimetric technique which follows the rate of heat production during enzymatic catalysis.

Recent Kinetic Experiments

Figure 1 is a high resolution spectrophotometric curve at 375 nm resulting from $S_2O_4^{2-}$ utilization during nitrogenase-catalyzed hydrogen evolution. The curve has no linear portion which would be required if $S_2O_4^{2-}$ were saturating (i.e., if the reaction were zero order in $S_2O_4^{2-}$). Analysis of this curve by usual kinetic procedures indicates that the nitrogenase-catalyzed reaction has a well defined one-half order dependence on $S_2O_4^{2-}$ even at high $S_2O_4^{2-}$ concentrations. Increasing the Fe protein/MoFe protein ratio while maintaining the MgATP level constant and the $S_2O_4^{2-}$ at the levels shown in Figure 1 causes the rate of $S_2O_4^{2-}$ utilization to increase but it still obeys a one-half order reaction. The Fe protein/MoFe protein ratio in these experiments was varied from 0.5 to 12. The one-half order rate constants for $S_2O_4^{2-}$ utilization resulting from varying the Fe/MoFe ratio provides information about the Fe protein dependence in enzymatic catalysis. Figure 2 is a plot of enzymatic velocity against Fe protein concentration and the inset is a double reciprocal plot of these quantities. This plot shows simple hyperbolic saturation of the MoFe protein by the Fe protein. These plots give the K_m value for the Fe protein-MoFe protein interaction as 3.12×10^{-5} M and Vmax as 1.92×10^{-2} $M^{1/2}mg^{-1}min^{-1}$. These data can be combined to give equation (3) which describes the rate of $S_2O_4^{2-}$ utilization at a constant, saturating MgATP concentration. Extensive

$$rate = \frac{Vmax[Fe]}{K_m+[Fe]} \quad [MoFe][S_2O_4^{2-}]^{1/2} \tag{3}$$

studies of this type have also been made using the polarographic method for following $S_2O_4^{2-}$ utilization during nitrogenase catalysis. Strictly obeyed one-half order reactions in $[S_2O_4^{2-}]$ are consistently found when the Fe protein/MoFe protein ratio varied from 0.5 to 40, to give a rate law of the same form as (3).

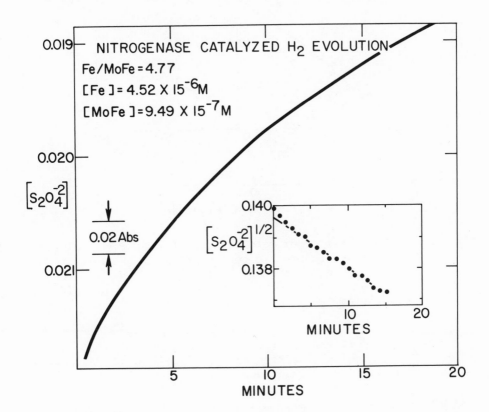

Fig. 1. $S_2O_4^{2-}$ utilization during the H_2 evolution reaction as catalyzed by recombined nitrogenase components. $S_2O_4^{2-}$ concentration was monitored at 375 nm in 1 cm. cells where the extinction coefficient is 71.   Inset is a plot of $[S_2O_4^{2-}]^{1/2}$ against time.

A calorimetric technique[17,18] was also employed to follow the rate at which heat is produced by the enzyme-catalyzed reaction, which in turn is related to the ΔH for the reaction and the rate law governing substrate utilization and product formation.   Because of ΔH value[17,18] for substrate reduction by nitrogenase is so large this method is highly sensitive for following the kinetics of nitrogenase catalysis.   The calorimetric method also has the advantage

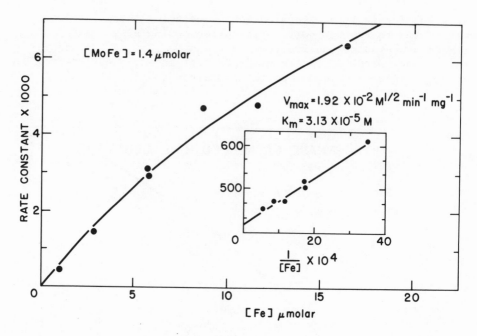

Fig. 2.   The one-half order rate constant for $S_2O_4^{2-}$ is plotted
           against the Fe protein concentration at a constant MoFe
           protein concentration of 1.4 μmolar. Inset is a double
           reciprocal plot of rate constant versus Fe protein con-
           centration.

Table I.   Equations describing the calorimetric response of a
           chemical reaction and predicted curves for several
           different reaction orders.

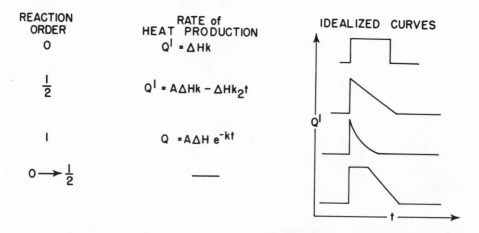

REACTION ORDER	RATE of HEAT PRODUCTION	IDEALIZED CURVES
0	$Q^l = \Delta Hk$	
$\frac{1}{2}$	$Q^l = A\Delta Hk - \Delta Hk_2 t$	
1	$Q = A\Delta H e^{-kt}$	
$0 \longrightarrow \frac{1}{2}$	—	

that the order of the reaction can be readily determined by direct inspection of the calorimetric curve. Table I contains the kinetic equations governing the heat production for reactions having different reaction orders, together with the idealized calorimetric curves expected from these various reaction orders. Further analysis of the curve in terms of the equations in Table I can provide additional kinetic details. The calorimetric curve in Figure 3a was obtained for the reaction where $S_2O_4^{2-}$ initially at 0.01M was limiting and MgATP was held constant while Figure 3b was obtained from limiting MgATP and excess $S_2O_4^{2-}$. Inspection of the two curves and reference to Table I shows that curve 3a is one-half order in $[S_2O_4^{2-}]$ and Figure 3b is first order in MgATP. The calorimetric method confirms the results obtained by the polarographic and optical methods by showing that $S_2O_4^{2-}$ is consumed in a one-half order reaction.

Conclusions

The previous kinetic studies allow us to conclude that $S_2O_4^{2-}$ is consumed in a strictly obeyed one-half order reaction during the nitrogenase-catalyzed hydrogen evolution reaction at $S_2O_4^{2-}$ concentrations up to 0.025M. This one-half order reaction order holds at Fe protein to MoFe protein ratios ranging from 0.5 to 40. We also find that at a given MoFe protein concentration and at fixed $S_2O_4^{2-}$ and MgATP concentrations, the rate of $S_2O_4^{2-}$ consumption (or the rate of H_2 evolution) follows simple Michaelis-Menten behavior in terms of Fe protein concentration the K_m value for this reaction is 3.12×10^{-5}M.

ELECTROCHEMICAL STUDIES OF NITROGENASE

Both the Fe protein and the MoFe protein as well as the nitrogenase complex itself are known to undergo redox reactions[1]. Measurement of the redox potentials and n values for these redox reactions gives the thermodynamic information which determines the direction of electron transfer during nitrogenase catalysis. In addition, measurements of the total electron transfer, the n values involved and the redox potential values, as well as evaluating which conditions of substrate, temperature, pH, etc. alter these, parameters are potentially useful in evaluating the numbers and types of metal clusters which occur in the nitrogenase component proteins. The redox states which have been reported[1,24] for the nitrogenase proteins will be briefly discussed followed by recent redox results which we have obtained on the MoFe protein.

Super-reduced MoFe Protein

The disappearance of the well known EPR signal associated with dithionite-reduced MoFe protein during active enzyme turnover has

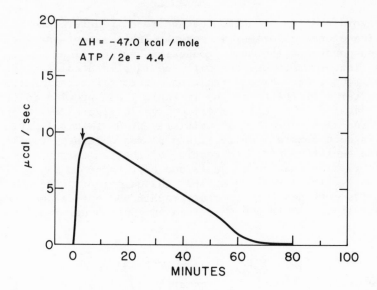

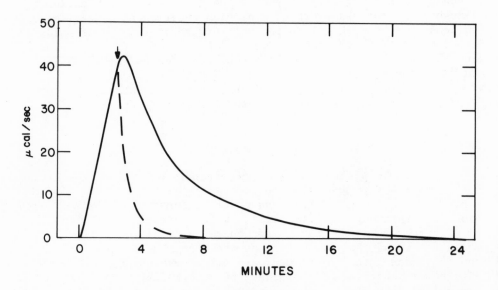

Fig. 3.   Calorimetric curves for nitrogenase catalysis. The top
curve is limiting $S_2O_4^{2-}$ in the presence of the MgATP
generating system and the bottom curve is limiting MgATP
in the presence of excess $S_2O_4^{2-}$. The arrows indicate the
point at which the reactants were added to the calorimeter
and the enzyme catalyzed reaction initiated. The reactions were carried out at 25.0° C.

been interpreted[25-27] as resulting from the formation of a meta-
stable, super-reduced, EPR-silent form of this protein. After it
is formed, this super-reduced enzyme state is thought to react
rapidly with reducible substrates (including the aqueous solvent in
the absence of other reducible substrates) giving products and
returning to its original EPR-active form. It seems unlikely, at
this time, that this enzyme form can be studied under equilibrium
conditions to obtain redox potentials and n values for the redox
reaction it undergoes because of its presumed short lifetime. However, Hageman and Burris[21] using low ratios of Fe protein/MoFe
protein suggest that the observed delay time before H_2 is evolved
from a competent, fixing system results from the formation of a
super-reduced MoFe protein state. The observed delay time of
several minutes indicates that the super-reduced form persists for
at least this time interval.

      In spite of its short lifetime, some redox information for the
super-reduced state has been reported. Walker and Mortenson,[28] in
a seemingly difficult-to-execute set of experiments, found that
four additional electrons were present in the MoFe protein under
turnover conditions than in MoFe protein controls not turning over.
This significant result, if verified, will definitely establish that
electron transfer has occurred to produce a further reduced form of
the MoFe protein.

Fe Protein

      Several studies of the redox properties of the Fe protein have
been reported[29,30]. A combined EPR-Potentiometric technique has been
used[29] in which a protein sample was brought to a desired potential
and samples withdrawn and frozen for EPR measurements. The signal
height at g = 1.87 was used to quantitate the ratio of oxidized to
reduced Fe protein at each set potential. A similar set of experiments has combined a potentiometric-spectrophotometric technique for
measuring the redox properties of the Fe protein[30]. Reduction
potentials of -294 ± 20 mv and -250 mv respectively were obtained
by these methods. When the two types of measurements were repeated
for Fe protein equilibrated with excess MgATP or MgADP, redox potentials near -400 mV were obtained by each technique. Burris and
Orme-Johnson[31] report slightly different numerical values for the
Fe protein and its MgATP-complexed form but conclude similarily
that the Fe proteins-$(MgATP)_2$ complex has a redox potential lowered
by 100 mV compared to free Fe protein.

MoFe Protein

      In addition to the presumed super-reduced state of this protein,
the resting, dithionite-reduced MoFe undergoes reaction with various

reagents $(O_2, Fe(CN)_6^{3-}$, and organic redox agents) resulting in oxidized forms of this protein, which in most cases remain catalytically active.[24] The reported redox potentials range from -30 mV to -450 mV versus the normal hydrogen electrode,[24,35] with in some cases, two detectable redox regions. The total number of electrons removed from the MoFe protein also varies, reports range from 1-12. Considerable confusion, thus exists in these redox measurements which must be eliminated by new measurements carried out under carefully controlled conditions.

More recently, quantitative thionine oxidation titations of the free MoFe protein, which follow the height of the EPR signal at $g = 3.65$ as a function of added thionine, show two redox regions.[24,32] One region requires 3-4 equivalents of thionine to be oxidized before the EPR signal is affected and the second requires another 3 electrons to abolish the EPR signal. Analysis of the Mössbauer spectra of the dithionite-reduced state, the state fully oxidized by six equivalents of thionine, and intermediate oxidation states of the MoFe protein provides a unique view of the various types of iron centers (four have been identified) in the MoFe protein. References 24 and 32 should be consulted for a thorough discussion of these interesting EPR and Mossbauer results.

Our studies[18,33] of the oxidation of dithionite-free MoFe protein show that several redox states are obtained by dye oxidation under a variety of conditions of salt concentration, pH and temperature. The number of electrons required to interconvert these redox states has been carefully measured using very precise electrochemical reduction techniques. Some of the spectral properties[33] associated with these redox states of the MoFe protein are known and are discussed below.

Recent Redox Experiments

A preliminary account[18,33] of the anaerobic preparation of the redox states to be discussed and the electrochemical techniques used for measuring the reducing equivalents transferred to them under anaerobic conditions has been described[34]. The experimental procedure which was used in the experiments to be described is essentially reductive in nature. The MoFe protein is oxidized in various ways and is then characterized by measuring the redox potentials and numbers of electrons required for its reduction.

Figure 4 is the result of reducing samples of oxidized MoFe protein at carefully controlled reducing potentials. A total of six electrons are required to fully reduce this oxidized protein at potentials more negative than -840 mV vs a saturated calomel reference electrode (-500 mV vs normal hydrogen electrode). Titrations with $S_2O_4^{2-}$ monitored polarographically confirm this result. The

six-electron reduction of this protein, occurring in two separate
redox regions, is numerically consistent with the oxidation studies
discussed above[24,32].   However, our results show that each redox
region accepts three electrons whereas Orme-Johnson et al.[24,32]
interpret their oxidation results as indicating that four electrons
are removed from one redox region and two from the other.   The
removal of the last two electrons from the second redox region causes
the EPR spectrum to decrease to near zero intensity and, according
to Orme-Johnson et al.,[24,32] is consistent with two distinct spin
centers per MoFe protein. We have measured the EPR intensity at
g = 4.3 and g = 3.65 for the EPR-silent MoFe protein (oxidized by
six electrons) which undergoes selective reduction at various
potentials along the curve in Figure 4.   EPR signals at these g
values grow in as the oxidized MoFe protein is reduced at potentials
corresponding to the most positive redox region.   The EPR intensity
reaches a maximum at the plateau between the two redox regions
(-640mV).   We conclude that the first redox region is associated
with the EPR-containing molybdenum cofactor center.

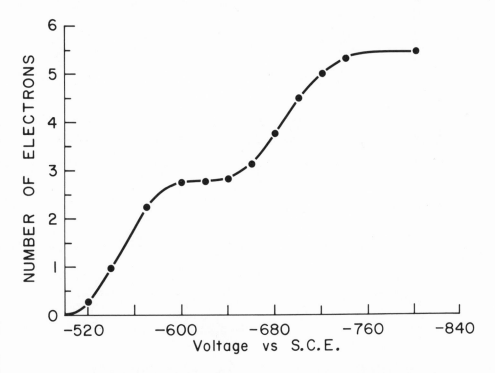

Fig. 4.   Controlled potential reduction of methylene blue-oxidized
           MoFe protein. Reduction occurred at pH 8.0 in 0.05 M Tris
           containing 0.1M NaCl and 5×10^{-5} M Methyl Viologen and
           Benzyl Viologen Mediators.

MoFe protein samples exhibiting redox activity in either of
the two redox regions shown in Figure 4 can be made separately by
selective oxidation of the MoFe protein under carefully controlled
oxidation conditions.[33] The state in which redox region I is
reduced and redox region II remains oxidized is prepared from the
oxidized MoFe protein shown in Figure 4 by selective electrochemical
reduction at the plateau between the two redox regions. In all
three cases, three electrons were taken up by each of these oxidized
MoFe protein preparations.

One other redox state (Figure 5) can be prepared but is only
partially characterized. This form of the oxidized MoFe protein
can accept a total of nine electrons as evidenced by both the
polarographic and controlled potential reduction techniques. The

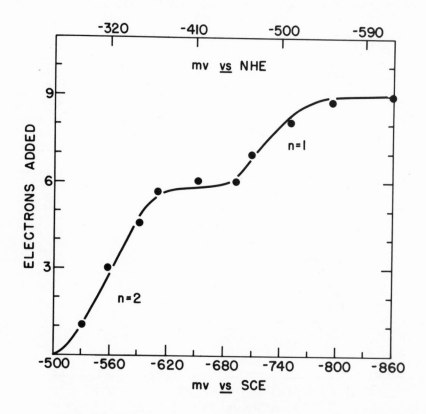

Fig. 5.   Controlled potential reduction of 2,6-dichlorophenolindo-
phenol-oxidized MoFe protein. Reduction occurred at pH 8
in 0.05M Tris containing 0.1M NaCl and 5×10^{-5} M Methyl
Viologen and Benzyl Viologen Mediators.

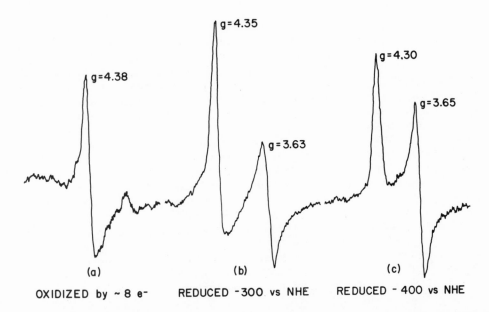

Fig. 6.   EPR spectra of the MoFe protein: a) oxidized as described
           in Figure 5; b) reduced at -300 mV; and c) reduced at
           -400 mV.

controlled potential reduction experiment indicates just two redox
regions, the more positive region accepting six electrons and the
more negative accepting three. Values of n = 2 and n = 1 were
observed for these two redox regions.

   Figure 6a shows the EPR spectrum of the MoFe protein oxidized
by ∿8 electrons. A single resonance is observed at g = 4.3. When
this protein is selectively reduced in the redox region that accepts
six electrons (Figure 5), the EPR signal at g = 4.3 is transformed
into the mixed signal shown in Figure 6b, which is composed of
Figures 6a and 6c. Reduction at the plateau transforms completely
the signal in Figure 6a into that shown in Figure 6c. At no point
during the controlled potential reductions is an EPR-silent form
of the protein observed. The EPR-silent form can be bypassed only
if a two-electron reduction step transforms signal 6a and 6c
directly. This view is consistent with n = 2 value observed for the
curve in Figure 5. The exact molecular origin of this signal
(Figure 6a) is not known but it can be shown to be an oxidized form
of the protein-bound molybdenum cofactor. This conclusion comes
from the results in Figure 5 and Figure 6 which show that reduction
of the six-electron region (with n = 2) of Figure 5 converts signal
6a into the cofactor signal 6c. The results of Newton, et al.
(these Proceedings) also demonstrate a nearly identical signal to

that of Figure 6a resulting from oxdiation of molybdenum cofactor in N-methylformamide solution.

The EPR signal in Figure 6a resembels the signal for Fe^{3+} ion and it might be argued that this signal arises from an Fe+ contaminant or that oxidation of the protein destroys certain clusters and produces Fe^{3+}. However, several lines of evidence run counter to this view. First, the oxidized protein still retains >75% of its original activity after production of this signal by oxidation. Second, the method[18,33] of preparation requires passing the oxidized protein through either P-2 polyacrylamide or G-25 sephadex columns which would remove contaminants of 25,000 or less. Free Fe^{3+} or aggregates of ferric hydroxide with molecular weights less than 25,000 would be easily separated. The final evidence is that electrolysis of the protein at -500 mV would quantitatively reduce any Fe^{3+} present but the results of Figure 5 and Figure 6 indicate that the center giving rise to the g = 4.3 signal responds in a well behaved manner to variation in potential. We conclude from this evidence that the g = 4.3 signal is from a naturally occurring, EPR-active state of the oxidized MoFe protein. Further study of the oxidized MoFe protein is presently underway.

REFERENCES

1. W.G. Zumft and L.E. Mortenson, The Nitrogen-Fixing Complex of Bacteria, Biochim. Biophys. Acta 1:416 (1975).
2. B.E. Smith, R.N.F. Thorneley, M.G. Yates, R.R. Eady and J.R. Postgate, Structure and Function of Nitrogenase from Klebsiella Pneumoniae and Azotobacter Chroococcum, in: "Proceedings of the First International Symposium on Nitrogen Fixation," W.E. Newton and C.J. Nyman, eds., Washington State University Press, Pullman, Washington (1976) Vol. I, p. 150.
3. G.D. Watt and A. Burns, Kinetics of Dithionite Utilization and ATP Hydrolysis for Reactions Catalyzed by the Nitrogenase Complex from Azotobacter vinelandii, Biochemistry, 16:264 (1977).
4. R.C. Burns, The Nitrogenase System from Azotobacter. Activation Energy and Divalent Cation Requirements, Biochim. Biophys. Acta 17:253 (1969).
5. R.N.F. Thorneley, M.G.Yates and D. Lowe, Nitrogenase of Azotobacter chroococcum: Kinetics of the Reduction of oxidized Iron-Protein by Sodium Dithionite, Biochem. J. 155:137 (1976).
6. M.G. Yates, R.N.F. Thorneley and D.J. Lowe, Nitrogenase of Azotobacter chroococcum: Inhibition by ADP of the Reduction of Oxidized Fe Protein by Sodium Dithionite, FEBS Letters 60:89 (1975).

7. R.N.F. Thorneley, Nitrogenase of Klebsiella: A Stopped Flow
    Study of Magnesium-Adenosine Triphosphate-Induced Electron
    Transfer Between the Component Proteins, Biochem. J. 145:
    391 (1975).

8. F.J. Bergersen and G.L. Turner, Kinetic Studies of Nitrogenase
    from Soya-Bean Root-Nodule Bacteroids, Biochem. J. 131:61
    (1973).

9. R. Silverstein and W.A. Bulen, Kinetic Studies of Nitrogenase-
    Catalyzed Hydrogen Evolution and Nitrogen Reduction Reactions,
    Biochemistry (:3809 (1970).

10. R.W.F. Hardy, R.C. Burns, and G.W. Parshall, The Biochemistry
    of N2 Fixation, in: "Bioinorganic Chemistry", R.F. Gould,
    ed. Advances in Chem. Series 100, ACS Publications,
    Washington, D.C. (1971) p. 219.

11. J.M. Rivera-Ortiz and R.H. Burris, Interactions Among Substrates
    and Inhibitors of Nitrogenase, J. Bacteriol 123:537 (1975).

12. L.C. Davis, V.K. Shah and W.J. Brill, Nitrogenase VII. Effect
    of Component Ratio, ATP and H_2 on the Distribution of Elec-
    trons to Alternative Substrates, Biochim. Biophys. Acta 403:
    67 (1975).

13. J.C. Hwang, C.H. Chen, and R.H. Burris, Inhibition of
    Nitrogenase-Catalyzed Reductions, Biochim. Biophys. Acta
    292:256 (1973).

14. R.N.F. Thorneley, R.R. Eady and D.J. Lowe, Biological Nitrogen
    Fixation by way of an Enzyme-Bound Dinitrogen-Hydrogen
    Intermediate, Nature 272:557 (1978).

15. W.A. Bulen, Nitrogenase from Azotobacter vinelandii and Reac-
    tions Affecting Mechanistic Interpretations, in: "Proceedings
    of the First International Symposium on Nitrogen Fixation,"
    W.E. Newton and C.J. Nyman, eds., Washington State University
    Press, Pullman, Washington (1979) Vol. I, p. 177.

16. W.E. Newton, W.A. Bulen, K.L. Hadfield, E.I. Stiefel and
    G.D. Watt, HD Formation as a Probe for Intermediates in N_2
    Reduction, in: "Recent Developments in Nitrogen Fixation",
    W.E. Newton, J.R. Postgate, and C. Rodriguez-Barrueco, eds.,
    Academic Press, London, New York and San Francisco (1977),
    p. 118.

17. G.D. Watt, W.A. Bulen, A. Burns and K.L. Hadfield, Stoichemis-
    try ATP/2e Values and Energy Requirements for Reactions Cata-
    lyzed by Nitrogenase from Azotobacter vinelandii,
    Biochemistry 14:4266 (1975).

18. G.D. Watt and W.A. Bulen, Calorimetric and Electrochemical
    Studies on Nitrogenase, in: "Proceedings of the First
    International Symposium on Nitrogen Fixation", W.E. Newton
    and C.J. Nyman, eds., Washington State University, Pullman,
    Washington (1976) Vol. I, p. 248.

19. M.C.W. Evans and S.L. Albrecht, Determination of the Applied
    Oxidation-Reduction Potential Required for Substrate
    Reduction by Chromatium Nitrogenase, Biochem. Biophys. Res.
    Comm. 61:1187 (1974).

20. R.V. Hageman and R.H. Burris, Kinetic Studies on Electron Trans-
     fer and Interaction between Nitrogenase Components from
     Azotobacter vinelandii, Biochemistry 17:4117 (1978).
21. R.V. Hageman and R.H. Burris, Nitrogenase and Nitrogenase
     Reductase Associate and Dissociate with each Catalytic Cycle,
     Proc. Natl. Acad. Sci. USA. 75:2699 (1978).
22. B.E. Smith, R.N.F. Thorneley, R.R. Eady and L.E. Mortenson,
     Nitrogenases from Klebsiella Pneumoniae and Clostridium
     Pasteurianum Kinetic Investigations of Cross Reactions as A
     Probe of the Enzyme Mechanism, Biochem. J. 157:439 (1976).
23. D.W. Emerich and R.H. Burris, Interactions of Heterologous
     Nitrogenase Components that Generate Catalytically Inactive
     Complexes, Proc. Natl. Acad. Sci. USA. 73:4369 (1976).
24. W.H. Orme-Johnson, L.C. Davis, M.T. Henzl, B.A. Averill, N.R
     Orme-Johnson, E. Munck and R. Zimmerman, Components and
     Pathways in Biological Nitrogen Fixation in: "Recent Develop-
     ments in Nitrogen Fixation", W.E. Newton, J.R. Postgate and
     C. Rodriguez-Barrueco, eds., Academic Press, London, New York
     and San Francisco (1977) p. 131.
25. L.E. Mortenson, W.G. Zumft and G. Palmer, Electron Paramagnetic
     Resonance Studies on Nitrogenase.III. Function of Magnesium
     Adenosine 5'-Triphosphate and Adenosine 5'Diphosphate in
     Catalysis by Nitrogenase, Biochim. Biophys. Acta 292:422
     (1973).
27. W.H. Orme-Johnson, W.D. Hamilton, T. Ljones, M.Y.W. Tso, R.H.
     Burris, V.K. Shah and W.J. Brill, Electron Paramagnetic
     Resonance of Nitrogenase and Nitrogenase Components from
     Clostridium Pasteurianum W5 and Azotobacter Vinelandii OP,
     Proc. Natl. Acad. Sci. USA 69:3142 (1972).
28. M.N. Walker and L.E. Mortenson, Evidence for the Existence of
     a Fully Reduced State of Molybdoferredoxin during the Func-
     tioning of Nitrogenase, and the Order of Electron Transfer
     from Reduced Ferredoxin, J. Biol. Chem. 249:6356 (1974).
29. W.G. Zumft, L.E. Mortenson and G. Palmer, Electron Parmagnetic
     Resonance Studies on Nitrogenase. Investigation of the Oxi-
     dation Reduction behavior of Azoferredoxin and Molybdoferre-
     doxin with Potentiometric and Rapid-Freeze Techniques, Eur.
     J. Biochem. 46:525 (1974).
30. L.E. Mortenson, M.N. Walker and G.A. Walker, Effect of Mag-
     nesium Di- and Triphosphates on the Structure and Electron
     Transport Function of the Components of Clostridial Nitrogen-
     ase, in: "Proceedings of the First International Symposium
     on Nitrogen Fixation", W.E. Newton and C.J. Nyman, eds.,
     Washington State University Press, Pullman, Washington (1975)
     Vol. I, p. 117.
31. R.H. Burris and W.H. Orme-Johnson, Mechanism of Biological N_2
     Fixation, in: "Proceedings of the First International Sym-
     posium on Nitrogen Fixation", W.E.Newton and C.J. Nyman,
     eds., Washing State University Press, Pullman, Washington.
     (1976) Vol. I p. 208.

32. R. Zimmerman, E. Munck, W.J. Brill, V.K. Shah, M.T. Henzl,
      J. Rawlings and W.H. Orme-Johnson, Nitrogenase X: Mossbauer
      and EPR Studies on Reversibly Oxidized MoFe Protein from
      Azotobacter vinelandii OP - Nature of the Iron Centers,
      Biochim. Biophys. ACta 537:185 (1978).
33. G.D. Watt, A. Burns and S. Lough, Redox Properties of Oxidized
      MoFe Protein, in: "Nitrogen Fixation", W.E. Newton and
      W.H. Orme-Johnson, eds., University Park Press, Baltimore
      (1979) Vol. I, in press.
34. G.D. Watt, An Electrochemical Technique for Measuring Redox
      Potentials of Low Potential Proteins by Microcoulometry,
      Anal. Biochem. In Press.
35. M. J. O'Donnell and B.E. Smith, Electron-Paramagnetic-Resonance
      Studies on the Redox Properties of the Molybdenum-Iron
      Protein of Nitrogenase between +50 and -450 mV, Biochem. J.
      173:831 (1978).

ELECTRON PARTITIONING FROM DINITROGENASE TO SUBSTRATE

AND THE KINETICS OF ATP UTILIZATION

R. H. Burris and Robert V. Hageman

Department of Biochemistry, College of Agricultural
and Life Sciences, University of Wisconsin-Madison,
Madison, Wisconsin 53706

INTRODUCTION

It now seems to be generally accepted that the flow of electrons[1,2] in the nitrogenase system is as shown in Figure 1. First let us clarify the nomenclature we will employ. The nomenclature suggested earlier by Hageman and Burris[3] has been modified somewhat based on suggestions we have received from a number of colleagues. The term nitrogenase, as it applies to the whole catalytic complex capable of reducing N_2, is retained in its accepted sense. As we are convinced that the Fe protein and the MoFe protein[3] associate and dissociate at each turn of the catalytic cycle[4] and that the MoFe protein binds and reduces substrates, we will utilize a nomenclature that reflects the function of the component proteins of nitrogenase. We will designate the MoFe protein as dinitrogenase (it really is the enzyme that reduces N_2 and other substrates) and the Fe protein as dinitrogenase reductase (its only function appears to be the reduction of dinitrogenase).[2,5]

By following the flow of electrons in Figure 1, you will observe that they originate in a reductant, such as reduced ferredoxin, reduced flavodoxin or $Na_2S_2O_4$ as used experimentally. These agents reduce dinitrogenase reductase. Before it can effect the reduction of dinitrogenase, it must bind two MgATP.[6,7] There is evidence that the dinitrogenase reductase is not reduced until after it binds the ATP. Upon binding MgATP, the dinitrogenase reductase changes in conformation so that its iron is subject to chelation, and its potential is lowered[8-10]. The dinitrogenase reductase and dinitrogenase bind momentarily (forming the nitrogenase complex), MgATP is hydrolyzed to reduce the potential of dinitrogenase reductase

23

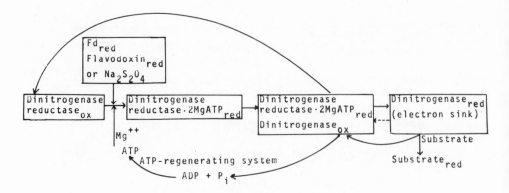

Fig. 1. Path of electron transfer in the nitrogenase system.

further, and it then transfers one electron to dinitrogenase.[11]   The
transitory complex dissociates, the dinitrogenase reductase returns
to its original oxidized state as shown, and the reduced dinitrogenase becomes a part of the electron sink. The transfer of the
electron from dinitrogenase reductase to dinitrogenase is accompanied by the hydrolysis of two MgATP to form ADP and P_i, and an
ATP-regenerating system must convert these back to ATP to sustain
the cycle. As reductions by dinitrogenase all require 2 or more
electrons, it is apparent that a dinitrogenase molecule must cycle
a second time to acquire a second electron before it can become
an effective reductant. Dinitrogenase molecules bearing at least
two electrons can bind and reduce substrate and then recycle to
acquire electrons again from dinitrogenase reductase.

There are a number of points in the scheme of Figure 1 that
we believe have been clarified by a kinetic examination of the
role of ATP and of the distribution of electrons among substrates
when three substrates have been made available simultaneously for
reduction. These are some of the issues we will discuss in this
communication.

DISSOCIATION OF THE NITROGENASE COMPLEX

What is the evidence that the two proteins of nitrogenase
associate and dissociate with each turn of the catalytic cycle?
As neither dinitrogenase nor dinitrogenase reductase has been
demonstrated to have catalytic activity by itself, it has been
customary to speak of the nitrogenase complex as the catalytic
entity. This concept has created a picture of a rather tight complex between dinitrogenase and dinitrogenase reductase that persists

during the catalytic cycling of nitrogenase. There never has been strong experimental evidence in support of this view; Hardy and Burns have expressed well deserved skepticism of the concept (p. 98 of ref.[12]). The observation of a peak in the ultracentrifuge corresponding to the combined weights of a molecule each of dinitrogenase and dinitrogenase reductase[13] has been interpreted to support a persistent complex between the components. The fact that the addition of $Na_2S_2O_4$ to the mixture destroyed the peak attributed to the complex in the ultracentrifuge cast doubt on the significance of the observed peak, because nitrogenase is functional in the presence of $Na_2S_2O_4$.

An alternative to the concept that the nitrogenase system functions with a persistent complex between its two proteins is that the two proteins dissociate after every electron transfer between the proteins. This suggestion carries with it the possibility that dinitrogenase is solely responsible for the reduction of the substrates. Current evidence strongly indicates both that the proteins dissociate after each electron transfer and that reduced dinitrogenase is responsible for reducing N_2 and other substrates.

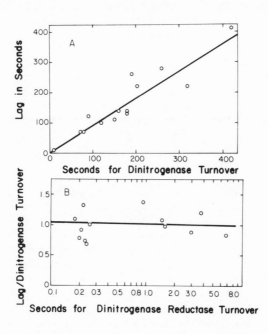

Fig. 2. Lag time as a function of protein turnover times. (A) Lag time plotted against nitrogenase turnover time; (B) lag time/nitrogenase turnover time plotted against nitrogenase reductase turnover time.

Hageman and Burris (see Fig. 1 of reference 3) have demonstrated the existence of a lag period before H_2 evolution occurs from the nitrogenase system[3]. What is particularly important is that the length of this lag is proportional to the turnover time of dinitrogenase but is unrelated to the turnover time of dinitrogenase reductase (Fig. 2). As you can see, the lag time increases linearly as the nitrogenase turnover time increases, whereas there is no increase in lag with an increase in the turnover time of nitrogenase reductase. The proportionality of the lag period to the average turnover time of dinitrogenase, even when dinitrogenase is present in large excess over dinitrogenase reductase, indicates that the electrons donated by dinitrogenase reductase are distributed randomly over the entire pool of dinitrogenase. As there is no evidence supporting the idea that transfer of electrons occurs between dinitrogenase molecules, this lag period demonstrates that the nitrogenase proteins dissociate after a single electron is transferred but before evolution of H_2 occurs. H_2 can be evolved only when dinitrogenase is in a two-electron reduced state; thus, the proteins must dissociate after each electron transfer between them and allow the dinitrogenase reductase to recycle to collect another electron to deliver to a molecule of dinitrogenase. The lag period represents the time necessary to charge the electron sink with molecules of dinitrogenase carrying two electrons each, for in this state they can function in the reduction of substrates.

As dinitrogenase reductase is a one-electron carrier, it is improbable that it can have a direct role in substrate reduction, because all of these reductions are two-electron processes. Also, if the lifetime of the nitrogenase complex is only long enough to transfer a single electron, the complex is unlikely to have a direct role in substrate reduction.

HYDROLYSIS OF ATP

As we[3] and others[22] have pointed out earlier, the hydrolysis of ATP accompanies the transfer of an electron from dinitrogenase reductase to dinitrogenase. We have shown that the lag in H_2 evolution by A. vinelandii nitrogenase is 4.3 minutes under the specific conditions employed (see Fig. 1 of reference 3). In contrast, there was no demonstrable lag in ATP hydrolysis of MgATP. MgATP is not hydrolyzed by dinitrogenase reductase nor by dinitrogenase alone. If MgATP were hydrolyzed when electrons were transferred from reduced dinitrogenase to substrate, there would be a lag in ATP hydrolysis comparable to the lag in H_2 evolution. The absence of such a lag supports the alternative explanation that MgATP is hydrolyzed when the electron is passed from dinitrogenase reductase to dinitrogenase.

ALLOCATION OF ELECTRONS AMONG SUBSTRATES

If we load the electron sink of nitrogenase with electrons so that dinitrogenase can effect two-electron reductions, how does it allocate electrons if more than one substrate is present? We have presented evidence[14] that the allocation is dependent upon the electron flux through dinitrogenase.

There are a number of ways to alter the electron flux through nitrogenase. One can change the balance between dinitrogenase and dinitrogenase reductase or change the level of reductant or MgATP. As indicated, the flux through dinitrogenase governs the allocation. Dinitrogenase serves as the electron sink. The sink fills one electron at a time, and the electrons go at random to the dinitrogenase molecules to reduce them first with one and then with another electron. The buildup of reducing activity under such conditions starting from the oxidized state of dinitrogenase is exponential until the steady state is reached. Overall reaction rates then are linear.

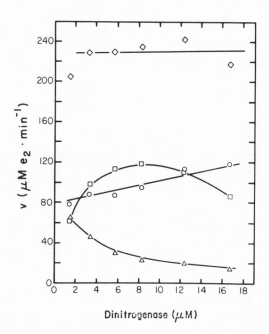

Fig. 3. Allocation of electrons with changing concentrations of dinitrogenase. Standard assay conditions were used, with 0.5% acetylene in nitrogen as a gas phase, 5 mM dithionite, 150 µM azotobacter flavodoxin, and 2.16 µM dinitrogenase reductase: (0-0) electron pairs to H_2 (µM min^{-1}); (□-□) electron pairs to ethylene; (Δ-Δ) electron pairs to ammonia; (◇-◇) total electron pairs transferred.

In an earlier publication,[14] we measured electron allocation
to substrates under conditions of essentially constant total electron flux through the nitrogenase system, as indicated in Figure 3.
The balance between dinitrogenase and dinitrogenase reductase was
altered. The acetylene level was 0.5%, dithionite concentration
5 mM, A. vinelandii flavodoxin concentration 150 μM, and the dinitrogenase was varied as shown on the x axis. Note that when the molar
concentration of dinitrogenase was about equivalent to that of di-
dinitrogenase reductase (constant at 2.16 μM), there was a near
equivalence of electrons passing to N_2 and to C_2H_2 and a somewhat
greater number passing to H^+. As the system was altered toward a
higher concentration of dinitrogenase relative to dinitrogenase
reductase, the proportion of electrons passing to H^+ increased
steadily, whereas the proportion to N_2 decreased rapidly and then
leveled off. The allocation to acetylene was more complex, as it
increased rapidly and then decreased at high levels of dinitrogenase.
At the terminal point, when the ratio of dinitrogenase to dinitrogenase reductase was approximately 8, the distribution of electrons
was about 54% to H^+, 39% to C_2H_2 and 7% to N_2. It is evident
that N_2 does not compete well with H^+ or C_2H_2 when dinitrogenase is
in excess so that the dinitrogenase is at a relatively low reduction
state when the electron flux through the system is constant. With
an approximate molar equivalence between dinitrogenase and dinitrogenase reductase, the dinitrogenase is in a more reduced state, and
N_2 competes quite effectively with C_2H_2 and H^+ for electrons. It
also is evident[16] that a change in substrate concentration, e.g.,
an increase in C_2H_2 concentration, will alter the specific response
from the response shown.

We will examine the response in electron allocation when the
total electron flux is varied by altering the MgATP or reductant
concentration or the ratio between nitrogenase components. The
response is shown in Figure 4. Part A of the figure shows the
effect of varying the concentration of MgATP and the component
ratio. The data are expressed as the pairs of electrons allocated
each minute per Mo (x axis), and the fraction of the electrons
going to each substrate (y axis). Instead of having an essentially
constant electron flux, as in Figure 3, the electron flux through
dinitrogenase increases as indicated on the logarithmic x axis.
Notice that the solid and open symbols nearly coincide, indicating
that electron allocation is much the same whether electron flux
through dinitrogenase is altered by changing the concentration of
MgATP or by changing the ratio of nitrogenase components. The
other prominent observation is that at lowest electron flux, N_2
does very badly, H^+ very well and C_2H_2 intermediate in the competition for electrons. At the highest electron flux, N_2 does very
well, C_2H_2 rather badly and H^+ intermediate.

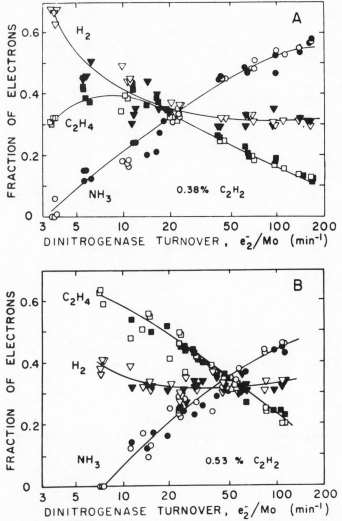

Fig. 4. Electron allocation as a function of the electron flux
through dinitrogenase. Standard assay conditions were used with
additions as indicated. o-o, e_2^- in NH_3; ∇-∇, H_2 evolved; and - ,
C_2H_4 formed. Figure 4A: 100 μM flavodoxin and 5 mM dithionite.
Open symbols, MgATP varied from 20 μM to 5 mM with 2.55 μM dinitrogenase and 14.2 μM dinitrogenase reductase. Closed symbols, component ratio varied from 1:5.57 to 13.4:1 dinitrogenase:dinitrogenase
reductase with 5 mM MgATP. Experiment carried out with 0.38% C_2H_2
in N_2 gas phase at 30°C. Figure 4B: 1.27 μM dinitrogenase and
2.37 μM dinitrogeanse reductase. Open symbols, MgATP varied from
40 μM to 5 mM with 5 mM dithionite and 100 μM flavodoxin as reductant. Closed symbols, reductant varied from 0.6 mM dithionite to
5 mM dithionite plus 100 μM flavodoxin with 5 mM MgATP. Experiments
carried out with 0.53% C_2H_2 in N_2 gas phase at 30°C.

For the experiment recorded in Figure 4B, the molar ratio of dinitrogenase reductase to dinitrogenase was kept constant at somewhat less than 2:1. The ATP and reductant levels were varied over a wide range. A. vinelandii flavodoxin was added to enhance electron transfer. Again the electron allocations were very similar whether electron flux through dinitrogenase was varied by changing the MgATP or reductant level. At the highest electron flux, the allocation was very similar to that shown in Figure 4A; the cross-over point for these curves was at about double the electron flux of Figure 4A (this resulted from an increased C_2H_2). The lowest electron flux was about twice that of Figure 4A, and in this instance, C_2H_2 was the most effective and H^+ the intermediate substrate. N_2 again was ineffective in capturing electrons under low electron flux but effective at high flux.

A comparison of the three methods of varying electron flux shows that neither dinitrogenase reductase nor MgATP alone have specific effects on the electron allocation to substrates, and these experiments furnish no evidence for a second role for either dinitrogenase reductase or MgATP beyond their roles in electron transfer to dinitrogenase.

Figure 4 also indicates the following qualitative effects of varying the electron flux through dinitrogenase. High electron flux through dinitrogenase favors N_2 reduction over reduction of C_2H_2 or H^+. As the electron flux is lowered, reduction of C_2H_2 becomes important. At very low electron flux through dinitrogenase, the reduction of H^+ becomes the dominant reaction. These observations support the results of Davis et al.[15] and Shah et al.[16] They made similar studies without quantitating the total electron flux through the system. The results of Bergersen and Turner[4], however, predict that C_2H_2 still should be an effective competitor with N_2 for electrons at very high electron fluxes through dinitrogenase. They found that the K_m's for C_2H_2 and N_2 varied in a parallel fashion when the ratio of the nitrogenase component proteins was varied. The parallel trend in the K_m's predicts that the relative efficiencies of these substrates as electron acceptors would remain the same, as the electron flux though dinitrogenase varied, a con-clusion at variance with our experimental data.

THE KINETICS OF SUBSTRATE REDUCTION

Figure 5 shows the dependency of the observed K_m for substrate reduction on the total electron flux through dinitrogenase. The total electron flux (x axis) for this figure was defined by H_2 evolution in the absence of added substrate. If the total electron flux is accepted as the flux at saturating substrate concentration, the shape of the curves is virtually the same as when the electron flux is observed under argon (Figure 5); the only change is a

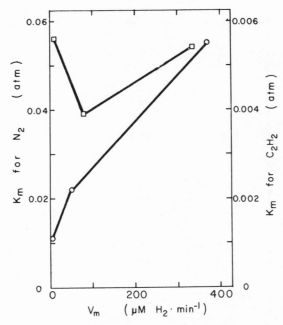

Fig. 5.   Dependency of the K_m's for C_2H_2 and N_2 on the total electron flux through dinitrogenase as described in the text.
o-o, K_m for C_2H_2 (right axis);   - , K_m for N_2 (left axis);
K_m expressed in atm.

horizontal shift in the curves. For C_2H_2, only the data below 5%
C_2H_2 were utilized to eliminate the uncertain interpretation of the
enhancement at high C_2H_2 concentrations. The observed change in the
K_m for C_2H_2 with changes in the electron flux is in general agreement with the results of Bergerson and Turner.[4] However, no definite
trend with changing electron flux in the K_m for N_2 was observed,
which contrasts with the results reported by Bergerson and
Turner.[4] The increased effectiveness of C_2H_2 as an inhibitor of N_2
reduction at low electron flux through dinitrogenase observed by
Davis et al.[15] would be consistent with little or no change in the
K_m for N_2 when the K_m for C_2H_2 decreases. The reason for the discrepancy between the results reported here and the results of
Bergerson and Turner[4] is unknown.

An initial lag in C_2H_2 reduction has been reported by other
workers,[17-21] with an accompanying burst in H_2 evolution to main-
tain a constant total electron flux.[18] A similar lag for N_2 reduction has been reported,[19] but it also has been denied.[20,21] The lag
in C_2H_2 reduction is much longer than the turnover times of the two
nitrogenase proteins. Data presented by Thorneley and Eady indicated
that the lag period might be proportional to the turnover time of

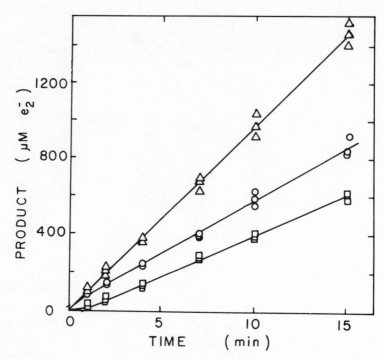

Fig. 6. Lag period before N_2 reduction. Standard assay conditions
were used with 20 mM dithionite as reductant in a N_2 gas
phase, 4.53 µM dinitrogenase (6.57 µM molybdenum) and
0.945 µM dinitrogenase reductase (0.50 µM Fe_4S_4) in each
assay. Assays were stopped at the indicated time by the
addition of HCl and analyzed for H_2 and NH_3 formed. $\square-\square$,
e_2^- in NH_3; o-o, H_2 evolved; and $\Delta-\Delta$, total e_2^- in products.

dinitrogenase, but not dinitrogenase reductase.[18] We have observed
a distinct lag phase before the onset of C_2H_2 reduction with a
complementary burst of H_2 evolution (data not shown), as well as a
lag before N_2 reduction begins (Figure 6). The lag before N_2
reduction is harder to demonstrate than the lag before C_2H_2 reduction, as a high rate of NH_3 formation is easy to measure but gives
a very short lag period, whereas a low rate of NH_3 formation gives
a longer lag, but is difficult to quantitate because of larger
percentage errors in the NH_3 determinations. The conditions
described in Figure 6 seem to be optimal for the system we were
using. A complementary burst of H_2 evolution was observed during
the lag in N_2 reduction such that the total electron flux remained
constant. Table 1 shows the effect of the electron flux through
dinitrogenase on the length of the lag for C_2H_2 reduction. The
ratio of the lag period to the turnover time shows no obvious trend,

Table 1. Effect of Electron Flux on the C_2H_2 Lag

v (μM e_2^-/min)	Dinitrogenase (μM)	Lag (minutes)	Dinitrogenase Turnover* (minutes)
C_2H_2 reduction			
6.00	9.59	22.6	2.32
7.15	4.79	14.5	0.97
14.3	4.79	4.1	0.49
48.2	4.70	2.3	0.141
93.9	4.60	0.7	0.071
N_2 reduction			
97.6	4.53	0.8	0.067

*Turnover time calculated by [molybdenum] x v^{-1} where [molybdenum] is the concentration of molybdenum in the assay as dinitrogenase, and in expressed on an e_2^- per Mo basis. Electron flux was varied by altering the component ratio.

and averages about 12. The data support a slow activation of dinitrogenase for C_2H_2 reduction; the activation occurs at a rate proportional to the rate of total electron transfer. Table 1 also shows that the N_2 lag period is approximately as long as the C_2H_2 lag period at the same electron flux.

HYDROLYSIS OF ATP IN THE PRE-STEADY STATE

Recently, Eady et al.[22] have shown that ATP is hydrolyzed during the initial electron transfer between the nitrogenase proteins of Klebsiella pneumoniae at 10°C, thus definitely establishing the link between MgATP and electron transfer. Although the time course of the ATP hydrolysis was indistinguishable from that of electron transfer, the stoichiometry of the hydrolytic reaction remained uncertain. They found that two ATP were hydrolyzed per dinitrogenase, and commented that this is less than half of the ATP hydrolyzed in the H_2 evolution reaction. They then proposed a second role for MgATP to account for the additional ATP hydrolyzed during H_2 evolution. However, while they did quantitate the amount of ATP hydrolyzed, they failed to quantitate the number of electrons transferred. The hydrolysis of two ATP per dinitrogenase corresponds to only 0.4 ATP per dinitrogenase reductase under their conditions, and this latter number does not give any reasonable stoichiometry.

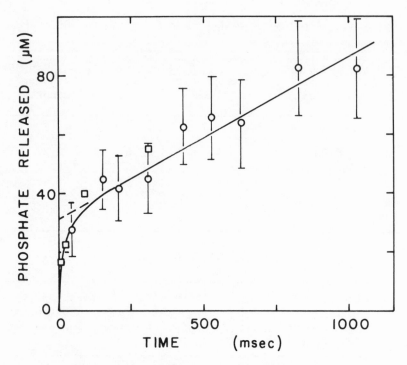

Fig. 7.   Pre-steady state hydrolysis of ATP by nitrogenase.   The
          chemical quench technique described in the text was used
          to measure phosphate release during the first second of the
          nitrogenase catalyzed reaction.   Syringe A contained 25 mM
          buffer, 5 mM dithionite and 4 mM ATP.   Syringe B contained
          25 mM buffer, 5 mM dithionite, 6 mM magnesium acetate,
          19.2 µM dinitrogenase (27.8 µM molybdenum) and 47.4 µM
          dinitrogenase reductase (25.1 µM Fe$_4$S$_4$ clusters).   Syringe
          C contained 25% perchloric acid.   The mixing ratio was
          5 A:5 B to initiate the reaction and approximately 2 C to
          stop the reaction.   Phosphate released was corrected for
          the background phosphate in the assay mixture.   The linear
          least squares fit was calculated from the data at times of
          90 msec and longer.   Two experiments are shown; o represents
          the data used in the least squares fit and □ represents
          a separate experiment.

     Eady et al.[22] also claim support for a second role for MgATP
from the uncoupling of ATP hydrolysis from H$_2$ evolution in the
steady state at 10°C.   This is a futile hydrolysis of ATP, and
Eady et al.[22] appear to confuse this with a second hydrolytic role
for ATP.   The hydrolysis of ATP upon electron transfer between the
two proteins does not require that all the ATP hydrolyzed at that
site support net electron transfer.   Hageman and Burris[14] have
proposed that the futile hydrolysis of ATP occurs in the complex of

oxidized dinitrogenase reductase with the dinitrogenase. Under these circumstances, ATP hydrolysis could occur at the site normally coupled to electron transfer without leading to net electron transfer. This would be futile hydrolysis, but not a second role for MgATP.

Figure 7 shows a similar pre-steady state experiment on ATP hydrolysis carried out at 30°C. An initial rapid release of phosphate is followed by a much slower steady state rate. The data were not adequate to determine the rate constant for the initial burst, but the results of Eady et al.[22] indicate that the rate constant would be the same as for electron transfer. The slow steady state rate of phosphate release in Figure 7 can be extrapolated to time zero to estimate the magnitude of the burst. The 32 µM phosphate released corresponds to 1.3 ATP hydrolyzed per dinitrogenase reductase molecule. This dinitrogenase reductase had been found to contain 0.53 Fe_4S_4 cluster per dinitrogenase reductase molecule. Ljones and Burris[10] have suggested that the Fe_4S_4 content represents the active protein concentration, and they have shown that there is one electron transferred from each Fe_4S_4 equivalent.[11] The ATP hydrolysis corresponds to 2.5 ATP hydrolyzed per Fe_4S_4 cluster or for each electron transferred. The ATP burst also is equivalent to 3.2 ATP hydrolyzed per dinitrogenase. Dinitrogenase from A. vinelandii contains 2.0 molybdenum per protein molecule and one spin per molybdenum.[23] The EPR-active center accepts one electron per molybdenum in a fast step. In Figure 7, the burst of phosphate would represent 2.2 ATP hydrolyzed per molybdenum in dinitrogenase (there were 1.45 molybdenums per dinitrogenase). These results strongly suggest that there are two ATP hydrolyzed upon the initial one electron transfer between the two proteins of the nitrogenase system of A. vinelandii. This value accounts for all of the ATP hydrolysis observed in the steady state under optimal conditions. There is no reason to postulate a second role for ATP hydrolysis, nor is there any excess ATP hydrolysis observed experimentally that could be used in such a role.

CONCLUSIONS

The data presented in this paper support the role of MgATP in electron transfer as the only role for MgATP in the nitrogenase reaction. Figures 3 and 4 show that electron allocation to substrates is controlled specifically by the electron flux through dinitrogenase, and not by the MgATP concentration or the dinitrogenase reductase concentration. Figure 7 shows that the ATP hydrolysis occurring in the steady state can be accounted for by the ATP hydrolysis coupled to electron transfer between the two proteins of the nitrogenase system. To correctly interpret the pre-steady state hydrolysis of ATP, it is necessary to compare the

hydrolysis of ATP to the number of electrons transferred during
the burst, and not to compare the ATP hydrolysis to the total
protein concentration.

REFERENCES

1. H. C. Winter and R. H. Burris, Nitrogenase, Ann. Rev. Biochem.
    34:409 (1976).
2. W. H. Orme-Johnson, W. D. Hamilton, T. Ljones, M.-Y. W. Tso,
    R. H. Burris, V. K. Shah, and W. J. Brill, Electron para-
    magnetic resonance of nitrogenase and nitrogenase components
    from Clostridium pasteurianum W5 and Azotobacter vinelandii
    OP, Proc. Natl. Acad. Sci. U.S.A., 69:3142 (1972).
3. R. V. Hageman and R. H. Burris, Nitrogenase and nitrogenase
    reductase associate and dissociate with each catalytic
    cycle, Proc. Natl. Acad. Sci. U.S.A., 75:2699 (1978).
4. F. J. Bergersen and G. L. Turner, Kinetic studies of nitro-
    genase from soya-bean root-nodule bacteroids, Biochem. J.,
    131:61 (1973).
5. B. E. Smith, D. J. Lowe, and R. C. Bray, Studies by electron
    paramagnetic resonance on the catalytic mechanism of
    nitrogenase of Klebsiella pneumoniae, Biochem J., 135:331
    (1973).
6. P. T. Bui and L. E. Mortenson, Mechanism of the enzymic re-
    duction of N_2: The binding of adenosine 5'-triphosphate
    and cyanide to the N_2-reducing system, Proc. Natl. Acad.
    Sci. U.S.A. 61:1021 (1968).
7. M.-Y. W. Tso and R. H. Burris, The binding of ATP and ADP by
    nitrogenase components from Clostridium pasteurianum,
    Biochem. Biophys. Acta 309:263 (1973).
8. W. G. Zumft, L. E. Mortenson, and G. Palmer, Electron para-
    magnetic resonance studies on nitrogenase. Investigation
    of the oxidation-reduction behavior of azoferredoxin and
    molybdoferredoxin with potentiometric and rapid-freeze
    techniques, Eur. J. Biochem., 46:525 (1974).
9. G. A. Walker and L. E. Mortenson, Effect of magnesium adenosine
    5'-triphosphate on the accessibility of the iron of clos-
    tridial azoferredoxin, a component of nitrogenase,
    Biochemistry, 13:2382 (1974).
10. T. Ljones and R. H. Burris, Nitrogenase: The reaction between
    the Fe protein and bathophenanthrolinedisulfonate as a
    probe for interactions with MgATP, Biochemistry, 17:1866
    (1978).
11. T. Ljones and R. H. Burris, Evidence for one-electron transfer
    by the Fe protein of nitrogenase, Biochem. Biophys. Res.
    Commun. 80:22 (1978).

12. R. C. Burns and R. W. F. Hardy, Nitrogen fixation in bacteria
    and higher plants, in: "Molecular Biology, Biochemistry,
    and Biophysics," Vol. 21., Springer, Berlin (1975).

13. R. N. F. Thorneley, R. R. Eady, and M. G. Yates, Nitrogenases
    of Klebsiella pneumoniae and Azotobacter chroococcum. Complex
    formation between the component proteins, Biochim. Biophys.
    Acta 403:269 (1975).

14. R. V. Hageman and R. H. Burris, Kinetic studies on electron
    transfer and interaction between nitrogenase components
    from Azotobacter vinelandii, Biochemistry, 17:4117 (1978).

15. L. C. Davis, V. K. Shah, and W. J. Brill, Nitrogenase. VII.
    Effects of component ratio, ATP and H_2 on the distribution
    of electrons to alternative substrates, Biochim. Biophys.
    Acta, 403:67 (1975).

16. V. K. Shah, L. C. Davis, and W. J. Brill, Nitrogenase. VI.
    Acetylene reduction assay. Dependence of nitrogen fixation
    estimates on component ratio and acetylene concentration,
    Biochim. Biophys. Acta, 384:353 (1975).

17. B. E. Smith, R. N. F. Thorneley, R. R. Eady, and L. E. Mortenson,
    Nitrogenases from Klebsiella pneumoniae and Clostridium
    pasteurianum. Kinetic investigations of cross-reactions as
    a probe of the enzyme mechanism, Biochem. J., 157:439 (1976).

18. R. N. F. Thorneley and R. R. Eady, Nitrogenase of Klebsiella
    pneumoniae. Distinction between proton-reducing and
    acetylene-reducing forms of the enzyme: Effect of tempera-
    ture and component protein ratio on substrate reduction
    kinetics, Biochem. J. 167:457 (1977).

19. B. E. Smith, R. R. Eady, R. N. F. Thorneley, M. G. Yates, and
    J. R. Postgate, Some aspects of the mechanism of nitrogenase,
    in: "Recent Developments in Nitrogen Fixation," W. Newton,
    J. R. Postgate, and C. Rodriguez-Barrueco, eds., Academic
    Press, New York (1977).

20. H. Berndt, D. J. Lowe, and M. G. Yates, The nitrogen-fixing
    system of Corynebacterium autotrophicum. Purification and
    properties of the nitrogenase components and two ferredoxins,
    Eur. J. Biochem. 86:133 (1978).

21. D. W. Emerich and R. H. Burris, Interactions of heterologous
    nitrogenase components that generate catalytically inactive
    complexes, Proc. Natl. Acad. Sci. U.S.A., 73:4369 (1976).

22. R. R. Eady, D. J. Lowe and R. N. F. Thorneley, Nitrogenase of
    Klebsiella pneumoniae: A pre-steady state burst of ATP
    hydrolysis is coupled to electron transfer between the
    component proteins, FEBS Letters, 95:211 (1978).

23. E. Muenck, H. Rhodes, W. H. Orme-Johnson, L. C. Davis,
    W. J. Brill, and V. K. Shah, Nitrogenase. VIII. Moessbauer
    and EPR spectroscopy. Molybdenum-iron protein component from
    Azotobacter vinelandii, Biochim. Biophys. Acta, 400:32 (1975).

ACTIVE SITE PROBES OF NITROGENESE

C.E. McKenna, T. Nakajima, J.B. Jones, C. Huang,
M.C. McKenna, H. Eran and Alan Osumi

Department of Chemistry
University of Southern California
Los Angeles, California 90007, USA

INTRODUCTION

Nitrogenase possesses the ability, thus far unique, to catalyze efficiently an ATP-dependent reduction of N_2 to NH_3 under
physiological conditions. Nitrogenase is a versatile catalyst,
capable of reducing other small, unsaturated molecules, such as
C_2H_2, HCN, N_2O and N_3^-. Unlike the parent compounds, higher homo-
logs of these adventitious substrates generally are reduced poorly,
or not detectably, by the enzyme. This situation has limited
development of new nitrogenase reduction chemistry as a strategy
to elucidate active site properties. Not long ago, we introduced
the concept of using a strained ring system as a chemical probe
for nitrogenase. This approach is currently being implemented in
our laboratory by studies of product formation parameters, relative
reduction rates, inhibition kinetics, effects on H_2 evolution, and
deuterium labeling patterns in reduction products. The results
obtained are providing new information on nitrogenase substrate
binding, reduction stereochemistry, and other mechanistically
significant aspects of nitrogenase function. The enzymological
data also create new criteria to evaluate nitrogenase model chemistry and co-factor function.

CYCLOPROPENE AS NITROGENASE SUBSTRATE AND INHIBITOR

Experimental Considerations

Cyclopropene is most conveniently prepared in a single-step
synthesis from commerically available allyl chloride.[1] In some
preparations, we obtained improved reproducibility and product

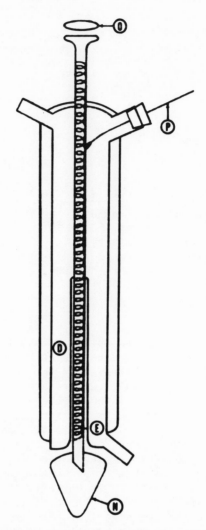

Figure 1.   Apparatus for low pressure distillation of cyclopropene.
             N₂ cooled in copper tubing immersed in liquid N₂ is
             passed through the vacuum-jacketed tube D at a rate
             sufficient to maintain the temperature in the upper,
             isothermal column segment at the desired value (measur-
             ed by thermocouple P). Due to the gap E, a temperature
             gradient exists in the lower half of the column. Sample
             is slowly distilled from bulb N into vacuum line traps
             via the O-ring joint O.

purity (less cyclopropane contamination) by substituting THF for
mineral oil as reaction solvent. Contaminating propene and methyl
acetylene levels can be decreased below 0.1-0.5% by a single

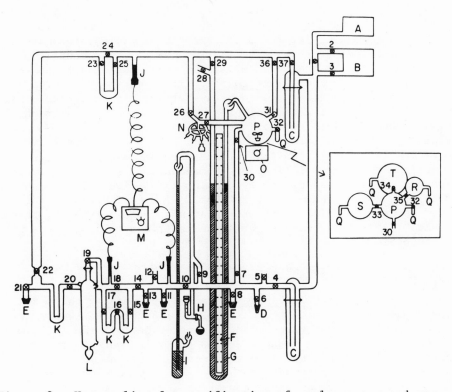

Figure 2. Vacuum line for purification of cyclopropene and pre-
           paration of cyclopropene assay mistures. A: mechanical
           pump, B: two-stage oil diffusion pump, C: pump trap,
           D: 10/30 inner joint inlet, E: 14/20 inner joint inlet
           F: doubly graduated meter stick, G: mercury manometer,
           H: Macleod gauge, I: mercury pressure release valve,
           J: thermocouple probe, K: collection trap, L: low temp-
           erature distillation column, M: thermocouple gauge,
           N: manifold, O: magnetic stirrer, P: 1 liter bulb con-
           taining a magnetically driven propeller, Q: freeze-out,
           R: 0.5 liter gas storage bulb, S: 1 liter gas storage
           bulb, T: 3 liter gas storage bulb. The distillation
           apparatus at L (cf. Fig.1) is used in conjunction with
           collection traps K (stopcocks 15-17). Distillation
           pressure is monitored with probe J near stopcock 18.
           Mixtures of cyclopropene in inert gases are prepared in
           bulb P (at a moderate overpressure) and used to fill
           assay vessels attached to manifold N. The vessels are
           pre-evacuated via stopcock 26. Recovery of unused cyclo-
           propene can be effected in the upper line trap K (stop-
           cocks 23-35). Cyclopropene is removed from the line for
           storage in a standard gas tube (in liquid nitrogen) via
           one of the lower line outlets D or E.

distillation of 2-10g of crude cyclopropene at -141° and 100-150
μ in the apparatus shown in Fig. 1 and as part L in the vacuum
line schematized in Fig. 2. Small amounts of cyclopropene are
readily purified by preparative gc (Porapak N). The distilled
cyclopropene may be stored indefinitely in liquid N_2 and can be
manipulated on a vacuum line equipped with a manifold interfacing
to conventional serum stoppered assay bottles, as depicted in Fig. 2
(manifold at N). At a partial pressure of < 0.25 atm in argon, He
or N_2 mixtures, cyclopropene is stable for hours at 25°. When exposed to certain buffers, it decomposes slowly on a time scale
considerably greater than that of standard nitrogenase assays. Very
rapid (30-60 sec) resolution of C_3H_6 reduction products derived
from cyclopropene can be achieved using $AgNO_3$-glycerol columns
containing a short segment of Porapak N; 1-3 μmol ethane can be
added as an internal standard to assay mixtures analyzed by this
column.[2] A detailed account of these experimental procedures will
be given elsewhere.

Reduction Products and Kinetics

Nitrogenase from A. vinelandii and K. pneumoniae catalyzed
ATP-dependent dithionite reduction of cyclopropene to two principal
hydrocarbon products, propene and cyclopropane.[2] Formally, these
represent two-electron reductions. A possible four-electron reduction product, propane, is not observed above the 1% level.
Neither propene nor cyclopropane is observably reduced by the
enzyme.[3] Under optimal assay conditions, the ratio of propene:
cyclopropane formation rates is approximatively 2:1.[2]

Similar reduction behavior is observed with this substrate
in vivo.[3,4] N_2-grown, but not NH_3-grown, A. vinelandii cultures
display O_2-dependent reduction of cyclopropene to a 2:1 mixture of
propene and cyclopropane. The reduction is inhibited by N_2 (1.0
atm) or CO (2 x 10^{-3} atm) consistent with nitrogenase as the effective catalyst.[4] The substrate Km for reduction to either product whether in vivo or in vitro, using highly purified components
combined in a ratio of 2 (Fe protein) to 1 (Mo-Fe protein), is
about 1 x 10^{-2} atm, corresponding to a molar Km value of about
10^{-4} (Fig. 3a,b).[4] This low value is equalled only by N_2 itself,
and perhaps C_2H_2, among known nitrogenase substrates.[3,4] The
structural isomers of cyclopropene, allene and methyl acetylene,
are much poorer substrates in vitro with Km values [3] that are larger
by 10^2-10^3. The in vivo and in vitro Vm (cyclopropene): Vm (C_2H_2)
ratios are also comparable, and correspond to a 43-60% maximal
electron flux to cyclopropene under enzyme-limiting conditions.[4]
Residual electron allocation largely appears as H_2 evolution
(H. Eran, unpublished). A 25% diversion of total electron flux
into H_2 formation has also been observed with N_2 as substrate,
while, at saturating pressures, C_2H_2 reduction can account for
100% of electron flux through the enzyme.[5]

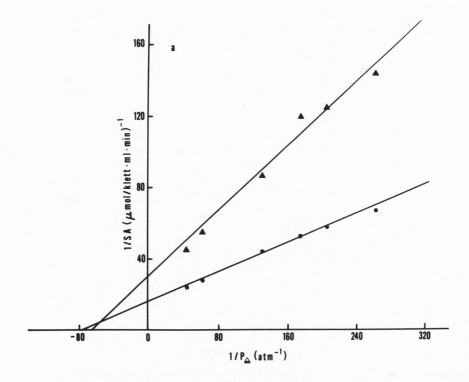

Figure 3a.   Double reciprocal (Lineweaver-Burk) plot of reduction
rate dependence on cyclopropene pressure with <u>A. vine
landii</u> cultured on N-free Burk's medium.   After addition of O_2 (0.1 atm) and 218 Klett-ml of log phase culture, time course plots for product formation (gc) were
made at each substrate pressure and the initial rate
($\overline{V}$) found from the linear portions.   V values were
divided by the product of the cell density (Klett units)
and assay culture volume to obtain specific activities
(SA values).   The data were fitted to a linear regres-
sion curve as shown.   Propene-●— , cyclopropane—▲—.
Data taken from ref. 4.

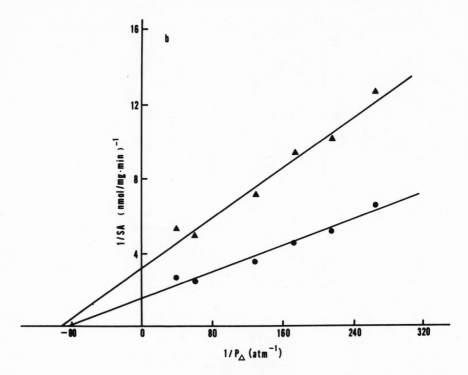

Figure 3b.   Double reciprocal (Lineweaver-Burk) plot of reaction
             rate dependence on cyclopropene pressure with purified
             nitrogenase components from A. vinelandii. The com-
             ponent ratio (Av1:Av2) was 1:2. Assay conditions were
             as described elsewhere.[2] Time course plots for product
             formation (gc) were made at each substrate pressure and
             the initial rates were found from the linear portion.
             V data were then converted to specific activity values
             (V/mg limiting protein-min). The data were fitted to
             a linear regression curve as shown.
             Propene –●– , cyclopropane –▲– . Data taken from ref. 4.

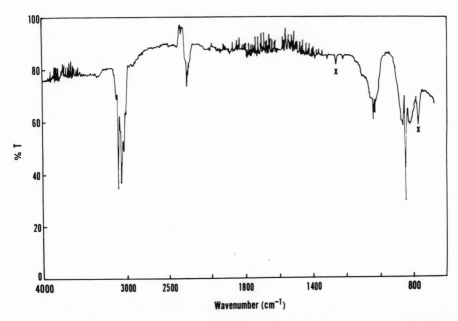

Figure 4. IR spectrum of cyclopropane isolated from nitrogenase-
           D_2O reduction of cyclopropene. The ir sample ($\sim$ 9 μmol)
           was run against air as blank using a 2x ordinate ex-
           pansion. Peaks marked "x" are assigned to cell back-
           ground. Data taken from ref. 6.

Inhibition Kinetics

In contrast to allene and methyl acetylene, cyclopropene is
an effective inhibitor of C_2H_2 reduction by nitrogenase.[3] Cyclopropene also strongly inhibits N_2 as substrate.[3] "Dissection" of
cyclopropene into its alkene (C_2H_4, propene) or saturated strained-
ring (cyclopropane) components respectively results in undetectable
and greatly diminished C_2H_2 inhibition activity.[3]

Deuterium Labeling Products in D_2O Nitrogenase

There are seven possible d_2-propenes and three possible
d_2-cyclopropanes (excluding the R/S stereoisomers of trans-1,
2-d_2-cyclopropane). Mass spectral analysis of the cyclopropane
formed by exposure of cyclopropene to active mixtures of nitrogenase (1Mo-Fe : 2 Fe) in D_2O-assay reagents indicate that the
predominant species is dideuterated.[6] The ir spectrum of this
product (Fig. 4) closely corresponds to that reported for cis-d_2-
cyclopropane and indicates that no more than 5-10% of the trans
isomer can be present.[6]

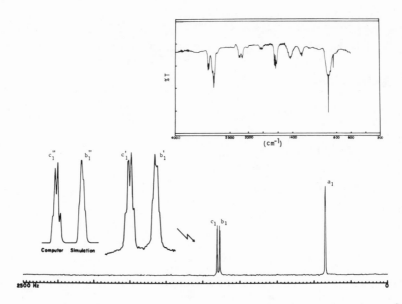

Figure 5.   IR spectrum versus air (inset, upper right) and ^{2}H-de-
            coupled 220 MHz FT ^{1}H nmr spectrum of propene isolated
            from nitrogenase-D_2O reduction of allene.   The full nmr
            spectrum (a_1, b_1, c_1) was obtained over a 2500 Hz sweep
            with the zero set 1643 Hz upfield from $CHCl_3$.   The ex-
            panded spectrum (b'_1, c'_1) used a sweep width of 250 Hz
            and a 2x ordinate scale expansion.   The simulated vinyl
            proton spectrum (b''_1, c''_1) was generated for weep width
            =250 Hz and a 1.3 Hz linewidth.   Data taken from ref. 6.

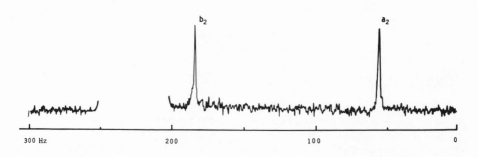

Figure 6. ^{1}H-coupled 30.7 MHz ^{2}H FT nmr spectrum of propene iso-
            lated from nitrogenase-D_2O reduction of allene (same
            sample as in Fig. 5).   The spectrum (a_2, b_2) was over a
            1000 Hz sweep with the zero set 222 Hz upfield from $CDCl_3$.
            Data taken from ref. 6.

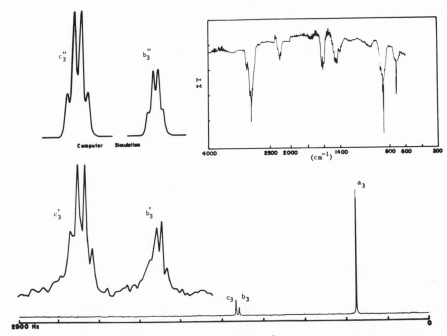

Figure 7.   IR (inset, upper right) and ^{2}H-decoupled 220 MHz FT ^{1}H
nmr spectra of propene isolated from nitrogenase-D$_2$O
reduction of methyl acetylene. The ir sample ($\sim$ 13 μmol)
was run against air as blank using a 2x ordinate expansion. For the full nmr spectrum (a$_3$, b$_3$,c$_3$), 32 transients were obtained on the sample in 0.4 ml CDCl$_3$ over
a 2500 Hz sweep width with the abscissa scale zero set
1658 Hz upfield from CHCl$_3$. To obtain the expanded
spectrum (b'$_3$, c'$_3$), a nominal sweep width of 100 Hz was
used with a 8.7x ordinate expansion. The simulated vinyl
proton spectrum (b"$_3$,c"$_3$) was generated for sweep width
= 100 Hz and a 1.1 Hz linewidth. Data taken from ref. 6.

     To facilitate identification of the deuterated propene prod-
uct(s), we adopted the strategy of first analyzing the deuterium-
labeling patterns in propene produced from the D$_2$O-nitrogenase
reduction of allene and methyl acetylene.[6] The ^{1}H and ^{2}H nmr
(Fig. 5 and 6) and ir spectra (Fig. 5) of the d$_2$-propene (by mass
spectrometery) from allene unequivocally establish that it is the
2,3-d$_2$-isomer. The ^{1}H nmr and ir spectra (Fig. 7) of the d$_2$-
propene (by mass spectrometry) from methyl acetylene suggest a
misture of the cis- and trans-1,2-d$_2$-isomers, with the cis form in
moderate excess ($\sim$ 65%).

     The ^{1}H nmr spectrum of propene derived from the D$_2$O-nitrogen-
ase reduction of cyclopropene is complex (Fig. 8). Analysis of the

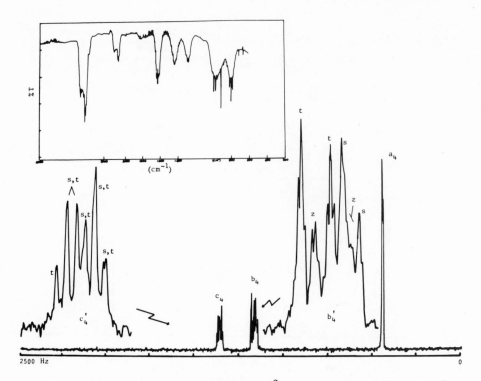

Figure 8. IR (inset, upper left) and ^{2}H-decoupled 220 MHz FT ^{1}H nmr
spectra of propene isolated from nitrogenase-D_2O reduction of cyclopropene. The ir sample ($\sim$ 20 µmol) was
run against air as blank. For the full nmr spectrum
(a_4, b_4, c_4), 16 transients were obtained on the sample
in 0.4 ml $CDCl_3$ over a 2500 Hz sweep width with the
abscissa zero scale set 1652 Hz upfield from $CHCl_3$. To
obtain the expanded spectrum (b'_4, c'_4) a nominal sweep
width of 250 Hz was used with a 4x ordinate expansion.
Assignments: s = cis-1,3-d_2-propene; t = trans-1,3-d_2-
propene; z = 2,3-d_2-propene. Data from ref. 6.

multiplet structure, aided by computer simulations, indicates that
protons are present in all four possible shift environments in the
molecule, corresponding to a 2:1 mixture of trans- and cis-1,3-d_2
isomers, with the 2,3-d_2-isomer present in an amount approximately
equal to the cis-1,3-d_2-isomer. This postulate predicts the observed methine: vinyl : mehtyl proton integration ratio of 1:1.7:2.7
and is strongly supported by the corresponding ^{2}H nmr spectrum
(Fig. 9), which shows deuteron resonances at the expected four
positions with the correct relative intensities. The present of
some 2,3-d_2-isomer is further confirmed by a characteristic sharp
band at 910 cm^{-1} in the ir spectrum of the product mixture.

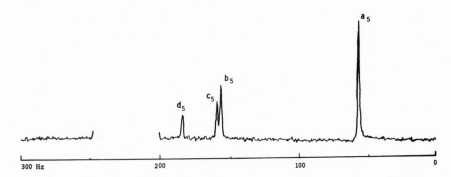

Figure 9. ^{1}H–decoupled 30.7 MHz ^{2}H nmr spectrum of propene isolated
            from nitrogenase–D$_2$O reduction of cyclopropene (same
            sample as in Fig. 8).   The spectrum (a$_5$, b$_5$, c$_5$, d$_5$) was
            plotted after 400 transients over a 1000 Hz sweep width
            with the abscissa scale zero set 222 Hz upfield from
            CDCl$_3$.   The spectral region from 200 to 250 Hz is ob-
            scured by solvent resonances.   Data from ref. 6.

CYCLOPROPENE REDUCTION AS A FUNCTIONAL CRITERION FOR NITROGENASE
MODEL CHEMISTRY

Artificial Model Systems

     Transition metal chemistry intended to imitate nitrogenase
catalysis has preoccupied many chemists over the past fifteen years.
The sole ability to reduce N$_2$ to NH$_3$ is not necessarily indicative
that the relevant model chemistry has been achieved.   Conversely,
some putative model systems, notably aqueous ones, have extremely
low turnover numbers for reduction of N$_2$ to NH$_3$, but display higher
activity for other substrates, such as C$_2$H$_2$.   The high reactivity
of C$_2$H$_2$ detracts from its significance as a model substrate, particularly because C$_2$H$_4$ is the only possible two-electron product
and is catalytically formed in such patently irrelevant processes
as hydrogenation of acetylene on supported Pd or Pt.   Cyclopropene
reduction by nitrogenase to a specific mixture of ring-opened and
ring-preserved products offers a more rigorous functional criterion
for model chemistry which, nevertheless, involves facile reductions.
So far, cyclopropene-transition metal chemistry is largely unex-
plored, but it has been reported[7] that bis(η^5-cyclopentadienyl)-
niobium chloride forms a cyclopropene complex which, on protonation,
yields only cyclopropane.   Newton, in collaboration with ourselves,
has found that three aqueous "model" systems - the heterogeneous
and homogeneous V^{2+} systems of Shilov and Schrauzer's molyb-
dothiol-borohydride reduce cyclopropene to cyclopropane alone.[8]
Thus, so far no model system has succeeded in duplicating the
reduction product stoichiometry displayed by the enzyme with cyclopropene.   A search for conditions under which existing models

fulfill this novel substrate criterion, or for new systems which can do so, may lead to a higher level of sophistication in functional modeling of nitrogenase.

The FeMo-co

The nitrogenase FeMo cofactor (FeMo-co) is presently characterized by its unique ability to restore N_2-fixing ability to extracts of certain mutants[9] and by its spectral properties, at least some of which (epr, EXAFS, Mossbauer) can be correlated with those of the Mo-Fe protein itself.[10] The functional role of FeMo-co in nitrogenase is not yet clear. Recently, it was reported[11] that uncomplemented FeMo-co catalyzed the reduction of C_2H_2 to C_2H_4 in a non-ATP-dependent process that could be inhibited by CO. This result was taken as evidence to support the idea that FeMo-co is the active site for N_2 reduction and binding in nitrogenase, although conversion of N_2 to NH_3 by cofactor-BH_4^- mixtures was not observed.[11] Our results[12] confirm that BH_4^--FeMo-co mixtures reduce C_2H_2 to C_2H_4 at an initial rate that depends on the FeMo-co concentration (determined over a range of 0-110 activity units). However, a dependence on the C_2H_2 concentration is observed at partial C_2H_2 pressures less than about 0.02 atm. Replacing the diluting gas (Ar) with He or H_2 has no effect, and dithionite alone did not support reduction. Based on estimated Mo content, the specific C_2H_2 reduction activity of our FeMo-co preparations was 5-10% relative to homogeneous Av1 (Table 1), which compares well with the relative specific activity of 8% obtained by Shah and Brill.[11] The reaction rate increases above 0° to a maximum at 25°-30°, then decreases, presumably due to thermal lability of the catalytic species or to accelerated decomposition of the BH_4^-. The maximum in the rate profile occuring at 25°-30° would explain the apparent invariability of the reaction rate to temperature reported previously.[11] The $t\frac{1}{2}$ for loss of C_2H_2 reduction activity in air-exposed FeMo-co at 25° was nearly 1 h, while the $t\frac{1}{2}$ for loss of reconstitutive activity was much shorter. C_2H_2 reduction activity was abolished by 0.12 atm CO or 1 mMKCN in the assay system. Besides C_2H_4, which is the major reduction product, we detected traces (up to 2%) of C_2H_6. A low level of background C_2H_2 reduction activity present in assay blanks did not account for this minor product. Subsidiary formation of C_2H_6 is not observed in ATP-dependent reductions of C_2H_2 to C_2H_4 by either intact or reconstituted nitrogenase with dithionite as electron donor, but is typical of BH_4^- reductions catalyzed by simple Mo-thiol complexes.[12]

DISCUSSION

Cyclopropene as a Nitrogenase Substrate

The high affinity of nitrogenase for C_2H_2 and cyclopropene, but not for C_2H_4, the linear C_3H_4 isomers, or cyclopropane, suggests

TABLE 1. Comparison of Nitrogenase and FeMo-co as Catalysts for C_2H_2 and Cyclopropene Reduction.[a]

Catalyst	Relative Activity (C_2H_2)	Reduction rate ratio (cyclopropene/C_2H_2)	Formation rate ratio (propene:cyclopropane)
Intact nitrogenase [b]	1.0[c]	0.58	2
Reconstituted nitrogenase [b]	–	0.56	2.5
FeMo-co	0.05 – 0.10	0.40	< 0.003
"Aproprotein Extract"	< 0.0001	–	–

[a] All data represent average values from two or more experiments. Data and experimental details are from ref. 12.

[b] Assayed as described in ref. 2.

[c] Reference activity of Av1 determined with 0.04 atm C_2H_2 and an optimal amount of Fe protein.

that both size (C_2H_2 vs methyl acetylene, allene) and electronic
orbital configuration (cyclopropene vs C_2H_4, cyclopropane) are important factors in determining substrate binding. As suggested by
the molecular profiles shown in Fig. 10, binding site topology
appears to restrict substrate size along the carbon-carbon multiple
bond axis.

Observation of similar kinetic parameters for cyclopropene
reduction in vitro under specified conditions and in vivo reinforce
the important postulate that purification, and use of an artificial
assay mixture, does not substantially alter the catalytic properties
of the enzyme.

Reduction Stereochemistry

Exposure of C_2H_2 to active nitrogenase-D_2O assay mixtures re-
portedly has resulted predominantly or exclusively in formation of
cis-1,2-d_2-C_2H_4.[13-15] This has been cited in support of a concerted transfer of two protons and two electrons to substrate, possibly involving a specifically positioned pair of acids derived
from the enzyme protein residues.[16] Because of its high symmetry
($D_{\infty h}$), C_2H_2 is a comparatively insensitive stereochemical probe.
A non-specific addition of exterior solvent protons to C_2H_2 bound
to an active site metal center, or other such mechanisms, could
arguably account for the observed preferential formation of the
cis-alkene isomer. For cyclopropene, reduction to cyclopropane
proceeds with rather high stereoselectivity, while reduction to
propene proceeds with rather low selectivity (scheme). With reduction of the cyclopropene double bond, protons, whether donated
by specific enzyme acids or by solvent, add to the ring syn fashion.
With cleavage of a cyclopropene C-C single bond, substantial addition of two protons across this bond is observed to yield, in
D_2O reductions, a 1,3-d_2-labeled product. However, the product
equivalent of more anti than syn addition is obtained, resulting
in an excess of trans 1,3-label. As most (75%) of the product is
1,3-labeled, the dominant mechanistic pathway cannot include pre-
reductive isomerization of cyclopropene to either allene or methyl
acetylene, since as we have proven, other products would then be
expected (scheme). The presence of a minor amount of 2,3-d_2-labeled
propene, with traces of 1,2-d_2-labeled propenes, nevertheless
makes such isomerization possible as a secondary mechanism. Per-
fectly concerted two-proton and two-electron transfer would require formation of a single 1,3-d_2-isomer; thus, this mechanism
cannot be involved in the major pathway to propene. Therefore,
alternative mechanisms are needed, in which protonation either lags
or precedes electron transfer, or is concerted with it but occurs
in discrete single-proton steps. Examples of the latter in which
syn or anti 1,2-addition of a nucleophilic metal and proton is
followed by reductive elimination concerted with protonation from
a specific side of the opening ring do not predict the observed

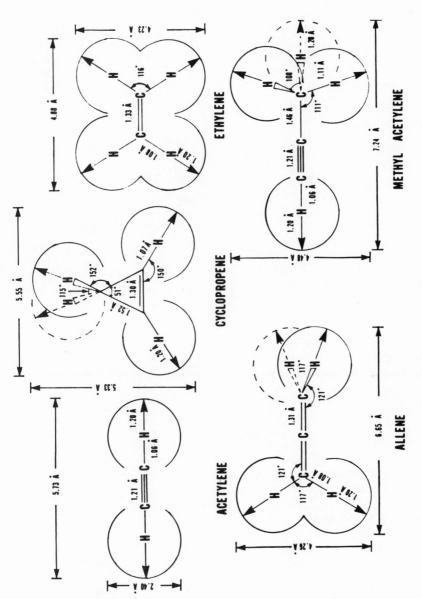

Figure 10. Molecular profiles of C_2H_2, C_2H_4 and C_3H_4 isomers. Distances indicated by arrows are Van der Waals radii.

$$H_2C=C=CH_2 \longrightarrow H_2C=CD-CH_2D$$

$$H_3C-C\equiv CH \longrightarrow \underset{1.8}{\overset{D}{\underset{H}{\text{C}}}=\overset{D}{\underset{CH_3}{\text{C}}}} + \underset{1.0}{\overset{H}{\underset{D}{\text{C}}}=\overset{D}{\underset{CH_3}{\text{C}}}}$$

$$\triangle \longrightarrow$$

$$\overset{H}{\underset{D}{\text{C}}}=\overset{H}{\underset{CH_2D}{\text{C}}} \quad \sim 1$$

$$\overset{D}{\underset{H}{\text{C}}}=\overset{H}{\underset{CH_2D}{\text{C}}} \quad \sim 2$$

$$H_2C=CD-CH_2D \quad \sim 1$$

and

$$\underset{D \quad D}{\triangle} \quad \geq 90\%$$

SCHEME.   Reduction of C_3H_4 Isomers by Nitrogenase in D_2O-Assay Mixtures.

products, however.   The interestingly low level of stereoselectivity we have observed with methyl acetylene reductions would tend to support these suggestions, although the large Km of this substrate makes direct comparison to C_2H_2 imperfect.

The reductions reported here were conducted under a specific set of assay conditions.   Variation of these parameters, and addition of other substrates or inhibitors may affect the reduction stereochemistry in an interpretively useful way.   The results obtained also present a relatively complex and subtle new challenge to biomimetic systems.

IMPLICATIONS FOR MODEL CHEMISTRY

It seems clear a priori that care must be exercised in drawing analogies between the catalytic behavior of BH_4^--FeMo-co and nitrogenase.   ATP-dependence, the reductant and the assay pH are quite different for the two systems, while the known instability of FeMo-co in anaerobic aqueous solution places in question the actual species responsible for substrate reductions by BH_4^--FeMo-co mixtures.   The striking differences in O_2-lability between the catalytic and reconstitutive FeMo-co activities may be relevant to this ambiguity, although it is possible that the catalytic and complementive functions of native cofactor involve distinct and non-interacting moieties. With these caveats in mind, the issue of appropriate substrate models can be addressed. Our experiments verify that on the basis of Mo content, FeMo-co added to aqueous BH_4^- is a highly effective, CO-sensitive catalyst for C_2H_2 reduction. However, C_2H_2 has been recognized to be a potentially misleading model substrate for nitrogenase, especially when used alone. By the more stringent cyclopropene criterion, BH_4^--FeMo-co is not functionally equivalent to nitrogenase. Interestingly, the inability of the cofactor to reduce cyclopropene to propene is shared by a number of transition metal complexes that also catalyze C_2H_2 reduction, including a Mo-thiol system. Incubation of FeMo-co with Kp 5058 "apoprotein" extract restored propene formation. To the extent that appropriate chemical equivalence exists among the FeMo-co catalyst in aqueous, alkaline BH_4^-, the spectroscopically defined FeMo-co species isolated in NMF, and those molybdenum and iron ions in intact nitrogenase assigned to the NMF cofactor, our findings are consistent with, but do not prove, the postulate that FeMo-co may represent only a portion of the active site responsible for binding and reducing N_2.

OTHER NEW SUBSTRATES

Variations on the basic cyclopropene structure could lead to additional new chemical probes for nitrogenase. Currently, we are interested in possible new heteroatom-containing substrates or inhibitors of the enzyme. In very recent work,[17] we have obtained evidence for the first fluorine-containing substrate of nitrogenase. Incubation of 3,3-difluorocyclopropene with an active Av1-Av2 mixture results in the formation of, at least, two reduction products, identified as 2-fluoropropene and propene, with approximately similar Km values. 2-Fluoropropene is not detectably reduced to propene. If the 3,3-difluorocyclopropene is directly reduced to either of these products, the reactions would be unprecedented in that reductive defluorination in effect has taken place. It is also noteworthy [18] that these reductions formally require four- and six electrons respectively.

CONCLUSION

Cyclopropene is a valuable active site probe for nitrogenase function and constitutes a novel and useful substrate criterion for the biomimetic chemistry of this enzyme.

ACKNOWLEDGEMENT

The authors wish to thank Dr. B.E. Smith for a gift of Kp 5058 extract and for providing preparative details on FeMo-co. This research was supported by grants from the Competitive Grants Research Office of the U.S. Department of Agriculture, the H.F. Frasch Foundation and the Petroleum Research Fund of the American Chemical Society.

REFERENCES

1. G. L. Closs and K. D. Krantz, A Convenient Synthesis of Cyclopropene, J. Org. Chem. 31:638 (1966).
2. C. E. McKenna, M-C. McKenna and M. T. Higa, Chemical Probes for Nitrogenase. 1. Cyclopropene. Reduction to Propene and Cyclopropane, J. Am. Chem. Soc. 98:4657 (1976).
3. C. E. McKenna, C. W. Huang, J. B. Jones, M-C. McKenna, T. Nakajima and H. T. Nguyen, Cyclopropenes: New Chemical Probes of Nitrogenase Active Site Interactions, "Nitrogen Fixation", W. E. Newton and W. H. Orme-Johnson, eds., University Press, Baltimore, (1979) Vol. 1. In press.
4. C. E. McKenna and C. W. Huang, In Vivo Reduction of Cyclopropene by A. vinelandii Nitrogenase, Nature 280:609 (1979).
5. R. C. Burns and R. W. F. Hardy, "Nitrogen Fixation in Bacteria and Higher Plants", A. Kleinzeller, G. F. Springer and H. G. Wiltman, eds., Molecualr Biology, Biochemistry and Biophysics, Vol. 21, Springer-Verlag, New York (1975).
6. C. E. McKenna, M-C. McKenna and C. W. Huang, Low Stereoselectivity in Methylacetylene and Cyclopropene Reductions by Nitrogenase, Proc. Nat. Acad. Sci. US 79:0000 (1979).
7. S. Fredericks and J. L. Thomas, Cyclopropene, Alkene, Alkyne, and Carbonyl Complexes of Cyclopentadienylniobium Species and Reduction of Cyclopropene , J. Am. Chem. Soc. 100:350 (1978).
8. C. E. McKenna, T. Nakajima, M-C. McKenna and W. E. Newton, Chemical Probes of Nitrogenase. 5. Cyclopropene: A Functional Criterion for Chemical Models of Biological N_2 Fixation. Submitted for publication.
9. V. K. Shah and W. J. Brill, Isolation of an Iron-Molybdenum Cofactor from Nitrogenase , Proc. Nat. Acad. Sci. US 74:3249 (1977).
10. J. Rawlings, V. K. Shah, J. R. Chisnell, W. J. Brill, R. Zimmerman, E. Munck and W. H. Orme-Johnson, Novel Metal Cluster in Iron-Molybdenum Cofactor of Nitrogenase , J. Biol. Chem. 253: 1001 (1978).
11. V. K. Shah, J. R. Chisnell and W. J. Brill, Acetylene Reduction by the Iron-Molybdenum Cofactor from Nitrogenase , Biochem. Biophys. Res. Comm. 81:232 (1978).

12. C. E. McKenna, J. B. Jones, H. Eran and C. W. Huang, Cyclopropene: A Functional Criterion for Evaluation of Active-Site Homology between Nitrogenase and its Molybdenum-Iron Cofactor Nature 280: 611 (1979).
13. M. J. Dilworth, Acetylene Reduction by Nitrogen-Fixing Preparations from Clostridium Pasteurianum , Biochim. Biophys. Acta 127:285 (1966).
14. R. W. F. Hardy, R. D. Holsten, E. R. Jackson and R. C. Burns, The Acetylene-Ethylene Assay for N_2 Fixation: Laboratory and Field Evaluation , Plant Physiol. 43:1185 (1968).
15. M. Kelly, Comparisons and Cross Reactions of Nitrogenase from Klebsiella Pneumoniae, Azotobacter Chroococcum and Bacillus Polymyxa, Biochim. Biophys. Acta 191:527 (1969).
16. E. I. Stiefel, Proposed Molecular Mechanism for the Action of Molybdenum in Enzymes; Coupled Proton and Electron Transfer Proc. Nat. Acad. Sci. US 70:988 (1973).
17. C. E. McKenna, T. Nakajima and H. Eran, A Fluorine-Containing Nitrogenase Substrate , In preparation.
18. R. H. Burris and W. H. Orme-Johnson, Mechanism of Biological N_2 Fixation , in: "Proceesings of the First International Symposium on N_2 Fixation", W. E. Newton and C. J. Nyman, eds., Washington State University Press, Pullman p. 208 (1976).

STUDIES ON THE SUBUNITS OF CLOSTRIDIAL MOLYBDENUM-IRON PROTEIN

Walter G. Zumft, Toshiharu Hase,[†] and Hiroshi Matsubara[†]

Department of Botany, Erlangen University, 8520 Erlangen
German Federal Republic
[†]Department of Biology, Faculty of Science, Osaka
University, Osaka, Japan

INTRODUCTION

In recent years we have witnessed a partial unveiling of the
puzzle that is nitrogenase, particularly with respect to its metal
centers. Most of this elegant work is described elsewhere in this
volume. Very little, however, is known about the role of the protein
moiety, although it is obvious that conformational changes of, and
subunit interactions between, the polypeptides, that is intrinsinc
protein interactions,[1] will take part in the overall reduction process. The general agreement about a $MgATP^{2-}$-induced conformational
change of the iron protein,[2-4] for instance, has perhaps prevented a
direct probing at the polypeptide level. The common knowledge of the
oligomeric nature of the molybdenum-iron protein,[5] has left the spa-
tial arrangement of the subunits largely a matter of conjecture, not
to mention any functional aspects of the quarternary structure.

Several questions pertain directly to the structural information
engraved in the polypeptide: what is the distribution of the metal
centers within the nitrogenase monomers; where is the binding site of
the molybdenum cofactor; and are there metal centers — and if so,
which ones — ligated jointly by two polypeptides? To gain the answers, the positions of the ligating groups, in particular the cysteinyl residues, will have to be identified and assigned to their
corresponding metal center. Also, their vicinal groups should be
known to gauge the role of the cluster or cofactor environment, e.g.,
in priming of the redox potential or in determining the direction of
electron flow. Eventually, some of these questions will be answered
by a complete three-dimensional model of nitrogenase from X-ray analysis. Until such data are forthcoming, investigation of the isolated

polypeptides will contribute to these problems too, even though only the apoproteins are amenable to isolation thus far.

SEPARATION OF THE MONOMERS OF THE MOLYBDENUM-IRON PROTEIN

Polyacrylamide gel electrophoresis in the presence of sodium dodecyl sulfate is frequently used with equivocal success to resolve the substructures of nitrogenase.[5] The Mo-Fe proteins of *Clostridium pasteurianum*[6] and *Klebsiella pneumoniae*[7] where shown by this method to contain two types of electrophoretically separable polypeptides. The difficulty in obtaining conclusive results with this procedure was elaborated by Kennedy et al.,[7] who tentatively resolved the Mo-Fe proteins of *Azotobacter vinelandii, Azotobacter chroococcum,* and *Rhizobium japonicum.* Previous investigations of the Mo-Fe proteins of these organisms had shown only one type of subunit. Recently, the monomers (subunits) of *A. vinelandii* Mo-Fe protein were separated by ion exchange chromatography on CM-cellulose[8] or sulfopropyl Sephadex[9] in the presence of urea, and were partially characterized.

The tetrameric structure of the Mo-Fe protein in solution might, therefore, inherently comprise two protomers with two polypeptides of different chemical composition, and the general structure of the Mo-Fe protein of many (all?) diazotrophs may be represented by $\alpha_2\beta_2$. As a matter of fact, no conclusive evidence to the contrary, i.e., a

Table 1. Procedures for monomer separation of the Mo-Fe protein

Procedure	Dissociating agent	Degree of separation	References
Electrophoretic methods			
PAGE[a]	SDS	Complete	6, 7, 11
PAGE	Urea	Complete	Unpublished
Isoelectric focusing	Urea	Complete	Unpublished
Molecular sieving	HCl	Incomplete	6
Ion exchange			
DEAE-cellulose	Protein derivatization	Incomplete	Unpublished
CM-cellulose	Prot. deriv. & urea	Incomplete	8
Sulfopropyl- Sephadex	Prot. deriv. & urea	Incomplete	9

[a]Polyacrylamide gel electrophoresis

Mo-Fe protein with only one type of monomer, is available yet from peptide mapping, analysis of terminal amino acid residues or terminal sequences. For convenience, we designate the larger subunit as the α-monomer and the smaller subunit as the ß-monomer where such easily distinguishable monomers exist, for instance, in *C. pasteurianum*. This nomenclature is formally identical to that used for *Azotobacter*, where the subunit with the lower electrophoretic mobility is the α-monomer. It remains, however, to be seen whether the polypeptides designated in this way are indeed homologous.

Table 1 summarizes the presently available procedures for monomer separation of the Mo-Fe protein. The most widely used method of detergent electrophoresis is never employed for preparative separation due to problems of scale-up and the additional complication of obtaining the proteins as their detergent complexes, which requires tedious procedures for detergent removal. This difficulty is not encountered with electrophoresis in the presence of 6 M urea, and excellent separation with high yields have been obtained with a commercially available preparative electrophoresis unit ("Ultraphor" made by Colora, Lorch, Germany; Zumft, unpublished results).

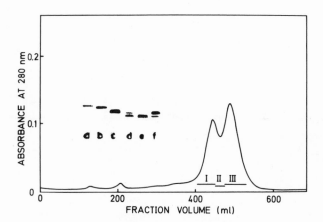

Fig. 1.   Separation of the monomers of derivatized Mo-Fe protein on DEAE-cellulose. Carboxymethylated and maleylated Mo-Fe protein was chromatographed on a DE-52 column (3 x 40 cm) with a linear gradient from 0.3 to 1.0 M NH_4HCO_3/CH_3COOH, pH 7.5; total gradient volume was 1 liter. Only two peaks were observed. The inset shows traces of SDS polyacrylamide gel electrophoresis for the α-monomer (a) and the ß-monomer (b) obtained from isoelectric focusing (see Fig. 2), and the pooled fractions I (c), II (d), and III (e). The difference in mobility of the monomers from the two separation procedures is due to maleylation. Trace (f) is a mixture of fractions I and III.

On a small scale, the monomers of clostridial Mo-Fe protein were
separated by gel filtration, using HCl as the dissociating agent.[6]
Amino acid composition and C-terminal amino acids were reported for
these monomers.[10] A drawback of this procedure was the incomplete dissociation of the Mo-Fe protein tetramer and a partial overlap of the
monomers due to the small difference in molecular weight.

A better separation method consists in protein derivatization by
carboxymethylation and maleylation with subsequent separation on
DEAE-cellulose. Carboxymethylation by itself was insufficient and did
not allow separation of the clostridial monomers on DEAE-cellulose
under a variety of conditions, including the addition of urea to the
eluant. Fig. 1 shows the elution profile of derivatized Mo-Fe protein
from a DEAE-cellulose column with a linear gradient of NH_4HCO_3/CH_3COOH.
As demonstrated in the inset of Fig. 1, electrophoretically homogeneous fractions of both monomers can be obtained. Although a complete
separation under the conditions employed here was not possible, the
method has great potential for scale-up and obtaining large quantities of both monomers.

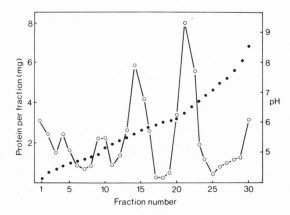

Fig. 2.   Isoelectric focusing of Mo-Fe protein by the flat-bed tech-
           nique with Sephadex G-75 as supporting matrix. Approx. 60 mg
           of Mo-Fe protein in 3 ml 0.05 M Tris-HCl buffer, pH 7.5, was
           made 6 M in urea, 1% in 2-mercaptoethanol, and adjusted to
           pH 2 with 1 M HCl. After 2 h incubation at room temperature,
           the protein was neutralized and applied to a "LKB multiphor"
           electrophoresis apparatus. The gel bed of Sephadex G-75,
           superfine, contained carrier ampholytes to create a pH gra-
           dient from 3.5 to 8, and was 6 M in urea and 1% in 2-mercap-
           toethanol. The dotted line shows the pH gradient after 16 h,
           when complete separation of the monomers was achieved. The
           α-monomer centered around a pH of 6.3; the ß-monomer around
           5.7. The temperature was kept at 6°C; the power supply out-
           put was 7 W. The protein was applied near the anode.

Isoelectric focusing in dextran gels as a supporting matrix and with urea as dissociating agent, combines complete separation of the monomers with high capacity for preparative runs. Fig. 2 shows the protein pattern of a flat-bed isoelectric focusing in Sephadex G-75, using a pH gradient from 3.5 to 8. The two monomers separated completely because of a substantial difference in their isoelectric points. After separation, the proteins were recovered in high yield by passage over a column of Sephadex G-100 and, at the same time, were efficiently freed from adhering ampholytes.

PROPERTIES OF CLOSTRIDIAL Mo-Fe PROTEIN MONOMERS

Several properties of the clostridial monomers are shown in Table 2. The molecular weight was re-determined by SDS polyacrylamide gel electrophoresis of the isolated monomers. Hydrodynamic methods proved unsuccessful because of the strong tendency of the monomers to aggregate. Their molecular weights were $60,670 \pm 970$ (S. D.) and $52,100 \pm 940$ (S. D.) in good agreement with previous determinations,[11,12] particularly to a determination of the unfolded polypeptide chain in guanidine·HCl.[6]

The N-terminal sequences show clearly the different nature of the two subunits. From native Mo-Fe protein four pairs of amino acids were cleaved in the following sequence: Ser/Met, Glu/Leu, Asn/Asp,

Table 2. Properties of the apoprotein monomers of clostridial molybdenum-iron protein

Property	α-Monomer	β-Monomer	Ref.
Mol. wt			
SDS electrophoresis	$60,670 \pm 970$	$52,100 \pm 940$	
Gel filtration in 6 M guanidine·HCl	62,000	52,500	6
SDS electrophoresis of native Mo-Fe protein	$59,500 \pm 1940$ / 60,000	$50,700 \pm 1950$ / 51,000	11 / 12
N-Terminal sequence[a]	Ser-Glu-Asn-Leu	Met-Leu-Asp-Ala	
Isoelectric point[b]	5.9 ± 0.1	5.3 ± 0.08	
Tryptophanyl residues[c]	3	0	
$E_{1cm}^{1\%}$ (λ_{max})[d]	8.9 (278 nm)	6.8 (277 nm)	

[a]Determined by manual Edman degradation and identification of phenylhydantoin derivatives of amino acids by thin-layer chromatography.[14]
[b]Determined in 6 M urea and corrected by 0.42.[13]
[c]Determined by the method of Spies and Chambers.[15]
[d]Solvent 10 mM NaCl, pH ∿ 7.5.

and Leu/Ala, which agrees fully with the sequence of the isolated
monomers. The acidity of the Mo-Fe protein is reflected in the isoelectric points of its monomers. According to the investigation by
Ui,[13] the isoelectric point (pI) of a protein measured in 6 M urea
should be decreased by 0.42; in addition, the pI of native and denatured protein might be different. Both monomers precipitate at pH
values close to their pI.

By two independent methods, the three tryptophanyl residues of
the Mo-Fe protein were exclusively found in the α-monomer. This
situation allowed a clear spectrophotometric distinction between the
two monomers (see below), without resorting to more time-consuming
methods of differentiation. We had previously suggested that tryptophan may play a role in forming an active nitrogenase complex and in

Table 3.   Amino acid composition of clostridial molybdenum-iron
            protein and its subunits

AMINO ACID	MoFe-PROTEIN	60,000 SUBUNIT	50,000 SUBUNIT	SUM OF 2 SUBUNITS
LYS	76	47 (47)	31 (43)	78
HIS	23	16 (16)	9 (12)	25
ARG	30	15 (16)	13 (16)	28
CM-CYS[a]	14	9 (9)	7 (8)	16
ASP	104	61 (59)	41 (42)	102
THR[b]	58	33 (32)	30 (27)	63
SER[b]	52	26 (25)	25 (22)	51
GLU	103	60 (53)	50 (45)	110
PRO	44	26 (23)	27 (26)	'53
GLY	97	53 (54)	44 (41)	97
ALA	71	39 (40)	33 (29)	72
VAL[c]	76	43 (42)	32 (27)	75
MET	31	16 (18)	12 (12)	28
ILE[c]	82	47 (42)	37 (34)	84
LEU	63	28 (33)	32 (30)	60
TYR	37	20 (20)	16 (17)	36
PHE	36	20 (19)	18 (16)	38
TRP[d]	3	3 (ND)	0 (ND)	3
TOTAL	1000	562	457	1019
MOL.WT	107,355	61,913	49,546	111,459

Proteins were reduced with 2-mercaptoethanol and carboxymethylated
with iodoacetic acid.[16] They were dialyzed against 0.01 M NH_4HCO_3
and lyophilized. Protein samples were hydrolyzed with 6 N HCl for
24 and 72 h; the amino acid composition was determined according to
the method of Spackman et al.[17] Calculations were based on molecular
weights of 110 000, 60 000, and 50 000 for the Mo-Fe protein,
α-monomer, and ß-monomer, respectively. The values in parentheses
are from ref. 10.
[a]As carboxymethyl-cystein.
[b]Values extrapolated to zero time.
[c]Values from 72 h hydrolysis.
[d]Values from hydrolysis with 3 N mercaptoethanesulfonic acid.[18]

determining perhaps the incompatibility with a heterologous Fe protein.[5] Although there is no evidence for that yet, it remains highly interesting to note that, of the three clostridial nitrogenase polypeptides, only one, i.e., the α-monomer, contains tryptophan. The amino acid composition of the Mo-Fe protein monomers is shown in Table 3. The sum of amino acid residues of the monomers agrees very well with the composition of the native Mo-Fe protein. A high degree of agreement is also found in the composition of the α-monomer with that of a previous determination[10] which, however, does not extend to the ß-monomer. The similarity of the monomers in their amino acid composition was pointed out previously with difference indices below ten.[9,10]

A tryptic digestion of a carboxymethylated and maleylated sample of the α-monomer separated on a column of Bio-Gel P 30 (3 x 240 cm) into a minimum of 13 peaks. Two peptides thereof were purified and their amino acid composition was as follows: peptide A, Arg (1), Thr (2), Pro (1), Gly (1), Ala (1), Val (1), and Ile (1); peptide B, Arg (1), Asp (2), Glu (1), Gly (1), and Tyr (1).

IMMUNOREACTION AND INHIBITION OF NITROGENASE REACTIONS BY ANTISERA AGAINST THE Mo-Fe PROTEIN MONOMERS

Antibody-antigen interactions might prove a useful tool in elucidating part of the conformational requirements of particular nitro-

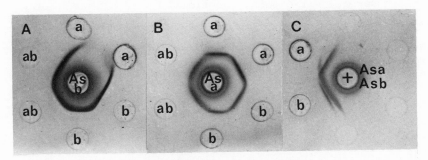

Fig. 3.   Immuno double diffusion patterns of Mo-Fe protein, its monomers, and rabbit gamma globulin against the monomers. Symbols: ab, Mo-Fe protein; a, α-monomer; b, ß-monomer; Asa and Asb are the gamma globulin fractions against the α-monomer and the ß-monomer, respectively. For each monomer, two white New Zealand rabbits were immunized with a protein emulsion in Freund's complete adjuvans. The sera of the two rabbits were mixed and the gamma globulin fraction was isolated by repeated $(NH_4)_2SO_4$ precipitation. The agar wells were charged with 10 µg antigen in 5 µl; the center well contained 10 µl immunoglobulin or 5 µl of each globulin in part C. Diffusion time was 30 h at 25°C; staining with 0.2% acid fuchsin.

genase reactions. For that purpose, rabbit antisera were obtained for each monomer. The antigens used for immunization were pure subunits with no complementary subunit present by the SDS electrophoresis criterion. The antiserum against the ß-monomer was antigen specific in immuno double diffusion tests (Fig. 3A); only the ß-monomer and native Mo-Fe protein were precipitated. The antiserum against the α-monomer, however, did not exhibit the analogous reaction. This antiserum precipitated the Mo-Fe protein and both monomers (Fig. 3B). A pattern of fusion indicated immunological relatedness between the two monomers which is interesting in the light of the above mentioned similarity in their amino acid composition. Immuno double diffusion of the monomers against a mixture of the two antisera showed the presence of two precipitin systems for the ß-monomer, one of which gave an intersecting pattern with the precipitin system of the α-monomer (Fig. 3C).

The antisera were also tested for their capability to inhibt partial reactions of nitrogenase. As shown in Table 4, the antisera inhibited ATP hydrolysis with three different substrates and concomitant product release to the same extent. When antisera were used in 15-fold excess over the equivalence point, the antiserum against the α-monomer inhibited nitrogenase reactions by approx. 40%, whereas the antiserum against the ß-monomer inhibited 20% or less. Only the anti-

Table 4. Effect of antisera against the Mo-Fe protein monomers on several nitrogenase reactions

Substrates & products		% Inhibition of product release in the presence of antiserum against the	
		α-monomer	ß-monomer
C_2H_2	P_i	38.2 ± 2.0 (3)	17.6 ± 1.3 (3)
	C_2H_4	40.9 ± 6.6 (3)	18.7 ± 3.9 (3)
H^+	P_i	40.2 ± 4.5 (3)	12.0 ± 3.4 (3)
	H_2	33.8 (1)	18.0 (1)
N_2	P_i	47.9 ± 8.4 (2)	20.5 ± 1.1 (2)

The reaction mixture contained 0.4 mg Mo-Fe protein and 0.7 mg Fe protein in a total volume of 1.5 ml. Purified gamma globulin was added in a 15-fold excess over the equivalence point to the Mo-Fe protein and incubated together for 50 min. The reaction was started by the addition of the Fe protein. The atmosphere was Ar or N_2; all gasses were measured by gas chromatography; other analytical details are as described in ref. 22. ( ) Number of experiments.

serum against the α-monomer showed a clear concentration dependency of its inhibitory action. Its inhibition could not be enhanced by addition of the antiserum against the ß-monomer. Possible explanations of the differential inhibition of the two antisera might be a localization of the binding site for the Fe protein on the α-monomer, or a greater proximity of the antibody-binding site on the α-monomer to the catalytical site. This latter argument is based on the premise

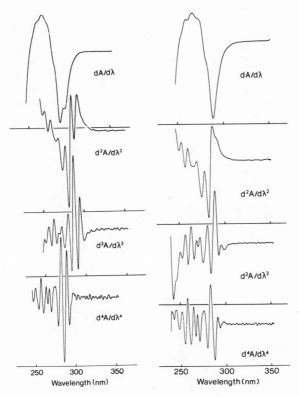

Fig. 4.   Ultraviolet derivative spectra of the α and ß-monomers of clostridial Mo-Fe protein. From top to bottom, first, second, third, and fourth derivative spectrum; the left part shows the α-monomer, the right part the ß-monomer. Note that the spectra in the left part of the figure are not vertically aligned with respect to the abscissa. The marker on each axis refers from left to right to 250, 300, and 350 nm. The derivative spectra were obtained with a dual wavelength spectrophotometer, equipped with a microprocessor for electronic analog differentiation. Differentiation was done on a time basis with constant scan speed (30 nm/min) and fixed band pass (2 nm). All spectra were recorded in cuvettes with 1 cm light path, which were thermostated at 21°C. For all recordings Δλ was 5 nm.

that the substrate-binding and -reducing site would be more susceptible to conformational restrictions induced by antibody binding.
Within the margin of experimental error, it was not possible to show
a shift of the $P_i/2e$ ratio under the action of antibodies as had
been indicated with an antiserum against the Mo-Fe protein from
Klebsiella.[19]

DERIVATIVE SPECTROPHOTOMETRY OF HIGHER ORDER OF THE Mo-Fe PROTEIN AND ITS SUBUNITS

Spectrophotometry has proven of limited use in studies of nitrogenase due to the featureless absorption spectrum of the enzyme. Even
redox changes of the protein show only subtle changes in absorbance
which were, however, exploited for several kinetic studies with the
Fe protein by Thorneley and colleagues.[20] The absence of specific
features extends to the ultraviolet part of the protein spectrum,

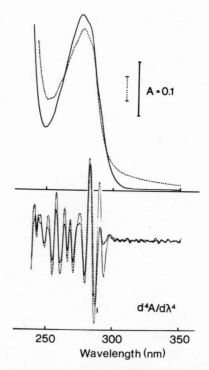

Fig. 5.   Comparison of ultraviolet absorption spectra and their
          fourth derivative of the clostridial Mo-Fe protein monomers.
          Solid line, α-monomer; dotted line, ß-monomer. The protein
          concentration, estimated by the Lowry procedure, was 0.465
          mg per ml for the α-monomer, and 0.8 mg per ml for the
          ß-monomer; solvent was 0.01 M HCl. Spectrophotometer set-
          tings as for Fig. 4.

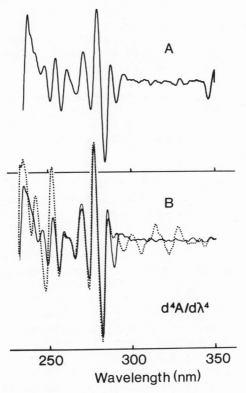

Fig. 6.   Effect of denaturing agents on the Mo-Fe protein. A) Spec-
          trum of native Mo-Fe protein in 15 mM anaerobic citric acid,
          83 mM in NaCl, and 8.3 mM in Tris-HCl; the pH was 2.7. The
          spectrum was recorded 10 min after protein addition to the
          solvent. B) Spectra of native Mo-Fe protein in 6 M urea,
          50 mM Tris-HCl, pH 7.5; dotted line, record of the spectrum
          10 min after protein addition to the solvent; solid line,
          the same solution after 8 h incubation at 23°C. All runs
          were made under anaerobic conditions; protein concentration,
          0.26 mg per ml; scan speed 60 nm/min; all spectra show the
          fourth derivative.

which consists of a broad band due to absorbance by tryptophan and
tyrosine and, to a lesser extent, by phenylalanine. The absorption of
the latter amino acid strongly overlaps with those of the two other
aromatic amino acids.

     In contrast, derivative spectra of higher order result in a much
better resolution of small absorption maxima and shoulders, although
the information in the original spectrum is not increased. Inflexion
points are converted to maxima in derivative spectra of odd order,
with overlapping bands, thus, being resolved. The number of extrema

rises by n+1 with the derivative order. As a result, the protein
spectrum is gradually transformed into a series of bipolar signals, a
process which is virtually complete with the fourth derivative.[21]
These effects are shown for both Mo-Fe protein monomers in Fig. 4.
Derivative spectrophotometry also allows an easy discrimination of
the two Mo-Fe protein monomers. In Fig. 5, their ultraviolet spectra
are compared with their fourth derivative spectra. The contribution
of tryptophan to the absorption spectrum of the α-monomer is barely
noticeable as a small shoulder, whereas the derivative spectrum possesses a separate band which is absent from the ß-monomer.

     Application of derivative spectrophotometry of higher order to
follow solvent perturbation of the ultraviolet spectrum of the Mo-Fe
protein indicates at the present state of experimentation the possibility of detecting redox and solvent-dependent conformations, which
might eventually also be followed kinetically. It was also of considerable interest to compare the denaturing effect of urea with that of
acid, since the latter has not only been used to dissociate the
Mo-Fe protein,[6] but also to liberate the molybdenum from the protein.[23] Hydrochloric acid and 6 M urea influenced the spectrum of the
Mo-Fe protein nearly indistinguishable. However, the spectrum in
citric acid was qualitatively different from that in 6 M urea. A
short citric acid treatment of the Mo-Fe protein and organic solvent
extraction has yielded a molybdenum-iron complex with acetylene-reducing activity,[24] whereas HCl treatment yielded derivatives of thio-
molybdates.[23] Only a prolonged incubation of the Mo-Fe protein in
urea converted the spectrum to that instantaneously obtained with
citric acid. We, therefore, suggest that citric acid specifically unfolds domains of the Mo-Fe protein which are relevant to binding of
the molybdenum cofactor, and these domains are rather resistent to
denaturation by urea. We are further tempted to suggest that the band
at 290 nm seen immediately in the citric acid-treated protein, which
is initially absent from the urea-treated protein, is due to tryptophan. Immunoreactivity, tryptophan distribution, and solvent perturbation have thus all singularized the α-monomer; whether this points
to a particular role in catalysis has yet to be examined with other
techniques.

ACKNOWLEDGEMENTS

     W. G. Z. wishes to thank Dr. W. Fekl of Pfrimmer & Co. for pro-
viding access to their animal test station, and Mr. G. Hünnebeck for
his help with the immunization procedure. Thanks are extended to Miss
S. Mümmler and Mrs. H. Edel for excellent technical assistance and to
Mrs. E. Weitemeyer for taking the photographs. The work of W. G. Z.
was supported by a grant from the Deutsche Forschungsgemeinschaft.

REFERENCES

1. I. M. Klotz, D. W. Darnall, and N. R. Langerman, Quarternary structure of proteins, in:"The Proteins," H. Neurath, R. L. Hill, and C.-L. Boeder, eds., vol. 1, p. 294, Academic Press, New York (1975).
2. W. G. Zumft, G. Palmer, and L. E. Mortenson, Electron paramagnetic resonance studies on nitrogenase. II. Interaction of adenosine 5'-triphosphate with azoferredoxin, *Biochim. Biophys. Acta*, 292:413 (1973).
3. W. H. Orme-Johnson, W. D. Hamilton, T. Ljones, M.-Y. W. Tso, R. H. Burris, V. K. Shah, and W. J. Brill, Electron paramagnetic resonance of nitrogenase and nitrogenase components from *Clostridium pasteurianum* W5 and *Azotobacter vinelandii, Proc. Nat. Acad. Sci. USA*, 69:3142 (1972).
4. B. E. Smith, D. L. Lowe, and R. C. Bray, Studies by electron paramagnetic resonance on the catalytic mechanism of nitrogenase of *Klebsiella pneumoniae, Biochem. J.*, 135:331 (1973).
5. W. G. Zumft, The molecular basis of biological dinitrogen fixation, *Structure and Bonding*, 29:1 (1976).
6. T. C. Huang, W. G. Zumft, and L. E. Mortenson, Structure of the molybdoferredoxin complex from *Clostridium pasteurianum* and isolation of its subunits, *J. Bacteriol.*, 113:884 (1973).
7. C. Kennedy, R. R. Eady, E. Kondorosi, and D. K. Rekosh, The molybdenum-iron protein of *Klebsiella pneumoniae* nitrogenase. Evidence for non-identical subunits from peptide 'mapping', *Biochem. J.*, 155:383 (1976).
8. R. H. Swisher, M. L. Land, and F. J. Reithel, The molecular weight of, and evidence for two types of subunits in, the molybdenum-iron protein of *Azotobacter vinelandii* nitrogenase, *Biochem. J.*, 163:427 (1977).
9. D. J. Lundell and J. B. Howard, Isolation and partial characterization of two different subunits from the molybdenum-iron protein of *Azotobacter vinelandii* nitrogenase, *J. Biol. Chem.* 253:3422 (1978).
10. J.-S. Chen, J. S. Multani, and L. E. Mortenson, Structural investigation of nitrogenase components from *Clostridium pasteurianum* and comparison with similar components of other organisms, *Biochim. Biophys. Acta,* 310:51 (1973).
11. G. Nakos and L. E. Mortenson, Molecular weight and subunit structure of molybdoferredoxin from *Clostridium pasteurianum* W5, *Biochim. Biophys. Acta*, 229:431 (1971).
12. M.-Y. W. Tso, Some properties of the nitrogenase from *Clostridium pasteurianum*. Molecular weight, subunit structure, isoelectric point, and EPR-spectra, *Arch. Microbiol.*, 99:71 (1974).
13. N. Ui, Isoelectric points and conformations of proteins. I. Effect or urea on the behaviour of some proteins in isoelectric focusing, *Biochim. Biophys. Acta,* 229:567 (1971).

14. A. Niederwieser, Thin-layer chromatography of amino acids and derivatives, *in*:"Methods in Enzymology", C. H. W. Hirs and S. N. Timashedd, eds., vol. 25, p. 60, Academic Press, New York (1972).

15. J. R. Spies and D. C. Chambers, Chemical determination of tryptophan in proteins, *Anal. Chem.*, 21:1249 (1949).

16. A. M. Crestfield, S. Moore, and W. H. Stein, The preparation and enzymatic hydrolysis of reduced and S-carboxymethylated proteins, *J. Biol. Chem.*, 238:622 (1963).

17. D. H. Spackman, W. H. Stein, and S. Moore, Automatic recording apparatus for use in the chromatography of amino acids, *Anal. Chem.*, 30:1190 (1958).

18. B. Penke, R. Ferenczi, and K. Kovács, A new acid hydrolysis method for determining tryptophan in peptides and proteins, *Anal. Biochem.*, 60:45 (1974).

19. R. J. Rennie, A. Funnell, and B. E. Smith, Immunochemistry of nitrogenase as a probe for the enzyme mechanism. Evidence for multiple enzyme forms and an $MgATP^{2-}$ binding site on the MoFe protein, *FEBS Letters*, 91:158 (1978).

20. R. R. Eady, D. J. Lowe, and R. N. F. Thorneley, Nitrogenase of *Klebsiella pneumoniae*: A pre-steady state burst of ATP hydrolysis is coupled to electron transfer between the component proteins, *FEBS Letters*, 95:211 (1978).

21. G. Talsky, L. Mayring, and H. Kreuzer, Feinauflösende UV/VIS Derivativspektrophotometrie höherer Ordnung, *Angew. Chem.*, 90:840 (1978).

22. W. G. Zumft and L. E. Mortenson, Evidence for a catalytic centre heterogeneity of molybdoferredoxin from *Clostridium pasteurianum*, *Eur. J. Biochem.*, 35:401 (1973).

23. W. G. Zumft, Isolation of thiomolybdate compounds from the molybdenum-iron protein of clostridial nitrogenase, *Eur. J. Biochem.*, 91:345 (1978).

24. V. K. Shah, J. R. Chisnell, and W. J. Brill, Acetylene reduction by the iron-molybdenum cofactor from nitrogenase, *Biochem. Biophys. Res. Commun.*, 81:232 (1978).

HD FORMATION BY NITROGENASE: A PROBE FOR N_2 REDUCTION
INTERMEDIATES

Barbara K. Burgess, Scot Wherland, Edward I. Stiefel
and William E. Newton

Charles F. Kettering Research Laboratory
150 East South College Street
Yellow Springs, Ohio 45387, USA

INTRODUCTION

Nitrogenase is a complex of two separately purifiable proteins,
the molybdenum-iron protein [MoFe] and the iron protein [Fe].[1,2]
In addition to catalyzing the reduction of N_2 to ammonia, nitrogenase has ATP-hydrolyzing activity,[3] ATP-dependent H_2-evolution
activity,[4] and supports a reaction between D_2 and protons (from
H_2O) to form HD.[5] In the absence of other substrates, all the
reductant consumed is used to reduce protons to H_2. When N_2 is
added as a substrate, an apparent maximum of 75% of the electrons
reduce N_2 while the remainder still reduce protons.[6]

Early studies on the inhibition of nitrogenase-catalyzed
reactions identified H_2 as a specific competitive inhibitor of N_2
reduction.[7-10] Formation of HD from D_2 and protons by excised
soybean nodules was first demonstrated by Hoch et al. and was found
to be dependent on N_2.[5] This result was confirmed and extended to
other N_2-fixing organisms.[11] Later, HD formation was observed in
cell-free extracts of Azotobacter vinelandii and was shown to
require N_2, ATP and $Na_2S_2O_4$. In addition, CO, an inhibitor of N_2
fixation but not of H_2 evolution, was shown to be a potent inhibitor
of HD formation.[12] These results were confirmed using crude, cell-
free preparations from several organisms[13-14] with only the N_2
requirement for HD formation being in dispute.[15,16]

Preliminary data from this laboratory, obtained using purified
azotobacter nitrogenase complex, confirmed the requirement for N_2
and showed that HD formation is a one-electron requiring
process.[9,17,19] To explain these observations, it has been

suggested that H_2 inhibition of nitrogen fixation, and HD formation under D_2 and N_2 are manifestations of the same molecular process in which a bound diimide-level species is postulated as an intermediate in the reduction of N_2 to ammonia.[20,21] This mechanism suggests hydrazine as the next intermediate in N_2 reduction, and in fact, hydrazine has been identified as a substrate of nitrogenase in preliminary studies.[17]

Further experiments on HD formation and H_2 inhibition using the purified component proteins of nitrogenase are reported here. These new data establish the following: (i) HD formation is a property of the purified component proteins; (ii) HD formation is largely but not entirely dependent on N_2; (iii) D_2 does not exchange with H_2O; (iv) one electron is required for each HD formed; and, (v) hydrazine is a substrate of nitrogenase which is not sensitive to H_2 inhibition and does not lead to HD formation under D_2. Finally, the effect of varying the ratio of [Fe] to [MoFe] on HD formation has been examined and is discussed.

HD FORMATION BY PURE NITROGENASE COMPONENTS

[Fe] and [MoFe] were purified from A. vinelandii by a method developed in our laboratory.[22] Both components are homogeneous by the criterion of SDS gel electrophoresis and are devoid of uptake hydrogenase activity. The specific activity of [Fe] is 1950-2100 nmoles H_2 evolved/min/mg[Fe] and of [MoFe] is 2800-3000 nmoles H_2/min/mg[MoFe]. Our previously reported experiments using pure [Fe] and [MoFe] have shown that HD is a major product of nitrogenase turnover under 50% D_2:40% N_2:10% Ar and that HD formation follows a similar smooth variation with changes in the ratio of [Fe] to [MoFe] as do H_2 evolution and NH_3 formation.[23] We have now shown that no HD formation occurs with either [Fe] or [MoFe] alone or with both components present if either MgATP or $Na_2S_2O_4$ is absent. These results firmly establish HD formation as a reaction which occurs during nirogenase turnover and confirm that previous results were not due to contaminants in nitrogenase preparations.

ELECTRON REQUIREMENT FOR HD FORMATION

In previous studies,[24] nitrogenase turnover was measured as the rate of either $Na_2S_2O_4$ utilization or ATP hydrolysis and was shown to be independent of the substrate(s) being reduced (at least for N_2 and C_2H_2 reduction and H_2 evolution). In Table 1, we show that the rates of $Na_2S_2O_4$ utilization, measured at the molar ratio [Fe]/[MoFe] of 5, are the same under H_2-evolving conditions, N_2-fixing conditions or H_2-inhibited N_2-fixing conditions. Thus, if electron flow through nitrogenase under D_2-inhibited N_2-fixing conditions is measured as a function of the partial pressure of

Table 1.   Dithionite Utilization Studies.[a]

GAS	μmoles e$^-$/min/mg[b]
100% Ar	1.20 ± 0.06
100% N$_2$	1.04 ± 0.10
40% N$_2$:60% Ar	1.05 ± 0.096
40% N$_2$:10% Ar:50% H$_2$	1.05 ± 0.12

[a]Dithionite utilization was measured using a
polarographic technique.[24]   The 2.5 ml reaction
mixture contained 38mM TES/KOH, pH 7.4,
2.5mM ATP, 5mM MgCl$_2$, 30mM creatine phosphate,
20mM neutralized Na$_2$S$_2$O$_4$, 20mM NaCl and 0.313mg
creatine phosphokinase. Reactions were run at
30°C for 10 min following equilibration with
the appropriate gas phase.
[b]2.5mg total protein ([MoFe] plus [Fe] at a 5:1
molar ratio).

N$_2$(pN$_2$), a constant total electron flow is expected as pN$_2$ increases.
Figure 1 shows that if two electrons are allocated for each H$_2$
evolved and three electrons are allocated for each ammonia formed
but any electron requirement for HD formation is ignored, total
electron flow decreases as pN$_2$ increases. If, however, one electron
is allocated for each HD formed, then the total electron flow is
shown to be constant as it must be to agree with the Na$_2$S$_2$O$_4$ utilization data. These electron balance studies confirm the conclusion
drawn from our earlier work on the azotobacter complex[18] as well as
on purified [Fe] and [MoFe][23] namely, one electron is required for
every HD formed.

The electron requirement for HD formation rules against a reversible exchange process (equation 1) as a possible mechanism for
HD formation.

$$D_2 + H_2O \rightleftharpoons HD + HOD \tag{1}$$

Our results lead to the prediction that while HD is formed in the
gas phase HOD is not formed concomitantly in the aqueous phase. A
more direct way to test this prediction of the absence of a simple
exchange mechanism involves the study of nitrogenase turnover
under 50% H$_2$ (labeled with T$_2$):40% N$_2$:10% Ar wherein T$^+$ incorporation

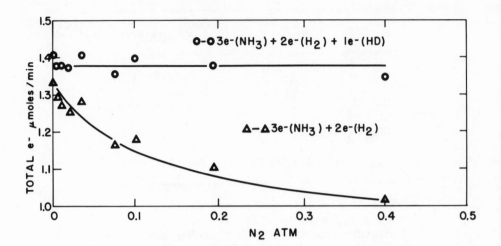

Fig. 1. Plots of total electron flow under 0.5 atm D_2 as a function
of N_2 pressure with one electron (O-O) and zero electrons
(Δ-Δ) assigned for each HD formed. Assay conditions are
as described in the legend to Table I using a 1ml
reaction mixture. HD and H_2 were measured on a
Finnigan Mass Spectrometer and ammonia by the Chaykin
method. Data are taken from Fig. 2 and points are
means of sextuplicate determinations with all products
measured on the same reaction vial.

into the aqueous phase can be monitored. When this experiment was
performed at [Fe]/[MoFe] of 5, only 5.7 ± 1.0 nmoles H^+ (labeled with
T^+)/min/mg total protein were found to be incorporated into the
H_2O. Using identical assay conditions, but under 50% D_2: 40% N_2:
10% Ar, 266 ± 14 nmoles HD/min/mg total protein were formed. Thus,
>98% of the HD formed is attributable to a mechanism other than
a reversible exchange involving the solvent.

N_2 DEPENDENCE OF HD FORMATION

Figure 2 shows the rates of H_2 evolution and ammonia and HD
formation as functions of pN_2. As pN_2 increases, H_2 evolution
decreases while both ammonia and HD formation increase. This figure
shows that although the majority of the HD formed is N_2 dependent,
there is a rather significant amount of HD formed even in the absence
of N_2. The N_2-independent HD formation was studied further by
measuring HD formation under 100% D_2 or 50% D_2: 50% Ar as a function
of the ratio [Fe]/[MoFe]. Twenty ratios ranging from 0.1 to 130
were used. We observed that, although both H_2 evolution and HD
formation show a component ratio dependence in these experiments,

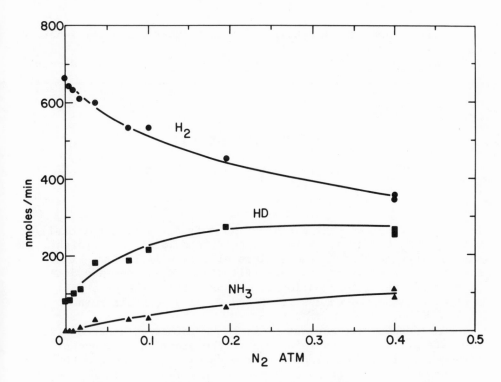

Fig. 2.   Plots of the rates of NH₃ (▲), HD (■) and H₂ (●) produc-
            tion as a function of varying N₂ pressure. Assay
            conditions etc. are as in the legend to Fig. 1 with
            a 1 ml reaction mixture under 50% D_2, balance N_2 plus Ar.

the percentage of total electron flow that is used to form HD is not
dependent on the [Fe]/[MoFe] ratio, but is constant. For 50% D_2:
50% Ar, it is 6%. This observation is in contrast to similar
experiments under 50% D_2: 40% N_2: 10% Ar, where the percentage of
total electron flow for N_2-dependent HD formation showed a marked
dependence upon component ratio.

        These findings lead us to hypothesize that, at a given level of
D_2, a constant percentage of the total electron flow destined to form
H_2 is intercepted by that D_2 to form HD* (N_2-independent HD formation) and a constant percentage of total electron flow destined to
form ammonia is intercepted by D_2 to form HD** (N_2-dependent HD
formation). This hypothesis, along with the other quantitative
results that we have observed, leads directly to formulation of
equation 2.

If $HD_{total} = HD* + HD**$

and $F = HD*/[HD* + 2(H_2)]$, $F' = HD**/[HD** + 3(NH_3)]$ (where F

and F' are expressed as fractions of the appropriate electron flow)

and $\varepsilon = HD_{total} + 2(H_2) + 3(NH_3)$

(where ε is the total electron flow)

then
$$\frac{HD_{total}}{\varepsilon} = \frac{H_2}{\varepsilon} 2\frac{(F-F')}{1-F} + F' \qquad (2)$$

Equation 2 predicts that a plot of HD(total)/ε versus H_2(total)/ε, using the N_2-dependence data in fig. 2, should give a straight line with an intercept of F' and a slope of $2[(F-F')/(1-F)]$. Figure 3 shows a least squares computer fit of our N_2-dependence data plotted according to equation 2. As predicted, the data fall on a straight line and values for F and F' have been determined.

Our results are consistent with the formation of HD by two separate pathways. Under 50% D_2, in the minor N_2-independent pathway, D_2 redirects electrons from H_2 evolution to form HD, with 6% of the electrons being utilized in this manner. In the major, N_2-dependent pathway under 50% D_2, D_2 redirects electrons from ammonia formation to form HD. Under these conditions, 50% of the electrons destined to form ammonia are lost to HD formation. In other experiments, H_2 evolution and ammonia formation were monitored as functions of the [Fe]/[MoFe] ratio under N_2-fixing conditions (60% Ar: 40% N_2) and $D_2(H_2)$-inhibited conditions (10% Ar; 40% N_2: 50% $D_2(H_2)$). These data showed that 50% less ammonia is formed under H_2-inhibited N_2-fixing conditions when compared to N_2-fixing conditions for the same pN_2 in the absence of H_2. A comparison of the results from these sets of experiments reveals that N_2-dependent HD formation and H_2 inhibition of nitrogen fixation occur by the same stoichiometric mechanism.[25]

MECHANISM FOR N_2-DEPENDENT HD FORMATION

The stoichiometry of N_2-dependent HD formation (H_2 inhibition) is:

$$E + N_2 + 2H^+ + 2e^- + D_2(H_2) \longrightarrow E + N_2 + 2HD(2H_2) \qquad (3)$$

Combining the terms $[E + N_2 + 2H^+ + 2e^-]$ implies the presence of a

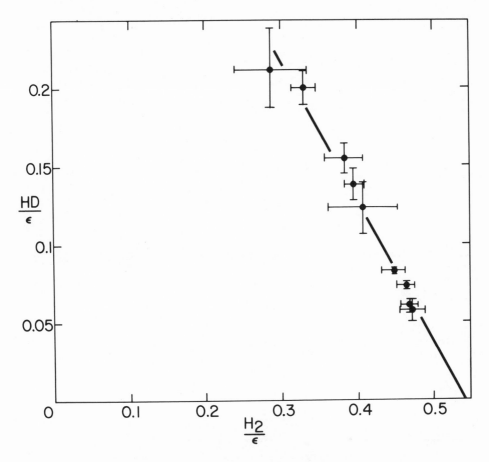

Fig. 3. Least squares computer fit of data shown in Fig. 2 plotted according to equation 2. Calculated values for F = 0.062±.01 and F' = 0.49±.03.

bound diimide-level species which reacts with D_2 or H_2 during the reduction of N_2 to ammonia. Our new data are consistent with this hypothesis, which was first formulated from earlier experiments on the azotobacter complex.[17,18,20,21] Thus, under H_2-inhibited conditions, one H_2 molecule reacts with a diimide-level species, e.g., E-N_2H_2, to form two H_2 molecules and reform N_2. Because one H_2 molecule is consumed in this reaction and two H_2 molecules are formed, the net gain is one H_2 molecule formed from the two protons and two electrons initially in the diimide-level species. Thus, we actually measure two electrons per one H_2 evolved with the H_2 formed in this manner being indistinguishable from normal ATP-dependent H_2 evolution. Under D_2-inhibited conditions, D_2 reacts with the same intermediate to generate 2HD and N_2, but now, as both molecules of product (HD) are recognizably different from the reactant (D_2) and from H_2 evolved by normal ATP-dependent hydrogen evolution, we actually measure two electrons per two HD molecules formed. Equations 4-8 illustrate this point.

For H_2 inhibition, we have

$$E + N_2 + 2H^+ + 2e^- \longrightarrow E\text{-}N_2H_2 \tag{4}$$

$$E\text{-}N_2H_2 + H_2 \longrightarrow E + N_2 + 2H_2 \tag{5}$$

and by combining equations (4) and (5) and cancelling, we have

$$2H^+ + 2e^- \longrightarrow H_2 \tag{6}$$

that is, two electrons are supplied by nitrogenase for each net H_2 evolved.

For D_2 inhibition (i.e., HD formation), we have

$$E + N_2 + 2H^+ + 2e^- \longrightarrow E\text{-}N_2H_2 \tag{4}$$

$$E\text{-}N_2H_2 + D_2 \longrightarrow E + N_2 + 2HD \tag{7}$$

$$\overline{2H^+ + 2e^- + D_2 \longrightarrow 2HD} \tag{8}$$

that is, two electrons are supplied by nitrogenase to produce two molecules of HD. It may well be significant from an efficiency standpoint that, under 50% $H_2(D_2)$ atmosphere, 50% of the electrons that should be used to reduce N_2 are intercepted in this fashion.

HYDRAZINE REDUCTION BY NITROGENASE

The reduction of N_2 to a diimide-level intermediate implies a mechanism of reduction involving two-electron steps. Carrying this argument one step further leads to the postulation of a hydrazine-level species as the next reduction intermediate. In fact, in pre-liminary studies, Bulen demonstrated that hydrazine is a substrate of nitrogenase and is reduced to ammonia.[17] We have confirmed hydrazine as a substrate of nitrogenase by monitoring hydrazine reduction and H_2 evolution as functions of the [Fe] to [MoFe] ratio under 100% Ar with 30mM hydrazine present. In addition, we have shown that hydrazine is not reduced by either [Fe] or [MoFe] alone, nor with both components present if either MgATP or $Na_2S_2O_4$ is absent and that 30mM hydrazine does not effect total electron flow through nitrogenase. Like N_2 reduction but unlike H_2 formation, hydrazine

Table 2. Hydrazine Reduction Under H_2 and D_2

CONDITIONS[a]	NH_3[b]		HD[b]	
100% Ar + 30mM HYDRAZINE	82.1	5.2	–	
100% H_2 + 30mM HYDRAZINE	81.7	5.3	–	
50% D_2: 50% Ar			83	1.7
50% D_2: 50% Ar + 30mM HYDRAZINE			81	3.2
50% D_2: 10% Ar: 40% N_2			266	14.0

[a]1mg total protein ([MoFe] plus [Fe] at a 5:1 molar ratio) in a 1 ml reaction volume. Assay conditions are described in the legend to Table I. Ammonia was determined with the Chaykin method after microdiffusion. HD was measured by a Finnigan Mass Spectrometer.
[b]nmoles produced per min per mg total protein

reduction is completely inhibited by 1% CO.

These data might suggest that hydrazine, as an intermediate in the reduction of N_2 to ammonia, could possibly be responsible for H_2 inhibition and HD formation from D_2, in contrast to the original hypothesis which suggested a bound diimide-level intermediate as being responsible for these reactions. However, Table 2 shows that hydrazine reduction to ammonia is not inhibited by H_2 and does not effect HD formation under D_2. Thus, if hydrazine is an intermediate in the N_2 to ammonia pathway, then H_2 inhibition and HD formation reactions must occur at an earlier reduction intermediate in this pathway, i.e., at the diimide-level.

CONCLUSIONS

The data presented here and in previous experiments[17,18,20,21] demonstrate unequivocally that a simple exchange mechanism cannot be evoked to explain HD formation and that all HD formed requires one electron per molecule. HD is formed by two separate pathways, a minor N_2-independent pathway and a major N_2-dependent pathway. H_2 inhibition of nitrogen fixation and N_2-dependent HD formation are manifestations of the same stoichiometric reaction which can be explained by the presence of a bound diimide-level intermediate in the reduction of N_2 to ammonia. This diimide-level intermediate suggests two-electron reduction steps which in turn implicates hydrazine as the next reduction intermediate. Hydrazine is, in fact, a substrate of nitrogenase reducible to ammonia in a reaction which is not sensitive to the presence of the isotopes of H_2 in the gas phase.

Recently, Thorneley et al. have produced independent evidence for a bound dinitrogen-hydride intermediate in the reduction of N_2 to ammonia.[26] It is important to note that the formation of this intermediate appears to be inhibited by H_2 and it will be interesting to discover the relationship between this intermediate and the ones we have postulated.

ACKNOWLEDGEMENT

S.W. is supported by a NSF National Needs Postdoctoral Fellowship. This manuscript constitutes Contribution No. 668 from the Charles F. Kettering Research Laboratory.

REFERENCES

1.   W.A. Bulen and J.R. LeComte, The nitrogenase system from
     Azotobacter: Two enzyme requirement for N_2 reduction,
     ATP dependent H_2 evolution and ATP hydrolysis, Proc.
     Acad. Sci. USA 56:979 (1966).
2.   L.E. Mortenson, Components of cell-free extracts of Clostridium
     pasteurianum required for ATP-dependent H_2 evolution from
     dithionite and for N_2 fixation, Biochim. Biophys. Acta
     127:18 (1966).
3.   M.J. Dilworth, D. Subramanian, T.O. Munson and R.H. Burris,
     The adenosine triphosphate requirement for nitrogen fixation
     in cell-free extracts of Clostridium pasteurianum, Biochim.
     Biophys. Acta 99:486 (1965).
4.   W.A. Bulen, R.C. Burns and J.R. LeComte, Hydrosulfite as elec-
     tron donor with cell-free preparations of Azotobacter
     vinelandii and Rhodospirillum rubrum, Proc. Nat. Acad. Sci.
     USA 53:532 (1965).
5.   G.E. Hoch, K.C. Schneider and R.H. Burris, Hydrogen evolution
     and exchange, and conversion of N_2O to N_2 by soybean root
     nodules, Biochim. Biophys. Acta 37:273 (1960)
6.   J.M. Rivera-Ortiz and R.H. Burris, Interaction among substrates
     and inhibitors of nitrogenase, J. Bacteriol. 123:537 (1975).
7.   P.W. Wilson and W.W. Umbreit, Symbiotic nitrogen fixation.
     III. Hydrogen as a specific inhibitor, Arch. Mikrobiol.
     8:440 (1937).
8.   G.W. Strandberg and P.W. Wilson, Formation of the nitrogen-
     fixing enzyme system in Azotobacter vinelandii, Can. J.
     Microbiol. 14:25 (1967).
9.   K.L. Hadfield and W.A. Bulen, Adenosine triphosphate require-
     ments of nitrogenase from Azotobacter vinelandii, Biochemistry
     8: 5103 (1969)
10. J.C. Hwang, C.H. Chen and R.H. Burris, Inhibition of nitro-
     genase catalyzed reductions, Biochim. Biophys. Acta 292:256
     (1973).

11. F.J. Bergersen, The relationship between hydrogen evolution,
     hydrogen exchange, nitrogen fixation and applied oxygen tension
     of soybean root nodules, Aust. J. Biol. Sci. 16:669 (1963).
12. E.K. Jackson, G.W. Parshall and R.W.F. Hardy, Hydrogen reactions
     of nitrogenase: Formation of the molecule HD by nitrogenase
     and an inorganic model, J. Biol. Chem. 19:4952 (1968).
13. R.C. Burns and W.A. Bulen, ATP-dependent hydrogen evolution
     by cell-free preparations of Azotobacter vinelandii, Biochim.
     Biophys. Acta 105:437 (1965).
14. G.L. Turner and F.J. Bergerson, The relationship between N_2-
     fixation and the production of HD from D_2 by cell-free extracts
     of soybean nodule bacteroids, Biochem. J. 115:529 (1969).
15. M. Kelly, Hydrogen-deuterium exchange reactions catalyzed by
     nitrogenase, Biochem. J. 109:322 (1968).
16. M. Kelly, Comparison and cross reactions of nitrogenase from
     Klebsiella penumoniae, Azotobacter chroococcum and Bacillus
     polymyxa, Biochim. Biophys. Acta 191:527 (1969).
17. W.A. Bulen, Nitrogenase from Azotobacter vinelandii and
     reactions affecting mechanistic interpretation, in "Proc.
     1st Int. Symp. on Nitrogen Fixation", W.E. Newton and
     C.J. Nyman, eds., Washington State University Press, Pullman,
     Wash. (1976) p. 177.
18. W.E. Newton, W.A. Bulen, K.L. Hadfield, E.I. Stiefel and
     G.D. Watt, HD formation as a probe for intermediates in N_2
     reduction, in "Recent Developments in Nitrogen Fixation",
     W.E. Newton, J.R. Postgate and C. Rodriguez-Barrueco, eds.
     Academic Press, London, New York. (1978) p. 119.
19. W.A. Bulen and J.R. LeComte, Nitrogenase complex and its
     components, in "Methods in Enzymology", A. San Pietro, ed.,
     Academic Press, New York, London. (1972) Vol. XXIV, Part B
     p. 456.
20. W.E. Newton, J.L. Corbin and J.W. McDonald, Nitrogenase:
     Mechanism and Models, in, "Proc. 1st Int. Symp. on Nitrogen
     Fixation", W.E. Newton and C.J. Nyman, eds., Washington
     State University Press, Pullman, Wash. (1976) p. 53.
21. E.I. Stiefel, W.E. Newton, G.D. Watt, K.L. Hadfield and
     W.A. Bulen, Molybdoenzymes: The role of electrons, protons
     and dihydrogen, in, "Bio-inorganic Chemistry II", K.N. Raymond,
     eds., Advances in Chemistry Series No. 162, American Chemical
     Society Publications, Washington D.C. (1977) p. 353.
22. B.K. Burgess, D. Jacobs and E.I. Stiefel, Azotobacter vinelandii
     nitrogenase: Large-scale, anaerobic, high-yield, high-
     activity preparation using modified Schlenk techniques,
     submitted for publication.
23. E.I. Stiefel, B.K. Burgess, S. Wherland, W.E. Newton, J.L. Corbin
     and G.D. Watt, Azotobacter vinelandii Biochemistry: $H_2(D_2)/N_2$
     relationships of nitrogenase and some aspects of iron metabo-
     lism, in "Nitrogen Fixation", W.E. Newton and W.H. Orme-Johnson,
     eds., University Park Press, Baltimore (1979) in press.

24. G.D. Watt and A. Burns, Kinetics of dithionite ion utilization
     and ATP hydrolysis for reactions catalyzed by the nitrogenase
     complex from Azotobacter vinelandii, Biochemistry 16:265
     (1977).
25. E.I. Stiefel, The mechanisms of nitrogen fixation in, "Recent
     Developments in Nitrogen Fixation", W.E. Newton, J.R. Postgate
     and C. Rodriguez-Barrueco, eds., Academic Press, London
     (1978) p. 69.
26. R.N.F. Thorneley, R.R. Eady and D.J. Lowe, Biological nitrogen
     fixation by way of enzyme-bound dinitrogen-hydride inter-
     mediate, Nature 272:557 (1978).

SPECTROSCOPIC AND CHEMICAL EVIDENCE FOR THE NATURE AND ROLE OF METAL CENTERS IN NITROGENASE AND NITRATE REDUCTASE

W.H. Orme-Johnson, N.R. Orme-Johnson, C. Touton,
M. Emptage, M. Henzl, J. Rawlings, K. Jacobson,
J.P. Smith, W.B. Mims[a], B.H. Huynh[b], E. Münck[b],
and G.S. Jacob[c]

College of Agricultural & Life Sciences, Department of
Biochemistry, University of Wisconsin, Madison, WI
53706, USA
[a]Bell Laboratories, Murray Hill, NJ 07974, USA
[b]The Freshwater Biology Research Institute, University
of Minnesota, Navarre, MN 55392
[c]IBM Watson Laboratories, Yorktown, NY 10598, USA

INTRODUCTION

With the advent of methods for the determination of FeS cluster type and concentration in proteins,[1] at least within the framework of known cluster systems, and with the recognition that a molybdenum-containing iron-sulfur FeS cluster can be dissociated from nitrogenase,[2,3] much recent work has been directed at the metal centers of nitrogenase. It turns out that combined epr and Mössbauer spectroscopy has been crucial to our understanding of this problem. In particular, nitrogenase studies have shown the power of the Mössbauer technique to analyze an extremely complex iron-containing protein, under favorable circumstances. We have repeatedly reviewed studies on the structure and function of the enzyme;[4-6] here, we will discuss recent work and attempt to assemble a speculative picture of the functioning enzyme.

IRON PROTEIN

This entity contains a Fe_4S_4 core[7] ligated to a protein composed of two identical subunits. The protein binds two MgATP/ molecule, changing its conformation, according to changes in the epr spectrum.[9,10] The core structure becomes accessible to

bipyridyl[11] in the presence of MgATP but not MgADP, suggesting
that binding ATP may expose the core to the solvent. Examination
of the redox potential of Fe_4S_4 cores at various degrees of exposure to water[12] shows that increasing the proportion of water in
the system raises the potential, not lowers it as is expected
during energy transduction. Therefore, the significance of the
MgATP effect may be to expose the core to groups on the Mo-Fe
protein when the complex between the two proteins is formed. The
transducing event may be of an entirely separate nature. For
instance, the fact that substitution of exosulfur ligands on synthetic $(RS)_4Fe_4S_4$ clusters by acyl groups raises the potential[13]
suggests that the ATP (bound to the Fe protein) phosphorylates a
cluster in the Mo-Fe protein, raising its potential so that electron
transfer from the Fe protein cluster takes place spontaneously.
Hydrolysis of the phosphorylated cluster or displacement by mercap-
tide would then cause the potential to fall, conserving the hydrolysis energy as a low potential electron on the cluster, which
subsequently reduces the other prosthetic groups and then the
substrates of nitrogenase. This process is illustrated in a
speculative figure at the end of the paper. Since the Fe protein
alone binds MgATP but does not cause its hydrolysis, the hydrolytic
event is triggered by binding to the Mo-Fe protein.

Mo-Fe PROTEIN

    The Mo-Fe protein from A. vinelandii (Av1) seems to contain
ca. 32 Fe and 2 Mo per molecule, arranged in four subunits of the
$\alpha_2\beta_2$ form. The epr spectrum reveals two S = 3/2 center per
molecule and either oxidation by dyes or reduction by the Fe
protein results in the disappearance of the signal. In the latter
case, reduction occurs within the turnover time of electron transfer to substrates. Mössbauer studies show that only about 45% of
the iron in the Mo-Fe protein is in these magnetic centers,[14-16]
but epr spectra, taken during CO inhibition[17] or under suboptimal
turnover conditions,[18] give evidence of both -1 and -3 oxidation
states[19] of iron-sulfur clusters. Oxidative titrations[16] show
that six electrons can easily be removed but that only two of the
electrons come from the epr-active clusters. Electrochemical
studies[20,21] suggest that the epr clusters have a rather high
(near 0) standard potential. In Figure 1, the effects of solvent
perturbation on the epr spectrum are shown. N-methylformamide
(NMF) unfolds the protein and converts the S=3/2 spectrum (around
g = 4) into the broad form characteristic of the Fe-Mo cofactor.[3]
At higher solvent concentrations, a g = 1.94 signal characteristic
of iron-sulfur centers in the -3 oxidation state appears. The
differential effect suggests two separare populations of metal
centers, and also suggests that the two types of centers are in
different domains of the protein. In particular, the removal of
the M centers (the epr moiety of the native protein) does not

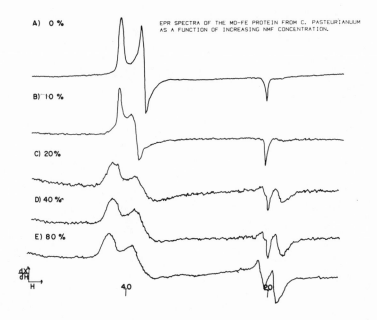

EPR SPECTRA OF THE MO-FE PROTEIN FROM C. PASTEURIANUUM
AS A FUNCTION OF INCREASING NMF CONCENTRATION.

A) 0 %

B) 10 %

C) 20%

D) 40 %

E) 80 %

Fig. 1.   Solvent perturbation of the epr spectrum of Cpl with
          graded amounts of NMF.

expose the P centers (vide infra) which require a higher solvent
concentration in order to appear in their epr-active form.

The M Centers

     The subpopulation of iron atoms in the Mo-Fe protein which
form the S = 3/2 epr entity were shown to be present in the
cofactor[2] extracted with NMF, according to epr and Mössbauer
spectroscopy.[3]   The cofactor can reactivate the apo-Mo-Fe protein
present in extracts of the A. vinelandii mutant UW-45.   By this
criterion and according to the preservation of the S = 3/2 epr
spectrum, the following reagents had no effect on the cofactor;
10 mM thiophenol, 1 mM bypyridyl, 5 M imidazole.   The first
reagent would be expected to displace[19] Fe_4S_4 or Fe_2S_2 cores from
derivatives, whereas the last two have been found to disrupt iron-
sulfur cores of either type, when present in either low molecular
weight derivatives or unfolded proteins.   These observations
suggested[3] the presence of a novel cluster structure outside present
concepts of biological iron-sulfur systems.   The Fe-Mo cofactor is
sensitive to RHgX, consistent with the presence of acid labile

sulfur.[2] Carbon monoxide binding was demonstrated when the strong
reductant system, 5-deazaflavin plus EDTA plus light, was used and
this CO binding was reversed in the dark.[3] The reversibility of
this reaction has been questioned.[22] We have found[3] that thiophenol appears to coordinate the cofactor according to changes
(narrowing) of its epr. We would also suggest that the special
property of NMF as an extractant of the cofactor (N-dimethylforma-
mide doesn't work) is due to the ability of NMF to mimic the protein
(peptide) milieu.

The Mössbauer studies[13],[15] on the Mo-Fe protein have recently
been focussed on the M-centers in situ.[20] Subtractive computer
techniques, allowing the removal of the contributions of the other
(P-centers, see below) iron atoms, has led to the following detailed
picture. The spectra of the S = 3/2 centers have been decomposed
into six subcomponents. Three of these shift appreciably in applied
magnetic fields, and three do not. Calculation of the effective
magnetic field suggests that, for hyperfine constants A > 0, terms
will add yielding an appreciable effect of external field, whereas
for A < 0, terms will cancel so that the Mössbauer features are
relatively unaffected in applied field. The observed result shows
that both kinds of atoms are present in the S = 3/2 cluster. We
cannot readily exclude 5 or 7 iron atoms on the basis of integra-
tions of Mössbauer spectra, but 6 is the most probable number. In
the Fe_2S_2 cores of chloroplast ferredoxins, the spin-coupled atoms
in the paramagnetic (-3) oxidation state are characterized by
hyperfine coupling constants of opposite signs, and the present
results can only mean that the S = 3/2 centers in nitrogenase are
spin-coupled structures.

The foregoing discussion concerns the iron in the cofactor.
What can one say concerning the molybdenum? EXAFS experiments[23]
suggest that 2 to 3 of the iron atoms are coupled by sulfur bridges
to the molybdenum atom. We have begun to examine the question using
pulsed epr spectroscopy.[24] When one measures the spin-echo amplitude as a function of the delay between echo-generating pulses, at
a suitably low temperature such that the spin phase memory time is
long ($\sim$ microseconds), the decay curve is found to be modulated by
nuclear frequencies. This technique has been applied to the study
of metal environments in heme and copper proteins.[25],[26] For purely
dipolar coupled spins (i.e, nuclear magnetic spins interacting
through space with the electron spin), the depth of the modulations
will scale as r-6 (I) (I+1), where r is the electron-nucleus
distance and I is the nuclear spin, at least for moderate interactions. Initial studies with [95]Mo- and [96]Mo-enriched Mo-Fe
protein from C. pasteurianum (Cp1) have been undertaken. The
echo decay curves are characterized by high frequency modulations
(due to nearby protons) and by lower frequency disturbances which
are emphasized in the [95]Mo spectrum. We wish to be extremely
cautious in interpreting these results at this early stage, but

three points are worth noticing. (a) The substitution of ^{95}Mo (I =
5/2) for ^{96}Mo (I = 0) does affect the spectrum both in detail and
in the rate of decay. The ^{95}Mo case rapidly decays, suggesting
that ^{95}Mo is indeed a component of the S = 3/2 system. (b) The
low frequency modulations are likely to be in part due to^{14} N (I = 1)
either from bound substrate or from peptide ligands. (c) The effects
noted are field-dependent. At different parts of the field, the
spin-echo experiment interrogates different subpopulations of the
(frozen) sample, depending on the orientation of the molecules.
Thus, the effects seen are dependent on the relative orientation
of the molecules and the magnetic field, suggesting anisotropy in
the underlying complex, not surprisingly. We are continuing experimentation on this matter using ^{15}N and ^{2}H labels.

The P-clusters

The perturbation experiments (Figure 1) clearly suggest the
presence of normal iron-sulfur clusters in the Mo-Fe proteins.
Mössbauer analysis[15] clearly demonstrates a minimum subcomponent
accounting for ∿ 6.5% of the non-cofactor iron. This result
independently suggests that the number of iron atoms in this class
is 16, agreeing with the analysis that half the iron atoms are of
non-cofactor type (i.e., 16 of ca. 32) based on integration of the
Mössbauer spectrum. During oxidative titrations, the protein
yields four electrons from this class and the iron atoms go from a
diamagnetic to a highly (S ≥ 3/2) magnetic state. These observations suggest that four spin-coupled clusters, each containing
four iron atoms, are present and are capable of being oxidized by
one electron each to a magnetic (but epr-silent) state. These we
term P-clusters and we have shown that, when the cluster transfer
technique[1] is applied to Av1 or Cp1, 4 ± 0.2 Fe_4S_4 cores are displaced. Why do these clusters show no epr ascribable to normal
iron-sulfur clusters - at least under normal enzymatic conditions?
A partial clue can be gleaned from the following experiment. We
examined the epr of Av1 samples prepared in a potentiostat at
fixed potentials in the presence of mediators.[21] To our surprise,
epr near g = 2 appeared transiently in the region of -200 to -280
mv. As the potential is raised, signal due to a -3 oxidation state
appears and then is quenched. This signal occurs within the potential range where the four P-clusters are undergoing a total of a
four-electron oxidation. This result suggests that, in native
Mo-Fe protein, the P-clusters sit in the $Fe_4S_4{}^{4-}$ state (i.e., "all
ferrous"). This interpretation accounts for the fact that we have
only observed oxidations of P-clusters, whereas the Mössbauer
spectrum of the steady state (where each M center is reduced by
one electron) shows no reduced P-cluster and no evidence for such
species has yet been recognized in epr studies. We suggest that
these observations are consistent with the notion that the reduction
of the Mo-Fe protein by the Fe protein involves production of the

Nitrogenase P

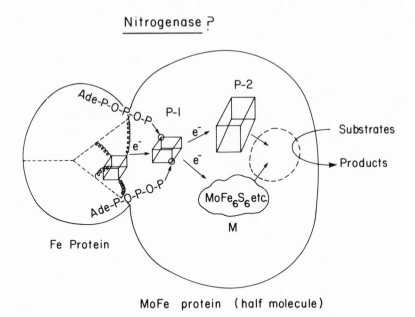

MoFe protein (half molecule)

Fig. 2.   Depiction of relationship of P and M clusters in MoFe
           protein, as well as action of Fe protein, showing a
           current hypothesis about energy transduction in this
           system.

$Fe_4S_4^{5-}$ state in the P-clusters by the mechanism of phosphorylation/
dephosphorylations alluded to above. This mechanism would conserve
the approximately 0.7 V of energy available from the hydrolysis of
two moles of MgATP under physiological conditions, and would lower
the potential from the -0.4 V input level[21] to a -1.1 V level.
Thus, the low potential form of the -5 oxidation state would be
produced which would be unstable with respect to either H^+ reduction
or, via a second P-cluster and a M center operating in tandem, the
reduction of substrates. This highly speculative picture is in
part depicted in Figure 2. One must note that no epr spectrum is
seen when both P-clusters (in a half-molecule) are oxidized to the
-3 state, though this state is paramagnetic according to the
Mössbauer effect. This result means that the P-clusters are grouped
in pairs or, in any event, are strongly linked. Thus, in the
$P_1(4-)P_2(3-)$ case, one observes an S = 1/2 magnetic state while a
spin-state change upon the oxidation of P_1 leaves the $P_1(3-)P_2(3-)$
case in the S ≥ 3/2 state. The reason that no epr is seen seems to
be that the g-values are extremely anisotropic.[20] If their proximity

gives rise to rapid relaxation of the epr, this would also complicate observation of the signals, by broadening them.

Mo OXIDATION STATES IN NITRATE REDUCTASE

We have previously[27] discussed the oxidation states used during nitrate reduction by the enzyme from Neurospora crassa, and Bray (this volume) has reported studies on the E. coli enzyme. We had observed transient epr signals during anaerobic reductive titrations with NADPH and during back titrations of reduced enzyme with NO_3^-, which we had ascribed to Mo(V) (g $\simeq$ 1.88) and Mo(III) (g = 4.15, 3.96, 2.01). A signal identical to this latter was subsequently described by Salerno et al.[28] who produced it by treating [Fe(II)EDTA] with nitric oxide (NO). Although nitrate reductase is not thought to generate NO, we suspected nonetheless that our enzyme, which contained about 0.3 mole Fe/mole enzyme and was kept in EDTA-containing buffer, might somehow be generating this type of compound. We have subsequently found that catalytic amounts of nitrate reductase will generate this signal from [Fe(II)EDTA] in the presence of TPNH. We are currently investigating the nature of this material, but feel that it is unlikely that nitrate reductase utilizes the Mo(III) state. Whatever the nature of the [Fe(II)EDTA]-derived species is, it is a powerful enough reductant to denote electrons to NO_3^- in the presence of the enzyme, regenerating ultimately [Fe(III)EDTA].

REFERENCES

1. W.H. Orme-Johnson and R.H. Holm, Identification of Iron-Sulfur Clusters in Proteins, in: "Methods in Enzymology", Vol. LIII, S. Fleisher and L. Packer, eds., Academic Press, New York, (1978) p. 268.
2. V.K. Shah and W.J. Brill, Isolation of an Iron Molybdenum Cofactor from Nitrogenase, Proc. Natl. Acad. Sci. USA 74:3249 (1977).
3. J. Rawlings, V.K. Shah, J.R. Chisnell, W.J. Brill, R. Zimmermann, E. Münck, and W.H. Orme-Johnson, Novel Metal Cluster in the Fe-Mo Cofactor of Nitrogenase: Spectroscopic Evidence, J. Biol. Chem. 253:1001 (1978).
4. W.H. Orme-Johnson and L.C. Davis, Current Topics and Problems in the Enzymology of Nitrogenase. in: "Iron-Sulfur Proteins", W. Lovenberg, ed., Vol. III, Chapter 2, Academic Press, New York (1976).

5. W.H. Orme-Johnson, L.C. Davis, M.T. Henzl, B.A. Averill,
    N.R. Orme-Johnson, E. Münck and R. Zimmermann, Components
    and Pathways in Biological Nitrogen Fixation, in: "Recent
    Developments in Nitrogen Fixation," W.E. Newton, J.
    Postgate, and C. Rodgriguez-Barrueco, eds., Academic
    Press, London (1977) p. 131.

6. W.H. Orme-Johnson, E. Münck, R. Zimmermann, W.J. Brill,
    V.K. Shah, J. Rawlings, M.T. Henzl, B.A. Averill,
    and N.R. Orme-Johnson, On the Metal Centers in
    Nitrogenase, in; "Mechanisms of Oxidative Enzymes",
    T.P. Singer and R. Ondarza, eds., Elsevier Press, New
    York, (1978) p. 165.

7. R.H. Holm and J.A. Ibers, Synthetic Analogues of the Active
    Sites of Iron-Sulfur Proteins, in: "Iron Sulfur Proteins",
    W. Lovenbert, ed., Vol. 3, Chapter 7, Academic Press,
    New York (1976).

8. C. Kennedy, R.R. Eady, and E. Konderosi, The Molybdenum-Iron
    Protein of Klebsiella pneumoniae Nitrogenase: Evidence
    for Non-Identical Subunits from Peptide Mapping, Biochem.
    J. 155:383 (1976).

9. W.H. Orme-Johnson, W.D. Hamilton, T. Ljones, M.-Y.W. Tso,
    R.H. Burris, V.K. Shah and W.J. Brill, Electron Para-
    magnetic Resonance (EPR) of Nitrogenase and Nitrogenase
    Components from Clostridium pasteurianum W5 and
    Azotobacter vinelandii OP, Proc. Natl. Acad. Sci. USA
    69:3142 (1972).

10. W.G. Zumft, W.C. Cretney, T.C. Huang, L.E. Mortenson and
    G. Palmer, On the Structure and Function of Nitrogenase
    from Clostridium pasteurianum W5, Biochem. Biophys. Res.
    Commun. 48:1525 (1972).

11. G.A. Walker and L.E. Mortenson, Effect of Magnesium Adenosine
    5'-Triphosphate on the Accessibility of the Iron of
    Clostridial Azoferredoxin, a Component of Nitrogenase,
    Biochemistry 13:2382 (1974).

12. C.L. Hill, J. Renaud, R.H. Holm and L.E. Mortenson, Synthetic
    Analogues of the Active Sites of Iron-Sulfur Proteins.
    Comparative Polarographic Potentials of the
    $[Fe_4S_4(SR)_4]^{2-,3-}$ and Clostridium pasteurianum Ferredoxin
    Redox Couples, J. Am. Chem. Soc. 99:2549 (1977).

13. B.E. Smith and G. Lang, Mössbauer Spectroscopy of the
    Nitrogenase Proteins from Klebsiella pneumoniae, Biochem.
    J. 137:169 (1974).

14. E. Münck, H. Rhodes, W.H. Orme-Johnson, L.C. Davis, W. Brill
    and V.K. Shah, Nitrogenase VII. EPR and Mössbauer
    Spectroscopy. Properties of the MoFe Protein of A.
    vinelandii Nitrogenase, Biochem. Biophys. Acta 400:32
    (1975).

15. R. Zimmermann, E. Münck, W.J. Brill, V.K. Shah, M.T. Henzl, J. Rawlings, and W.H. Orme-Johnson, Nitrogenase X: Mössbauer and EPR Studies on Reversibly Oxidized MoFe Protein from Azotobacter vinelandii OP, Biochem. Biophys. Acta 537:185 (1978).

16. D.J. Lowe, R.R. Eady, and R.N.F. Thorneley, Electron-Paramagnetic-Resonance Studies on Nitrogenase of Klebsiella pneumoniae. Evidence for Acetylene- and Ethylene-Nitrogenase Transient Complexes, Biochem. J. 173:277 (1978).

17. W.G. Zumft, L.E. Mortenson and G. Palmer, Electron-Paramagnetic-Resonance Studies on Nitrogenase Investigation of the Oxidation-Reduction Behavior of Azoferredoxin and Molybdo Ferredoxin with Potentiometric and Rapid-Freeze Techniques, Eur. J. Biochem. 46:525 (1974).

18. M.J. O'Donnell and B.E. Smith, Electron-Paramagnetic-Resonance Studies on the Redox Properties of the Molybdenum-Iron Protein of Nitrogenase between +50 and -450 mV, Biochem. J. 173:831 (1978).

19. B.A. Averill, J.R. Bale, and W.H. Orme-Johnson, Displacement of FeS Clusters from Ferredoxins and other FeS Proteins, J. Am. Chem. Soc. 100:3034 (1978).

20. B.H. Huhnh, E. Münck and W.H. Orme-Johnson, Nitrogenase XI: Mössbauer Studies on the Cofactor Centers of the MoFe Protein of A. vinelandii OP, Biochem. Biophys. Acta 527:192 (1979).

21. G.D. Watt, this volume.

22. W.E. Newton, B.K. Burgess, and E.I. Stiefel, this volume.

23. T.D. Tullius, S.D. Conradson, J.M. Berg, and K.O. Hodgson, this volume.

24. W.B. Mims, Measurement of the Linear Electric Field Effect in EPR Using the Spin Echo Method, Rev. Sci. Instrum. 45:1583 (1974).

25. W.B. Mims and J. Peisach, Assignment of a Ligand in Stellacyanin by a Pulsed Electron Paramagnetic Resonance Method, Biochemistry 15:3863 (1976).

26. W.B. Mims, J. Peisach, and J.L. Davis, Nuclear Modulation of the Electron Spin Echo Envelope in Glossy Materials, J. Chem. Phys. 66:5336 (1977).

27. W.H. Orme-Johnson, G.S. Jacob, M.T. Henzl and B.A. Averill, Molybdenum in Enzymes, in: "Bioinorganic Chemistry-II," K.N. Raymond, ed., Adv. Chem. Series, Vol. 162, American Chemical Society, Washington, D.C. (1977), p. 389; also G.S. Jacob and W.H. Orme-Johnson, On the Prosthetic Groups of Nitrate Reductase, in: "Molybdenum-containing Enzymes," M.P. Coughlan, ed., Pergamon Press, Oxford (1979). In press.

28. P.R. Rich, J.C. Salerno, J.S. Leigh, and W.D. Bonner, A Spin
     3/2, Ferrous-Nitric Oxide Derivative of and Iron-Containing
     Moiety Associated with Neurospora Crassa and Higher Plant
     Mitochondria, FEBS Lett. 93:323 (1978).

C_2H_2 REDUCTION AND $^{15}N_2$ FIXATION IN BLUE-GREEN ALGAE

Masayuki Ohmori and Akihiko Hattori

Ocean Research Institute, University of Tokyo

Nakano, Tokyo 164, Japan

INTRODUCTION

The capacity for nitrogen fixation in microorganisms is controlled by the intracellular level of nitrogenase and by the supply of reductant and ATP necessary for the operation of the nitrogenase system[1,2] Combined nitrogen invariably represses the formation of nitrogenase and inactivation of the nitrogenase complex by oxygen is also commonly observed[1,2] Because of its simplicity, inexpensiveness and high sensitivity, the acetylene reduction technique has been widely used to investigate the mechanism of biological nitrogen fixation. However, certain features of nitrogen fixation are not always identical to those of acetylene reduction[3] In this communication, we describe data obtained in our laboratory concerning nitrogen fixation and acetylene reduction in a blue-green alga *Anabaena cylindrica* with emphasis on the effects of ammonium and oxygen. A marked difference between nitrogen fixation and acetylene reduction appears with respect to the sensitivity to oxygen. Experimental details are presented elsewhere.

EFFECTS OF AMMONIUM ON ACETYLENE REDUCTION

The growth of *A. cylindrica* was equally supported by nitrate, ammonium and molecular nitrogen(Fig. 1). Molecular nitrogen is reduced to ammonium by nitrogenase, and reduction of nitrate to ammonium is mediated by nitrate reductase and nitrite reductase. Ammonium, either produced within cells or provided exogenously, is assimilated to yield glutamine by the action of glutamine synthetase[4] The nitrogenase activity of *A. cylindrica* assayed by acetylene reduction varied to a considerable extent during growth(Fig.2).

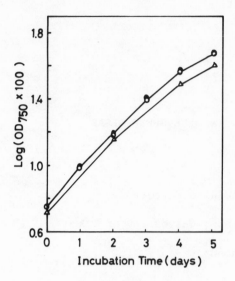

Fig. 1. Effect of nitrogen sources on growth of *A. cylindrica*. Cells were grown at 28°C in the light(5,000 lux) using the medium of Allen and Arnon[5] O, N_2; ●, NO_3^- (2 x 10^{-2} M); Δ, NH_4^+ (1 x 10^{-3}M on each day). (After Ohmori and Hattori[6])

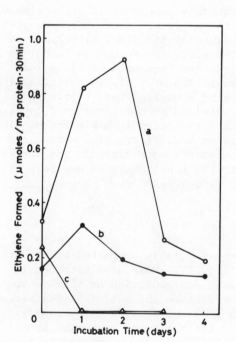

Fig. 2. Changes in acetylene-reducing activity during growth of *A. cylindrica*. (a), Cells grown on N_2; (b), cells grown on NO_3^-; (c), cells grown on NH_4^+. In (a) and (c), cells grown on N_2 and in (b), cells grown on NO_3^- were used as inocula. (After Ohmori and Hattori[6])

In the absence of combined nitrogen, the activity increased rapidly, attained a maximal level after 1 to 2 days, and then declined. Variation of the activity during growth in the presence of nitrate was relatively small, and the activity was invariably lower than that in the absence of nitrate. When 1 mM ammonium was added to the culture medium, the acetylene-reducing activity was diminished within 24 hr. Exposure of algal cells to nitrate or ammonium for less than 1 hr brought about no significant change in the acetylene-reducing activity.[6] Hence, the variation of the activity observed during growth can be attributed to the variation of the nitrogenase level, which is subject to the genetic control.

A different effect of ammonium was observed when the supply of ATP was limited. At a low light intensity of 400 lux, the acetylene-reducing activity was immediately suppressed by 85% by 5 mM of ammonium.[7] In the presence of 5 µM of cabonyl cyanide m-chlorophenylhydrazone(CCCP), a potent uncoupler of phosphorylation, in the light (22,000 lux), ammonium exhibited a similar inhibitory effect.[7] Although ammonium is known to uncouple photophosphorylation in the broken chloroplast system[8], we have found that this is not the case with intact cells of A. *cylindrica*. When ammonium (50 µM or 5 mM) was added in the dark, the intracellular level of ATP was reduced by 40% within 1 min, but restored after 5 min(Fig.3). On the other hand,

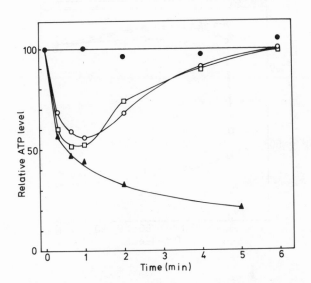

Fig. 3. Effects of ammonium and CCCP on the ATP pool of A. *cylindrica* in the dark. ●, no addition; ○, NH₄Cl was added to 50 µM; □, NH₄Cl was added to 5 mM; ▲, CCCP was added to 10 µM. (After Ohmori and Hattori[9])

the ATP level simply decreased when the algal cells were treated
with 10 μM of CCCP. Upon illuminating the algal cells 1 min after
the addition of 50 μM or 5 mM of ammoinum in the dark, the ATP level
rapidly increased and attained the same level as that of the control
without ammonium(Fig. 4). In the presence of CCCP, however, the ATP
level continued to decrease after the dark-to-light transition. Ammonium was ineffective for uncoupling phosphorylation at a concentration as high as 5 mM.

Both nitrogen fixation(acetylene reduction) and glutamine
synthesis require ATP.[1] When glutamine synthesis was inhibited by
methionine sulfoximine(MSX), the ammonium-induced change in the ATP
level disappeared(Fig. 5), indicating a rapid consumption of ATP
associated with ammonium assimilation to glutamine. We conclude
that ammonium not only represses the biosynthesis of nitrogenase but
also inhibits the nitrogenase activity by lowering the ATP supply to
the nitrogenase system. Nitrate uptake by *A. cylindrica*, which is
also an ATP-dependent process was inhibited by ammonium,[10] a finding
that supports this inference.

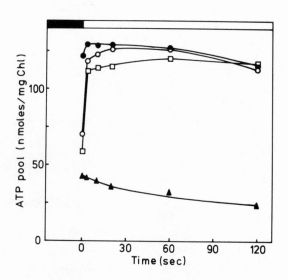

Fig. 4. Changes in the ATP pool of *A. cylindrica* after dark-to-light
transition. NH₄Cl or CCCP was added in the dark 1 min before the
light was turned on. Light intensity, 50 mW cm^{-2}; temperature, 25° C;
●, no addition; ○, NH₄Cl was added to 50 μM; □, NH₄Cl was
added to 5 mM; ▲ , CCCP was added to 10 μM. (After Ohmori and
Hattori[9])

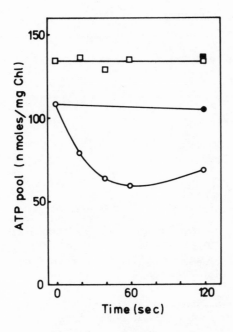

Fig. 5. Effect of MSX on ammonium-induced change in the ATP pools of *A. cylindrica*. Cells were preincubated for 4 hr at 25°C with or without MSX. □, MSX(2 mg ml^{-1}) treated, NH₄Cl(50 μM) added; ■, MSX treated, NH₄Cl not added; O, MSX not treated, NH₄Cl added; ●, MSX not treated, NH₄Cl not added. (After Ohmori and Hattori[9])

VARIATION OF C_2H_2/N_2 RATIO

Nitrogen fixation requires six electrons for the reduction of N_2 to NH_3, whereas reduction of acetylene to ethylene requires only two electrons. A factor of three is conventionally used to convert the activity of acetylene reduction to that of nitrogen fixation. Peterson and Burris[3] have measured acetylene reduction and nitrogen fixation simultaneously, and obtained values of 3 to 7 for the C_2H_2/N_2 reduction ratio. Ohmori and Hattori[6] have shown that acetylene reduction by *A. cylindrica* increased with time and attained a constant rate after about 30 min. Such a lag phase was not observed for nitrogen fixation assayed by an ^{15}N tracer technique. Thus, the C_2H_2/N_2 ratio increased with reaction time. The lag phase for acetylene reduction was eliminated when algal cells were preincubated in the light under argon atmosphere,[6] and the values of 6.1 to 7.9(average 7.2) were obtained for the C_2H_2/N_2 ratio(Table 1).

Nitrogenase catalyzes both nitrogen fixation and hydrogen evolution.[11] Peterson and Burris[12] have shown that hydrogen evolution

Table 1. N_2 fixation and C_2H_2 reduction by *A. cylindrica*.

Exp.	N_2 fixation (nmoles mg prot.$^{-1}$hr^{-1})	C_2H_2 reduction (nmoles mg prot.$^{-1}$hr^{-1})	C_2H_2/N_2
1	92	687	7.5
2	100	610	6.1
3	72	570	7.9
4	101	722	7.2

Cells were incubated under $^{15}N_2$ (95 atom%)-Ar (0.5 : 0.5) or C_2H_2-Ar (0.2 : 0.8) atmosphere. For other experimental conditions, see legend to Fig. 6. (After Ohmori and Hattori[13])

by heterocysts isolated from *Anabaena sp.* is most active under an argon atmosphere and that it is reduced by ca. 70% when argon is replaced with N_2. It is apparent that electrons flow simultaneously from reductant to both N_2 and H^+ to produce NH_3 and H_2, respectively. Acetylene at a partial pressure of 0.17 atm. completely inhibits hydrogen evolution,[12] therefore, under standard assay conditions of acetylene reduction, electrons from reductant predominantly flow to acetylene. Observed values higher than 3 for the C_2H_2/N_2 ratio can thus, at least partly, be explained.

EFFECTS OF OXYGEN ON NITROGEN FIXATION AND ACETYLENE REDUCTION

Sensitivity of acetylene-reducing activity to oxygen varies depending upon the physiological conditions of the alga.[14] We observed that preincubation of *A. cylindrica* under an argon atmosphere in the light raises this sensitivity. When algal cells grown under the same conditions with the same culture age were used, nitrogen fixation was invariably less sensitive to oxygen than acetylene reduction(Fig. 6). At a pO_2 of 0.2 atm., nitrogen fixation was reduced by 50%. This varying sensitivity is another factor which might be responsible for the variation of the C_2H_2/N_2 ratio. In the absence of oxygen, the ratio was 7 and at pO_2 levels of 0.1, 0.2 and 0.3, the ratios were 6, 5 and 2, respectively.

The oxygen-induced inhibition of acetylene reduction was reduced by the introduction of hydrogen(Fig. 7). At a pH_2 of 0.2 atm., the rate of acetylene reduction in the presence of oxygen(pO_2 0.2 atm.) was the same as that in the absence of oxygen. Thus, a lowered

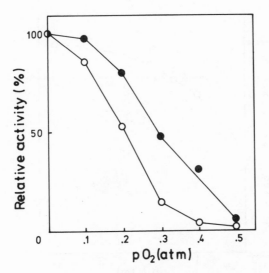

Fig. 6.   Effects of oxygen partial pressure(pO₂) on nitrogen fixation and acetylene reduction by *A. cylindrica*. Three ml portions
of algal suspension were placed in serum bottles(13 ml) with Venoject
caps and preincubated for 30 min in the light(10,000 lux) under argon
atmosphere.   Then 3-(3,4-dichlorophenyl)-1,1-dimethylurea(5 x 10⁻⁵M)
was added to each bottle in order to prevent photosynthetic oxygen
evolution.   Acetylene redution was determined by measuring C₂H₄ production by gas chromatography.[6] Nitrogen fixation was determined by
a ¹⁵N technique. The gas space was filled with gas mixture of ¹⁵N₂:Ar
=0.5:0.5, or ¹⁵N₂:O₂:Ar=0.5:p:(0.5-p), with p being the partial
pressure of O₂.   The incubation was conducted at 28°C and 10,000 lux
with shaking.   At 15 min intervals, algal cells were collected on
Reeve Angel 984H glass fiber filters and dried in a vacuum desiccator.
Cellular nitrogen was converted to N₂ by a modified Dumas method[15]
and ¹⁵N content was determined by emission spectroscopy using a JASCO
NIA-1 ¹⁵N analyzer.   ●, N₂-fixing activity; O, C₂H₂-reducing
activity. (After Ohmori and Hattori[13])

sensitivity of nitrogen fixation to oxygen would result from the
evolution of hydrogen associated with nitrogen fixation.

According to Bothe et al.[16] and Eisbrenner et al.[17] blue-green
algae contain uptake hydrogenase which catalyzes an oxyhydrogen reaction and protects nitrogenase from damage by oxygen. This uptake
hydrogenase differs from classical hydrogenase in the following respects:[12] (i) uptake hydrogenase cannot reduce ferredoxin; and (ii)
oxygen at low concentrations is very effective as an electron acceptor for the uptake hydrogenase both in vivo and in vitro.   Thus,

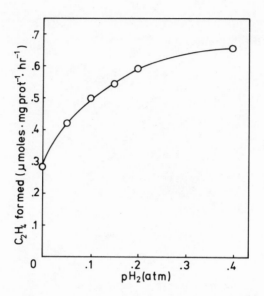

Fig. 7. Effect of hydrogen partial pressure (pH_2) on acetylene
reduction by *A. cylindrica* under aerobic conditions. Gas space
contained 0.2 atm. O_2, 0.2 atm. C_2H_2 and indicated amount of H_2,
and was balanced with Ar. For other experimental conditions, see
legend to Fig. 6. (After Ohmori and Hattori[13])

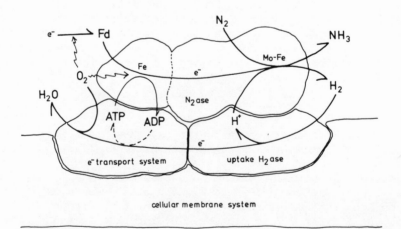

Fig. 8. A model of nitrogenase-hydrogenase relationship in the cells
of *A. cylindrica*.

hydrogen evolved by nitrogenase is quickly recaptured by uptake hydrogenase and is efficiently removed from the site of nitrogenase. In other words, uptake hydrogenase works as an oxygen scavenger in the cell. Two additional functions have been suggested for uptake hydrogenase:[18] (i) the production of ATP for nitrogenase coupled to the oxyhydrogen reaction; and (ii) to supply reductant for nitrogenase in the light in the absence of oxygen.

The uptake hydrogenase is tightly bound to the cellular membrane,[19] but nitrogenase is easily solubilized when the algal cells are disrupted.[20] In the *A. cylindrica* cell, nitrogenase is probably associated with uptake hydrogenase as illustrated in Fig. 8. The pathway of possible electron flow under aerobic conditions in the light is also included in this figure. Such a structural association will be beneficial for rapid recycling of hydrogen and for efficient removal of oxygen.

REFERENCES

1. M. J. Dilworth, Dinitrogen fixation. Ann. Rev. Plant Physiol. 25: 81 (1974).
2. W. D. P. Stewart, Nitrogen fixation by photosynthetic micro-organisms. Ann. Rev. Microbiol. 27: 283 (1973).
3. R. B. Peterson and R. H. Burris, Conversion of acetylene reduction rates to nitrogen fixation rates in natural populations of blue-green algae. Anal. Biochem. 73: 404 (1976).
4. B. J. Miflin and P. J. Lea, The pathway of nitrogen assimilation in plants. Phytochemistry 15: 873 (1976).
5. A. Watanabe, List of algal strains in collection at the Institute of Applied Microbiology, University of Tokyo. J. Gen. Appl. Microbiol. (Tokyo) 6: 283 (1960).
6. M. Ohmori and A. Hattori, Effect of nitrate on nitrogen-fixation by the blue-green alga Anabaena cylindrica. Plant Cell Physiol. 13: 589 (1972).
7. M. Ohmori and A. Hattori, Effect of ammonia on nitrogen fixation by the blue-green alga Anabaena cylindrica. Plant Cell Physiol. 15: 131 (1974).
8. N. E. Good, Activation of the Hill reaction by amines. Biochim. Biophys. Acta 40: 502 (1960).
9. M. Ohmori and A. Hattori, Transient change in the ATP pool of A. cylindrica associated with ammonia assimilation. Arch. Microbiol. 117: 17 (1978).
10. M. Ohmori, K. Ohmori and H. Strotmann, Inhibition of nitrate uptake by ammonia in a blue-green alga, Anabaena cylindrica. Arch. Microbiol. 114: 225 (1977).
11. R. H. Burris, Progress in the biochemistry of nitrogen fixation. Proc. Roy. Soc. B. 172: 339 (1969).
12. R. B. Peterson and R. H. Burris, Hydrogen metabolism in isolated heterocysts of Anabaena 7120. Arch. Microbiol.116: 125 (1978).

13. M. Ohmori and A. Hattori, Differential effects of oxygen on N_2 fixation and C_2H_2 reduction by <u>Anabaena</u> <u>cylindrica</u>. <u>Plant Cell Physiol</u>. (in press).

14. J. R. Benemann and N. M. Weare, Nitrogen fixation by <u>Anabaena cylindrica</u>. III. Hydrogen-supported nitrogenase activity. <u>Arch. Microbiol</u>. 101: 401 (1974).

15. E. Wada, T. Tsuji, T. Saino and A. Hattori, A simple procedure for mass spectrometric microanalysis of ^{15}N in particulate organic matter with special reference to ^{15}N-tracer experiments. <u>Anal. Biochem</u>. 80: 312 (1977).

16. H. Bothe, J. Tennigkeit and G. Eisbrenner, The utilization of molecular hydrogen by the blue-green alga <u>Anabaena cylindrica</u>. <u>Arch. Microbiol</u>. 114: 43 (1977).

17. G. Eisbrenner, E. Distler, L. Floener and H. Bothe, The occurrence of the hydrogenase in some blue-green algae. <u>Arch. Microbiol</u>. 118: 177 (1978).

18. H. Bothe, E. Distler and G. Eisbrenner, Hydrogen metabolism in blue-green algae. <u>Biochimie</u> 60: 277 (1978).

19. E. Tel-Or, L. W. Luijk and L. Packer, Hydrogenase in N_2-fixing cyanobacteria. <u>Arch. Biochem. Biophys</u>. 185: 185 (1978).

20. A. Haystead and W. D. P. Stewart, Characteristics of the nitrogenase system of the blue-green alga <u>Anabaena cylindrica</u>. <u>Arch. Mikrobiol</u>. 82: 325 (1972).

COMMON ANTIGENIC SITES FOR NITROGENASE FROM *Azotobacter* AND BACTERIAL NITRATE REDUCTASES

D. J. D. Nicholas

Agricultural Biochemistry Department
Waite Agricultural Research Institute
University of Adelaide
Glen Osmond, 5064, South Australia

INTRODUCTION

Nitrogenase is composed of two proteins, namely Component I containing Mo, Fe and S and Component II containing Fe and S.[1] Antibodies, raised in mice and rabbits, to Component I of nitrogenase from *Azotobacter* have been successfully used to follow the fate of this protein during its repression by ammonium ions.[2] Immuno-diffusion tests have also been used by Bishop et al.[3] to demonstrate cross-reacting material, in extracts of free-living *Rhizobium japonicum*, to antiserum prepared against Component I of nitrogenase from bacteroids of soyabean nodules.

About 25 years ago, nitrate reductase from *Neurospora crassa* was first identified as a NADPH-linked flavoprotein containing molybdenum and a mechanism of action was proposed for this enzyme.[4-6] Subsequent work with bacteria, fungi and green plants has amply con-firmed these early findings.[7-8] Later, Nason and his collaborators reported that a cytochrome b 557 was also involved in electron transfer to nitrate in *Neurospora*.[9,10] The role of various cyto-chromes as electron carriers in "nitrate respiration" is well documented.[11-13]

In this paper, results of experiments are reported for anti-bodies to purified Component I of nitrogenase from *Azotobacter* and to a purified nitrate reductase from *E.coli*. It will be shown that each antibody type not only cross-reacts with its corresponding antigen protein, but also with the other protein as well.

PREPARATION OF COMPONENT I FROM *Azotobacter*

Components I and II were purified from N_2-fixing cells by methods described previously.[2] Polyacrylamide gel techniques[2] and the C_2H_2 reduction method[2] were used to check that Component I was free from Component II before it was injected into a goat.

PREPARATION OF PURIFIED NITRATE REDUCTASE FROM *E.coli*

The enzyme was prepared from washed cells of *E.coli* B grown with nitrate (Table 1). The method used was a modification of that of Lund and DeMoss.[14] The cells (7g to 15 ml 0.025M Tris-HCl buffer pH 7.5 containing 5 mM $MgSO_4$, 5 mg/ml ribonuclease and 50 µg/ml DNAse) were ruptured at 2°C in an Aminco French Pressure Cell at

Table 1. Purification of Nitrate Reductase from *E.coli* B

Frac-tion	Volume	Total Protein mg.	Total Enzyme units	Specific Activity (nmoles NO_2^- produced/min/mg protein)	% Recovery Enzyme Activity	Purifi-cation
1	17.0	911	41,430	45.5	100	1
2	15.5	302	31,060	111	75	2.4
3	13.0	72	23,990	332	58	7.3

Nitrate reductase activity was assayed at 30°C. The following components of the reaction mixture (µmole) were contained in a final volume of 2 ml in 1.1 x 10 cm test tubes: sodium phosphate buffer (pH 7.5) 320, sodium nitrate 10, benzyl viologen 0.4. An appropriate aliquot of the enzyme was added and the liquid surface was layered with liquid paraffin. Then, 100 µl freshly prepared $Na_2S_2O_4$ (1.56 µmol) in 2% w/v sodium bicarbonate was added below the paraffin layer via a gas tight microsyringe. After a 10 min incubation period, 1 ml aliquots were dispensed into test tubes in air and 1 ml 1% (w/v) sulfanilimide in N.HCl and 1 ml 0.01% w/v α-naphthylethylenediamine HCl were added. After 10 min, the absorbance at 540 nm was determined in 1 cm cuvettes in a Shimadzu spectrophotometer. Protein was determined by a standard microbiuret method.

200 Pa.   The crude extract was centrifuged at 20,000 g for 30 min
in a Sorvall RC2B (SS 34 Rotor) and the supernatant (fraction 1)
retained.   Fraction I was centrifuged at 144,000 g for 120 min in a
Beckman type 65 Rotor in a Spinco ultracentrifuge and the pellet
washed repeatedly with 0.025 M Tris-HCl buffer (pH 7.5).   The pellet
(fraction 2) was then resuspended in 0.1 M Tris-HCl buffer (pH 8.3)
and placed in a waterbath at 60°C for 20 min.   The heat-treated
extract was then cooled in ice and again centrifuged at 120,000 g
for 90 min.   The supernatant contained the heat-released nitrate
reductase (fraction 3) and this was approximately 80% of the enzyme
present in fraction 2.   This enzyme was resolved into two major
proteins by non-dissociating polyacrylamide gel electrophoresis
(Fig.1a).   In duplicate gels, it was established that nitrate

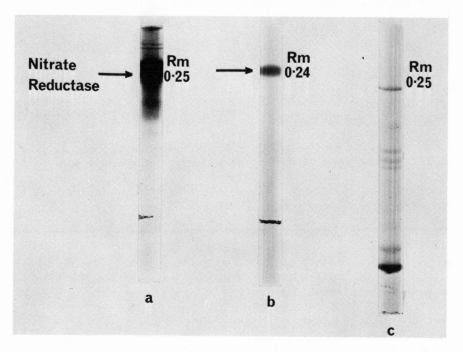

Fig.1   *Detection of Nitrate Reductase in Polyacrylamide Gels.*
         (a) Protein pattern of a heat-released nitrate reductase
         (250 μg of fraction 3 of Table 1 run in a non-dissociating
         polyacrylamide gel electrophoresis system and stained with
         Coomassie Brilliant Blue R250.   (b) Nitrate reductase
         activity of the heat-released enzyme (500 μg of fraction
         3 of Table 1 run in a non-dissociating polyacrylamide gel
         electrophoresis system.   (c) Protein pattern of a heat-
         released nitrate-reductase (100 μg of fraction 3 of Table
         1 run in an SDS-polyacrylamide disc gel electrophoresis
         system.   Nitrate reductase activity indicated by arrows.

reductase activity was associated with the protein band at Rm = 0.24-0.25 (Fig.1b). When fraction 3 was dissociated and reduced with SDS and 2-mercaptoethanol and run on SDS-polyacrylamide gel disc electrophoresis, although the protein band at Rm = 0.24-0.25 remained, it was less intense while other bands appeared at Rm = 0.42, 0.47 and 0.48 (Fig.1c). The most prominent band in the solubilized enzyme was a fast-moving component at Rm = 0.98 (Fig.1c).

Fraction 3 was separated on a Sepharose 6B gel column (2.5 x 100 cm) and 4 ml fractions collected (Fig.2). Nitrate reductase activity was located entirely in the first peak of the elution profile. An analysis of the proteins in this area by non-dissociating polyacrylamide gel electrophoresis showed that the protein composition changed across the peak (Fig.2). Nitrate reductase was associated with both major protein bands in fraction 70 (Fig.2b) which had the highest specific activity for the enzyme. Initial and late fractions of this peak, 65 (Fig.2a) and 75 (Fig. 2c), had lower specific activities and each contained one of the two major proteins present in fraction 70.

PREPARATION OF ANTISERA FOR COMPONENT I AND NITRATE REDUCTASE

Antiserum to purified Component I (see Methods in ref.2) was raised in a goat by a series of once-fortnightly subcutaneous injections over a period of eight weeks using aliquots of 3.5 mg purified Component I in Freund's complete adjuvant. The animal was challenged with the same dose of antigen three months later and serum obtained in the hyperimmune state. The IgG rich fraction of the antiserum was prepared by neutral salt fractionation.[15]

An antiserum to the heat-released nitrate reductase (fraction 3 of Table 1) was raised in a rabbit. Three subcutaneous injections of 1.5 mg protein emulsified in Freund's complete adjuvant were administered. This treatment was followed by two intravenous injections of 1.5 and 3 mg antigen respectively. All injections were done at fortnightly intervals. The animal was bled ten days after the final injection and the IgG fraction of the whole anti-serum prepared by neutral salt fractionation.[15]

IMMUNOLOGICAL REACTIONS OF COMPONENT I OF NITROGENASE FROM *Azotobacter*

Antisera to purified Component I from *Azotobacter* produced a characteristic double precipitin line after immunodiffusion against the purified immunogen as shown in Fig.3 and described and discussed previously.[2] Reactions of partial identity were also observed between antibody to Component I of nitrogenase and cell-free extracts

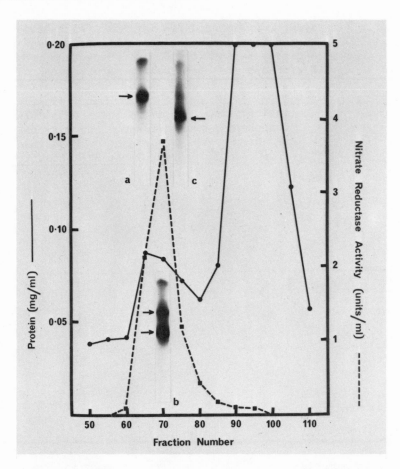

Fig.2   *Purification of a heat-released nitrate reductase (fraction*
         3 of Table 1) on a Sepharose 6B gel column equilibrated
         with 0.1M NaCl in 0.05M Tris-HCl, pH 7.5.   Nitrate reductase
         activity expressed as μmol NO$_2^-$ produced/ml and protein as
         mg/ml.   Protein patterns of eluted fractions run in a non-
         dissociating polyacrylamide gel electrophoresis system and
         stained with Coomassie Brilliant Blue R250.   Amounts loaded
         were 150, 150 and 145 μg for (a), (b) and (c) respectively.
         Nitrate reductase activity indicated by arrows.

of bacteria containing nitrate reductase, namely *E.coli* K12,
Thiobacillus denitrificans[16-17] (Fig.3) and *Agrobacterium tumefaciens*
(Fig.4) and Table 2.   Extracts of these bacteria did not have any
nitrogenase activity as determined by the C$_2$H$_2$ reduction method and
confirmed by gel electrophoresis.

    Cell-free extracts of the blue-green alga *Anabaena cylindrica*[18]

Table 2.   Results of Immunodiffusion Tests with Antibody
           to purified Component I from *A. vinelandii* and
           Antibody to Purified Nitrate Reductase from *E.coli*
           B with Antigens from various sources.

	Antibody to Component I from *A.vinelandii*	Antibody to Nitrate Reductase from *E.coli* B	BVH-linked Nitrate Reductase Activity	Ref.
Purified Component I from *A.vinelandii*	+	+	+	2
Purified Nitrate Reductase from *E.coli* B.	+	+	+	14
Crude Extracts of:				
Agrobacterium tumefaciens	+	+	+	14
E.coli K$_{12}$	+	+	+	14
Thiobacillus denitrificans	+	+	+	16,17
Anabaena cylindrica (grown with nitrate)	–	–	+	18
Neurospora crassa	–	–	+	7
A.vinelandii (grown with ammonia)	–	–	–	2
A.vinelandii (grown with nitrate)	+	+	+	2
*Aldehyde dehydrogenase (from yeast) EC 1.2.1.3	–	–	–	
*Xanthine oxidase (from buttermilk) EC 1.2.3.2	–	–	–	

Sigma Chemical Company, St. Louis, Missouri, U.S.A.

References listed refer to the growth conditions for the microorganisms as well as the preparation of cell-free extracts and the purified proteins.

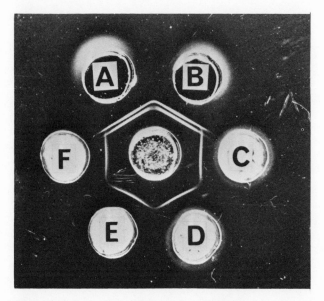

Fig.3   *Ouchterlony immunodiffusion assays in agar gel.*
        Comparison of the precipitin reaction of purified Com-
        ponent I (50 μg of Fraction 5 of Table 1 in ref.2)
        (A,B), 450 μg cell-free extract (S₃) of E.coli K₁₂ (C,D),
        and 400 μg cell-free extract (S₃) of T.denitrificans
        (E,F) against 50 μg of IgG fraction of antiserum to the
        same purified Component I from A.vinelandii.

and of the fungus *Neurospora crassa*,[7] both containing nitrate
reductase activity, did not react immunologically either with antibodies to Component I from *Azotobacter* (Fig.4) or those of nitrate
reductase from *E.coli* B (see also Table 2).   Cell-free extracts of
Azotobacter grown with ammonium salts did not react immunologically
with antiserum to either Component I of nitrogenase or nitrate
reductase from *E.coli* as shown in Table 2.   Neither nitrate
reductase activity nor nitrogenase was detected in extracts of
these cells of *E.coli*.

The purified nitrate reductase from *E.coli* prepared by passage
through a Sepharose 6B column (Fig.2b) gave a distinct precipitin
line when reacted against the IgG fraction of the antiserum to
purified Component I of nitrogenase from *Azotobacter* as illustrated
in Fig.5.

IMMUNOLOGICAL REACTIONS OF NITRATE REDUCTASE FROM *E.coli*

The reaction of the immunoglobulin to heat-released nitrate
reductase from *E.coli* (fraction 3 of Table 1) with its immunogen,

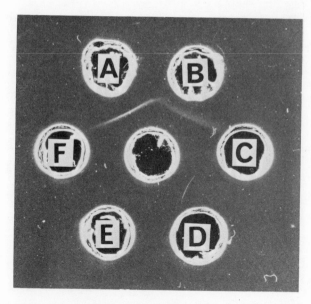

Fig.4   *Ouchterlony immunodiffusion assays in agar gel.*
The precipitin reaction of 200 µg cell-free extract (S3)
of Agrobacterium tumefaciens (wells A and B) with 50 µg
of IgG fraction of antiserum to purified Component I
from A. vinelandii (center well).   Wells C and D contain
200 µg cell-free extract of the blue green alga Anabaena
cylindrica grown with nitrate and wells E,F contain 200
µg cell-free extract of Neurospora crassa grown with
nitrate.

consisted of a heavy precipitin line (nearer the immunogen wells)
and a less discrete band near the three center wells (Fig.6, A,B,C
and G).   In immunodiffusion reactions where cell-free extracts of
Azotobacter grown with nitrate and heat-released nitrate reductase
from *E.coli* B were placed in adjacent wells (H and G respectively)
and reacted against the antibodies of nitrate reductase, a spurring
reaction of partial identity was observed (Fig.6).   The extract
of *Azotobacter* did not contain Component I of nitrogenase because,
when it was supplemented with purified Component II (see Methods in
ref.2), it did not reduce C_2H_2 to C_2H_4 and neither was Component I
detected by gel electrophoresis.

Purified aldehyde dehydrogenase from yeast as well as xanthine
oxidase from buttermilk, which contain Mo and non-heme Fe, did not
react immunologically with antiserum for either Component I or
nitrate reductase.   Nitrate reductases prepared from plant sources
did not form precipitin lines with either antibodies to Component I
or nitrate reductase from *E.coli*.

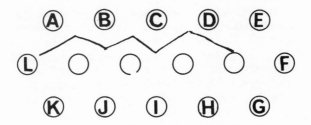

Fig.5   Ouchterlony immunodiffusion assays in agar gel.
         The precipitin reaction for a series of twofold
         dilutions of the purified nitrate reductase from
         E.coli after passage through Sepharose 6B (Fig.2b).
         Well A contains 60 µg, B 30 µg and so on through
         well L.   The four center wells contained 50 µg
         of the IgG fraction of antiserum to purified Component
         I of nitrogenase from A. vinelandii.   Nitrate
         reductase was not detected in wells E to L because
         it was too dilute.

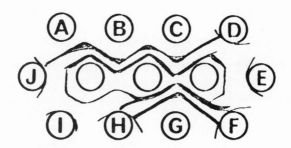

Fig.6   Ouchterlony immunodiffusion assays in agar gel.
         Comparison of precipitin reaction of (a) 60 µg heat-
         released nitrate reductase (fraction 3 of Table 1) in
         wells A,B,C and G and (b) 380 µg of a cell-free
         extract of A. vinelandii grown with nitrate (S105)
         in wells D,E,F,G,I and J.   500 µg of the IgG fraction
         of antiserum to heat-released nitrate reductase was
         contained in the three center wells.

CONCLUSIONS

The data indicate that Component I of nitrogenase from
Azotobacter shares common antigenic sites with a number of bacterial
nitrate reductases.   It is of interest that this relation does not
hold for nitrate reductases from either a blue-green alga *Anabaena
cylindrica* or higher plants.   Moreover, antibodies for Component I
did not cross-react with the molybdoenzymes, aldehyde dehydrogenase
and xanthine oxidase, from animal sources.   Thus, it is only in
bacteria that this antigenic similarity between two molybdenum-
containing proteins has been observed.   This result indicates that
during evolution both Component I of nitrogenase and bacterial
nitrate reductases could well have been derived from a common
ancestor protein.

The concept that there is a common cofactor for a range of
molybdoenzymes has been supported by genetic[19-21] and biochemical
[22-24] evidence.   Thus, Ketchum et al.[22] and Nason et al.[25]
demonstrated that an *in vitro* activation of nitrate reductase in
extracts of a mutant strain (nit-1) of *N.crassa* could be achieved
by adding various acid-treated molybdoenzymes.   They suggested that
a molybdenum moiety associated with a small organic molecule or polypeptide was the common factor for these enzymes.   Subsequently,
Shah and Brill[26] isolated a Fe-Mo cofactor from Component I of
nitrogenase of both *Azotobacter vinelandii* and *Clostridium
pasteurianum*, which reactivated inactive Component I from a mutant
strain UW45 of *Azotobacter*.   Pienkos et al.[27] isolated a molybdenum
cofactor from xanthine oxidase which also activated nitrate reductase
in extracts of *N.crassa* mutant strain (nit-1) but it had no effect
on an inactive nitrogenase in extracts of the *Azotobacter* mutant
strain UW45.   In view of these results,[26-27] the possibility cannot
be excluded that the antigenic similarities noted between Component
I of nitrogenase from *Azotobacter* and bacterial nitrate reductases
could be associated with the presence of a common contaminating
protein which is co-purified with either enzyme.   Further work is
required to establish whether this occurs.

Bishop et al.[3] have shown that antiserum prepared against the
Mo-Fe protein component of nitrogenase from soybean nodule bacteroids
cross-reacted with extracts of free-living *Rhizobium japonicum* grown
under various cultural conditions.   The most intense precipitin
bands resulted from cross-reaction of the antiserum with extracts of
cells grown anaerobically with nitrate or anaerobically with nitrate
and ammonia.   Our observations that antiserum to Component I of
nitrogenase cross-reacts with bacterial nitrate reductases and *vice
versa* are relevant to the interpretation of the data of Bishop et al.
because Cheniae and Evans[28] have shown that bacteroids from soybean
contain nitrate reductase, as do cells of *Rhizobium japonicum*
grown with nitrate.[29]   Thus, it would be of interest to establish

whether antibodies raised to a bacterial nitrate reductase would
also cross-react with extracts of bacteroids from soybean.

REFERENCES

1. R.H. Burris, Fixation by free-living micro-organisms: Enzymology,
    in: "The Chemistry and Biochemistry of Nitrogen Fixation",
    J.R. Postgate, ed., Plenum Press, London (1971), pp.326.
2. D.J.D. Nicholas and J.V. Deering, Repression, derepression and
    activation of nitrogenase in *Azotobacter vinelandii*, Aust.
    J. Biol. Sci., 29: 147-161 (1976)
3. P.E. Bishop, H.J. Evans, R.M. Daniel and R.O. Hampton,
    Immunological evidence for the capability of free-living
    Rhizobium japonicum to synthesize a portion of a nitrogenase
    component, Biochim. Biophys. Acta, 381: 248-256 (1975).
4. A. Nason and H.J. Evans, Triphosphopyridine nucleotide nitrate
    reductase in *Neurospora*, J. Biol. Chem. 202: 655-673 (1953)
5. D.J.D. Nicholas and A. Nason, Molybdenum and nitrate reductase.
    II. Molybdenum as a constituent of nitrate reductase,
    J. Biol. Chem. 207: 353-360 (1954).
6. D.J.D. Nicholas and A. Nason, Mechanisms of action of nitrate
    reductase from *Neurospora*, J. Biol. Chem. 211: 183-197 (1954)
7. D.J.D. Nicholas and A. Nason, Diphosphopyridine nucleotide-
    nitrate reductase from *Escherichia coli*, J. Bacteriol. 69:
    580-583 (1955).
8. D.J.D. Nicholas and A. Nason, Role of molybdenum as a constitu-
    ent of nitrate reductase from soybean leaves, Plant Physiol.
    30: 135-138 (1955).
9. R.H. Garrett and A. Nason, Involvement of a b-type cytochrome
    in the assimilatory nitrate reductase of *Neurospora crassa*,
    Proc. Nat. Acad. Sci. USA 58: 1603-1610 (1967).
10. R.H. Garrett and A. Nason, Further purification and properties
    of *Neurospora* nitrate reductase, J. Biol. Chem. 244:2870-2882
    (1969).
11. Y. Lam and D.J.D. Nicholas, A nitrate reductase from *Micrococcus
    denitrificans*, Biochim. Biophy. Acta 178: 225-234 (1969).
12. W.J. Payne, Reduction of nitrogenous oxides by micro-organisms,
    Bacteriol. Rev. 37: 409-452 (1973).
13. C.H. MacGregor, Anaerobic cytochrome b in *E.coli*: association
    with and regulation of nitrate reductase, J. Bacteriol.
    121: 111-1116 (1975).
14. K. Lund and J.A. DeMoss, Association-Dissociation behavior and
    subunit structure of heat-released nitrate reductase from
    Escherichia coli, J. Biol. Chem. 251: 2207-2216 (1976).
15. J.V. Ferrante and D.J.D. Nicholas, Use of immunoadsorbent affin-
    ity chromatography to purify Component I of nitrogenase from
    extracts of *Azotobacter vinelandii*, FEBS Lett.66:187-190(1976).
16. C.A. Adams, G.M. Warnes and D.J.D. Nicholas, A sulphite-

dependent nitrate reductase from *Thiobacillus denitrificans*, Biochim. Biophys. Acta 235: 398-406 (1971).

17. V. Sawhney and D.J.D. Nicholas, Sulphite and NADH-dependent nitrate reductase from *Thiobacillus denitrificans*, J. Gen. Microbiol. 100: 49-58 (1977).

18. P.F. Brownell and D.J.D. Nicholas, Some effects of sodium on nitrate assimilation and N_2 fixation in *Anabaena cylindrica*, Plant Physiol. 42: 915-921 (1967).

19. J.A. Pateman, J.D. Cove, B.M. Rever and D.B. Roberts, A common cofactor for nitrate reduction and xanthine dehydrogenase which also regulates the synthesis of nitrate reductase, Nature (London) 201: 58-60 (1964).

20. A. Kondorosi, I. Barabas, A. Svab, L. Orosz, T. Sik and R.D. Hotchkiss, Evidence for common genetic determinants of nitrogenase and nitrate reductase in *Rhizobium meliloti*, Nature (New Biology) 246: 153-154 (1973).

21. G. Scazzocchio, F.E. Holl and A.I. Foguelman, The genetic control of molybdoflavoproteins in *Aspergillus nidulans*. Allopurinolresistant mutants constitutive for xanthine dehydrogenase, Eur. J. Biochem. 36: 428-445 (1973).

22. P.A. Ketchum, H.Y. Cambier, W.A. Frazier, III, C.H. Madansky and A. Nason, *In vitro* assembly of *Neurospora* assimilatory nitrate reductase from protein subunits of a *Neurospora* mutant and the xanthine oxidizing or aldehyde oxidase systems of higher animals, Proc. Nat. Acad. Sci. USA 66: 1016-1023 (1970).

23. A. Nason, K.Y. Lee, S.S. Pan, P.A. Ketchum, A. Lamberti and J. DeVries, *In vitro* formation of assimilatory reduced nicotin-amide adenine dinucleotide phosphate: nitrate reductase from a *Neurospora* mutant and a component of molybdenum-enzymes, Proc. Nat. Acad. Sci. USA 68: 3242-3246 (1971).

24. G.M. Cheniae and H.J. Evans, Physiological studies of nodule-nitrate reductase, Plant Physiol. 35: 454-462 (1960).

25. A. Nason, A.D. An toine, P.A. Ketchum, W.A. Frazier, III and D K. Lee, Formation of assimilatory nitrate reductase by *in vitro* inter-cistronic complementation in *Neurospora crassa*, Proc. Nat. Acad. Sci. USA 65: 137-144 (1970).

26. V.K. Shah and W.J. Brill, Isolation of an iron-molybdenum co-factor of nitrogenase, Proc. Nat. Acad. Sci. USA 74: 3249-3253 (1977).

27. P.J. Pienkos, V.K. Shah and W.J. Brill, Molybdenum cofactors from molybdoenzymes and *in vitro* reconstitution of nitrogen-ase and nitrate reductase, Proc.Nat. Acad. Sci USA 74: 5468-5471 (1977).

28. G.M. Cheniae and H.J. Evans, On the relation between nitrogen fixation and nodule nitrate reductase of soybean root nodules, Biochim. Biophy. Acta 26: 654-655 (1957).

29. D.J.D. Nicholas, Y. Maruyama and D.J. Fisher, The effect of cobalt deficiency on the utilization of nitrate nitrogen in *Rhizobium*, Biochim. Biophys. Acta 56: 623-626 (1962).

E.P.R. AND THE ACTIVE CENTRES OF XANTHINE OXIDASE AND OF OTHER MOLYBDENUM-CONTAINING ENZYMES

Robert C. Bray

School of Molecular Sciences
University of Sussex
Falmer, Brighton, BN1 9QJ, U.K.

INTRODUCTION

This short review summarizes recent work and current hypotheses from the author's laboratory concerning the role of molybdenum in xanthine oxidase from milk and in nitrate reductase from E. coli. More limited studies on xanthine dehydrogenase from turkey liver and on sulfite oxidase from chicken liver are also referred to briefly. Much earlier work has been summarized[1] previously and a more complete review[2] will soon be available. Our work shows all these enzymes to have much in common. Further, in the case of xanthine oxidase, it provides direct evidence supporting Stiefel's hypothesis that the mechanisms of all molybdenum-containing enzymes involve coupled proton- and electron-transfers from (or to) the substrates, with the electrons being accepted (or donated) by the molybdenum atom of the enzyme and the protons being accepted (or donated) by one of its ligands, rendered more basic (or acidic) by the reduction (or oxidation) of the metal.[3] Our work serves further, however, to emphasize differences in the nature and reactions of the active centers of the various enzymes.

REDOX POTENTIALS

Data presently available on the redox potentials of molybdenum in enzymes, as determined by an e.p.r. method, are summarized in Table 1. Potentials for the relevant product/substrate complexes are also listed.

Several points are apparent.   Firstly, we note the very wide range of
potentials covered by molybdenum in the various enzymes, which span as
much as 700 mV   if the desulfo modification of xanthine oxidase as well as
the functional enzyme is included.   This large span emphasizes the
versatility of molybdenum in enzymes.   Secondly, we can see some
correlation between the potentials of the substrates on the one hand, and
those of the enzymic molybdenum on the other.   Thus, potentials relating
to nitrate reduction are much higher than those relating to the other
enzymes.   Such a correlation is in accord with expectations if a redox role
for molybdenum in the enzymes is assumed.   Lastly, and most
interestingly, we see that, in every case, the potential for the Mo(VI)/
Mo(V) couple is about the same (within 40 mV) as is that for the Mo(V)/
Mo(IV) couple.   As pointed out by Williams,[4] the fact that, starting with
Mo(VI), a second electron may be transferred to the metal as readily as is
the first, is consistent with the conclusion[2,5,6] that the enzymic reactions
all involve transfer of two electrons simultaneously.

Table 1.   Redox potentials for molybdenum in enzymes and for their
            substrates

Enzyme	Potential (mV) at pH 7		
	Prod./Subs.[a]	Mo(VI)/Mo(V)	Mo(V)/Mo(IV)
Aldehyde oxidase	−581[b]		
Sulfite oxidase	−454[c]		
Formate dehydrogenase	−420[d]		
Xanthine oxidase (functional)	−360[e]	−355[f,g]	−355[f,g]
Xanthine oxidase (desulfo)		−440[f,g]	−480[f,g]
Xanthine dehydrogenase	−360[e]	−350[f,h]	−362[f,h]
Nitrate reductase	+421[i]	+180[j]	+220[j]

a, Data from reference 7.   b, Potential for the couple acetate/
acetaldehyde.   c, Potential for the couple sulfate/sulfite.   d, Potential
for the couple CO_2/formate.   e, Potential for the couple uric acid/
xanthine.   f, Data at pH 8.   g, Data from reference 8.   h, Data from
reference 9.   i, Potential for the couple nitrate/nitrite.   j, Data from
reference 10.

Though the redox potential data establishes the thermodynamic reasonableness of a redox role for molybdenum in enzymes, nevertheless, the kinetic competence of particular reduced forms of the enzyme in the turnover processes is a separate question.   In fact, the first data on this point were obtained for xanthine oxidase as long ago as 1961 by the use of the rapid freezing method.   Mo(V) was found to appear and disappear again at rates compatible with the turnover rate during a single cycle of reduction of this enzyme by a substrate and reoxidation by oxygen.[11]

ANION- AND SUBSTRATE-BINDING SITES

It is reasonable to expect to find, in these enzymes, a ligand position capable of coordinating the substrate, or other molecules, directly to molybdenum.   In agreement, it has long been known[12,13] that substrates modify the e.p.r. signals from molybdenum in functional xanthine oxidase. In principle, however, there are problems in distinguishing the e.p.r. spectra of Michaelis complexes* from those of intermediates or product complexes.   In a number of cases, such problems have now been solved and considerable progress has been made recently in defining parameters of the complexes for all three types of enzyme.   For nitrate reductase, both nitrate and nitrite form well-defined complexes with the low-pH form of the enzyme.[16]   For sulfite oxidase, although we were unable to repeat all the work of Kessler and Rajagopalan,[17] at least one ion, phosphate, forms a complex having well-defined e.p.r. parameters (M.T. Lamy and R.C. Bray, unpublished work).   Finally, for xanthine oxidase, the nitrate ion and the substrates, xanthine, purine and 1-methylxanthine, all form well-defined complexes, giving signals whose parameters are scarcely distinguishable from one another.[13,18,19] Here, the ability to form complexes extends to desulfo xanthine oxidase as

---

*Many different reduced forms of xanthine oxidase are possible, bearing different numbers of electrons, and all of these, as well as the oxidized enzyme, can in principle form Michaelis complexes with substrate molecules.[5]   Though the complex of xanthine with the two-electron reduced enzyme is likely to have no more than a transient existence, on the other hand, for instance, the five-electron reduced enzyme is expected to yield, with xanthine, what is effectively a stable dead-end complex.   Molybdenum signals from these different reduced forms appear to be indistinguishable from one another[14] at temperatures where splitting due to interaction with reduced iron-sulphur centers[15] is not observed.

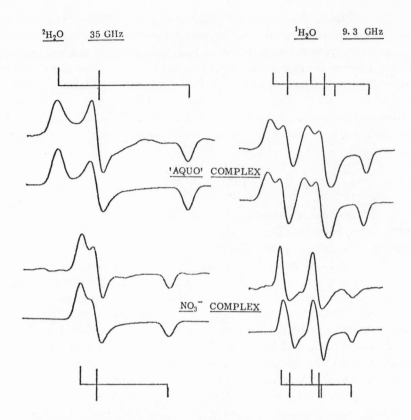

^2H_2O 35 GHz ^1H_2O 9.3 GHz

'AQUO' COMPLEX

NO_3^- COMPLEX

Fig. 1.   Evidence for an anion binding site: the Slow signal from
           desulfo xanthine oxidase.   The top half of the figure shows e.p.r.
           spectra in the absence, and the bottom half in the presence, of
           nitrate ions.   For each pair of spectra, the upper one is
           experimental, while the lower one is a computer simulation.
           The left-hand part of the figure shows spectra recorded in
           ^2H_2O at 35 GHz while the right-hand side shows spectra in
           ^1H_2O at 9.3 GHz.   Stick diagrams show the positions of the
           principle g-values, split by interaction with the strongly coupled
           protons, where these are present.   Parameters used in the
           simulations (splittings in mT) were as follows.   "Aquo"
           complex - $g_{1,2,3}$: 1.9719, 1.9671, 1.9551; $A(^1H)_{1,1;2,2;3,3}$: 1.66,
           0.16; 1.66, 0.16; 1.56, 0.16.   Nitrate complex - g_1, g_2, g_3:
           1.9689, 1.9666, 1.9571; $A(^1H)_{1,2,3}$: 1.50, 1.48, 1.52.
           Appropriate allowance was made for replacement of ^1H by ^2H,
           when ^2H_2O was employed.   The weakly coupled proton in the
           "aquo" complex affects only the linewidth; the possibility of there
           being a weakly coupled proton in the nitrate complex has not been
           investigated in detail.   (Modified from Gutteridge et al.[21]).

well as to the functional form of the enzyme.[*] Indeed, the lineshape of the
signal from the nitrate complex of the desulfo enzyme is a particularly
characteristic one (Fig. 1).   Though work on turkey liver xanthine
dehydrogenase[20] has been much less extensive than that on xanthine oxidase,
all the indications are that the molybdenum conters of these two enzymes are
virtually identical to one another.

A puzzling question concerning xanthine oxidase is the two types of
signal from complexes of the functional enzyme.   These are the type 1
Rapid signals, which are those just discussed, and the type 2 Rapid signals,
which, as we shall consider below, show interaction of molybdenum with two
equivalent protons.[**]   Though more work on this problem is needed,
probably the best hypothesis at the present time is that the type 1 and type 2
signals represent isomeric complexes having different ligand geometries.
Thus, it is suggested that the type 2 signal corresponds to a species in
which, say, an anion and a hydroxyl ligand have changed places with one
another relative to their positions in the corresponding type 1 species.

PROTONS COUPLED TO Mo(V)

It is now established that all three types of enzyme show
protons coupled to Mo(V), at least, in the e.p.r.-active species observed
at low pH values.   Table 2 summarizes the various proton coupling
constants.   As has been discussed elsewhere,[1,2,21] the best interpretation
of all, except possibly the smallest, of these splittings is that they are due
to Mo-OH or Mo-SH groups in the enzymes.   Unfortunately, to our
knowledge, no e.p.r. data is available on suitable low molecular weight
molybdenum compounds having such structures, which might serve as
"models" for the enzymes to check this conclusion.

---

[*]Reduced functional xanthine oxidase gives the signals named Rapid by Bray
and Vänngård,[13] whereas the desulfo modification gives the Slow signal.

[**]Type 2 Rapid signals from xanthine oxidase were originally reported in
the presence of xanthine[13] or uric acid.[18]   The latter workers[18] concluded
that the xanthine type 2 signal represented a non-productive complex of the
reduced enzyme with the substrate, presumably having the substrate wrongly
oriented, so that catalysis could not occur.   Type 2 signals have, however,
more recently been observed under other conditions in the absence of purines,
for example, in the presence of borate ions or at very high pH values
(S. Gutteridge, J.P. Malthouse and R.C. Bray, unpublished work).

Table 2.   Exchangeable protons coupled to Mo (V) in enzymes

Values of the hyperfine couplings constants, $A(^1H)$, are given in mT.

Enzyme	A_1	A_2	A_3	A_{av}
Xanthine oxidase (functional; type 1)[a]	1.30 0.40	1.39 0.30	1.40 0.20	1.36 0.30
Xanthine oxidase (desulfo)[b]	1.66 0.16	1.66 0.16	1.56 0.16	1.63 0.16
Xanthine oxidase (functional; type 2)[c]	1.07 1.07	0.98 0.98	0.93 0.93	1.01 1.01
Nitrate reductase[d]	1.11	0.84	0.81	0.92
Sulfite oxidase[e]	0.85	0.80	1.20	0.95

a, Data of Gutteridge et al.[19] for the 1-methylxanthine Rapid type 1 signal.   b, Data of Gutteridge et al.[21] for the "Aquo" Slow signal. c, Data of Bray et al.[18] for the xanthine Rapid type 2 signal. d, Data of Vincent and Bray[16] for the "Aquo" form of the low pH signal.   e, Unpublished data of M. T. Lamy and R. C. Bray for the "Aquo" form of the low pH signal.

It is interesting that, in xanthine oxidase, all forms of the enzyme listed in Table 2 show interaction of two protons with molybdenum, either with one strongly coupled and the other more weakly so, or with both strongly coupled.   On the other hand, sulfite oxidase and nitrate reductase each show one strongly-coupled proton only.   In the xanthine oxidase Rapid type 2 species, according to the hypothesis advanced in the previous section, what was originally the weakly coupled proton of the Rapid type 1 has become a proton strongly coupled to molybdenum because of a change of ligand geometry.

Deprotonated species containing terminal oxygen or sulfur ligands are presumably in all cases in equilibrium with the observed protonated species (see Gutteridge et al.[21]) and pK values for the exchangeable protons have been measured by two different procedures.   For sulfite oxidase and for nitrate reductase, relative proportions of the high- and low-pH signals were measured as a function of pH to give the pK values directly.[16, 22]   In contrast, for xanthine oxidase, hydrogen/deuterium exchange rate studies were used for indirect estimation of the pK values.[21]   In all cases studied, pK values found were of the order of 8, the one exception being for the very strongly coupled proton of desulfo xanthine oxidase, for which a pK of around 10 was reported.

Perhaps the most important feature of the proton coupling is that it has enabled us to demonstrate direct hydrogen transfer from substrates to xanthine oxidase.[19, 23]   When xanthine oxidase is treated with 1-methyl-xanthine in ^2H_2O, ^1H, which could only (apart from the methyl group) have come from the 8-position of the substrate molecule, appears in the enzyme site strongly coupled to molybdenum, but then exchanges out again into the medium, with a rate constant of 27 sec^{-1} at 12^0C (see Fig. 2).   Conversely, when 1-methyl[8-^2H-]xanthine is employed in ^1H_2O, then ^2H appears, initially, in this same site.[19, 24]

MECHANISMS OF ACTION AND DIFFERENCES BETWEEN
XANTHINE OXIDASE AND THE OTHER ENZYMES

The data summarized above seem to leave little room for doubt regarding certain aspects of the mechanism of xanthine oxidase and presumably of xanthine dehydrogenase, too.   The mechanism presumably involves abstraction of two electrons from the substrate by Mo(VI) and of a proton by a ligand of the metal, either sulfur or oxygen.   Rapid intra-molecular reoxidation of molybdenum would then give the Mo(V) seen by e.p.r.   Gutteridge et al.[21] assumed that the relatively low pK value for the strongly coupled proton of functional xanthine oxidase, in comparison with that of the corresponding proton of the desulfo enzyme, indicated the acceptor in the functional enzyme to be a terminal sulfur atom.

We suggest that, for sulfite oxidase, a reaction such as the one described above would yield the enzymic product, sulfate, directly from the substrate, sulfite.   Similarly, for nitrate reductase the reverse of this process would yield the product, when starting from nitrate.   In the case of xanthine oxidase, disposal of the residual carbonium ion produced by abstraction of the hydride equivalent from the substrate would be more of a problem.   To cope with this, it seems that the active center of this enzyme has evolved with a nucleophilic center capable of reacting with the carbonium ion.   Bray et al.[6] proposed that the product of such a reaction is detectable by e.p.r. as the Very Rapid signal.   According to this scheme,[6] hydrolysis of the Very Rapid species yields the product and

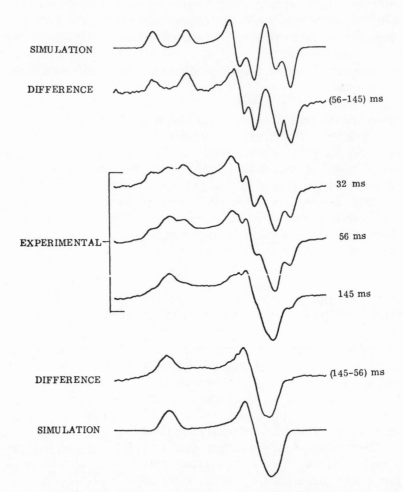

<u>Fig. 2.</u>   Evidence for direct transfer of hydrogen from 1-methylxanthine
to xanthine oxidase.   At long reaction times (>145 ms.),
treatment of xanthine oxidase with 1-methylxanthine in ^2H_2O
yields the Rapid signal with ^2H in both the strongly-coupled and
the weakly-coupled sites (see lower simulated spectrum).
However, at shorter reaction times, the spectrum is different.
By computer subtraction procedures, the spectrum may be
"extrapolated" back to what it would have been had there been no
exchange of protons with the medium.   Comparison of the upper
simulated spectrum with the difference spectrum [(56-145) ms.)
shows the latter to correspond to the Rapid signal with ^1H in the
strongly-coupled site and with ^2H in the weakly-coupled site.

(continued opposite)

the free reduced enzyme, which becomes reoxidized to complete the catalytic cycle (see Olson et al. [5]).

Accepting the above mechanisms, we are left with the problem of identifying the nucleophile. The active center of functional xanthine oxidase possesses two features, absent in both sulfite oxidase and nitrate reductase, which might be related to the nucleophile. Thus, its molybdenum shows interaction with a second exchangeable proton, and further, it possesses the cyanide-labile sulfur atom which is lost on conversion to the desulfo enzyme. [21,25] It is now tentatively suggested that it is the second proton which gives the clue to the nature of the nucleophile, and that this center is in fact the proton-bearing site in its deprotonated form, that is, a terminal oxygen or terminal sulfur ligand of molybdenum. Thus, we are suggesting (cf. Bray[2])* that both the proton-accepting group and the nucleophile of xanthine oxidase are either terminal oxygen or terminal sulfur atoms with their properties no doubt suitably varied by ligand geometry and the influence of neighbouring groups. **

---

*Williams[4] has independently come to a similar conclusion to the present one about the mechanism of action of xanthine oxidase.

**Bray et al.[6] and Gutteridge et al.[21], combining pK and other data, concluded that the cyanide-labile sulfur was present as a terminal sulfur ligand of molybdenum and that this constituted the proton-accepting group. They did not specifically identify the nucleophile. On the other hand, Olson et al.[5] earlier proposed that the sulfur (which they took to be a persulfide) was the nucleophile. In the absence of e.p.r. studies with [33]S and with [17]O, distinguishing between such schemes is still not easy. On the other hand, S. Tullius, K. Hodgson and S. Cramer (personal communication) have concluded, from preliminary EXAFS studies, that there is no terminal sulfur in functional xanthine oxidase. If this turns out to be correct, it could be consistent with the persulfide hypothesis,[5] leaving this group as the nucleophile, with the proton acceptor becoming a terminal oxygen rather than a terminal sulfur ligand.

Fig. 2. (continued)
The ^{1}H must have been transferred from the substrate; the rate constant for its subsequent exchange into the medium was estimated to be 27 sec^{-1}. Parameters used in the simulations (splittings in mT) were: $g_{1,2,3}$: 1.9893, 1.9703, 1.9657; for the upper simulation $A(^1H)_{1,2,3}$: 1.30, 1.39, 1.40; $A(^2H)_{1,2,3}$: 0.06, 0.05, 0.03; for the lower simulation $A(^2H)_{1,1;2,2;3,3}$: 0.20, 0.06; 0.21, 0.05; 0.22, 0.03. (Modified from Gutteridge et al.[19])

If such chemical likeness of the proton–accepting group and the nucleophile is accepted, then there is a further consequence.   The Inhibited series of signals[*] can then readily be explained[6] in terms of a "wrong way round" reaction, in which the nucleophile and the proton acceptor have exchanged roles.   Stability, in the Inhibited series, of signal-giving species bearing a Mo–coupled –RCO group from the substrate would then be due to the absence (or wrong location in the active center) of an auxiliary group.   This group would be the one which normally facilitates hydrolysis to yield the product in the corresponding Very Rapid species obtained in the "right way round" and catalytically effective normal enzymic reaction.

ACKNOWLEDGEMENTS

Work from the author's laboratory has been supported by a Programme Grant from the Medical Research Council.   Help from the colleagues on whose work the latest advances have depended is gratefully acknowledged.   Thanks are particularly due to Dr. S. Gutteridge and also to Dr. P. Malthouse and Miss M. T. Lamy.

REFERENCES

1. R. C. Bray, Molybdenum iron–sulphur flavin hydroxylases and related
        enzymes, in: "The Enzymes," P. D. Boyer, ed. , 3rd ed. ,
        Academic Press, New York (1975), Vol. 12, p. 299.
2. R. C. Bray, E. p. r. of molybdenum–containing enzymes, in:
        "Biological Magnetic Resonance," J. Reuben and L. Berliner,
        eds. , Vol. 2, Plenum Press, New York (1979).   In press.
3. E. I. Stiefel, Proposed molecular mechanism for the action of
        molybdenum in enzymes: coupled proton and electron transfer,
        Proc. Nat. Acad. Sci. U. S. A.   70: 988 (1973).
4. R. J. P. Williams, "The Biological Role of Molybdenum," Climax
        Molybdenum Company, London (1978).

---

[*] The Inhibited signal is obtained on treating xanthine oxidase with methanol or formaldehyde.[26]   It is also obtained, though without a proton splitting, on treatment with ethylene glycol[27] or acetaldehyde (J. P. Malthouse and R. C. Bray, unpublished work).   Similarly, the Desulfo Inhibited (or Resting 2) signal is obtained on treating the desulfo enzyme with ethylene glycol.[28]

5. L. Olson, D. Ballou, G. Palmer and V. Massey, The mechanism of action of xanthine oxidase, J. Biol. Chem. 249: 4363 (1974).

6. R. C. Bray, S. Gutteridge, S. J. Tanner and D. A. Stotter, The mechanism of action of xanthine oxidase: the relationship between the Rapid and Very Rapid molybdenum electron paramagnetic resonance signals, Biochem. J. 177: 357 (1979).

7. H. A. Sober, ed. , "Handbook of biochemistry: selected data for molecular biology," 2nd Ed. , C. R. C. Press, Cleveland (1970), p. J-34.

8. R. Cammack, M. J. Barber and R. C. Bray, Oxidation-reduction potentials of molybdenum, flavin and iron-sulphur centres in milk xanthine oxidase, Biochem. J. 157:469 (1976).

9. M. J. Barber, R. C. Bray, R. Cammack and M. J. Coughlan, Oxidation-reduction potentials of turkey liver xanthine dehydrogenase and the origins of oxidase and dehydrogenase behaviour in molybdenum-containing hydroxylases, Biochem. J. 163: 279 (1977).

10. S. P. Vincent, Oxidation-reduction potentials of molybdenum and iron-sulphur centres in nitrate reductase from Escherichia coli, Biochem. J. 177:757 (1979).

11. R. C. Bray, The chemistry of xanthine oxidase. 8. Electron-spin-resonance measurements during the enzymic reaction, Biochem. J. 81:196 (1961).

12. F. M. Pick and R. C. Bray, Complex-formation between reduced xanthine oxidase and purine substrates demonstrated by electron paramagnetic resonance, Biochem. J. 115: 735 (1969).

13. R. C. Bray and T. Vänngård, 'Rapidly appearing' molybdenum electron-paramagnetic-resonance signals from reduced xanthine oxidase, Biochem. J. 114: 725 (1969).

14. R. C. Bray, M. J. Barber, D. J. Lowe, R. Fox and D. Cammack, Mechanisms of electron transfer within multicomponent oxidative enzymes, illustrated by studies on xanthine oxidase, FEBS Symposium 40: 159 (1975).

15. D. J. Lowe and R. C. Bray, Magnetic coupling of the molybdenum and iron-sulphur centres in xanthine oxidase and xanthine dehydrogenases, Biochem. J. 169: 471 (1978).

16. S. P. Vincent and R. C. Bray, Electron-paramagnetic-resonance studies on the nitrate reductase from Escherichia coli K12, Biochem. J. 171: 639 (1978).

17. D. L. Kessler and K. V. Rajagopalan, Hepatic sulphite oxidase: effect of anions on interaction with cytochrome-c, Biochim. Biophys. Acta 370: 389 (1974).

18. R. C. Bray, M. J. Barber and D. J. Lowe, Electron-paramagnetic-resonance spectroscopy of complexes of xanthine oxidase with xanthine and uric acid, Biochem. J. 171: 653 (1978).

19. S. Gutteridge, S. J. Tanner and R. C. Bray, The molybdenum centre of native xanthine oxidase: evidence for proton transfer from substrates to the centre and for existence of an anion-binding site, Biochem. J. 175: 869 (1978).

20. M. J. Barber, R. C. Bray, D. J. Lowe and M. P. Coughlan, Studies by electron-paramagnetic-resonance spectroscopy and stopped-flow spectrophotometry on the mechanism of action of turkey liver xanthine dehydrogenase, Biochem. J. 153: 297 (1976).

21. S. Gutteridge, S. J. Tanner and R. C. Bray, Comparison of the molybdenum centres of native and desulpho xanthine oxidase: the nature of the cyanide-labile sulphur atom and the nature of the proton-accepting group, Biochem. J. 175: 887 (1978).

22. H. J. Cohen, I. Fridovich and K. V. Rajagopalan, Hepatic sulphite oxidase: a functional role for molybdenum, J. Biol. Chem. 246: 374 (1971).

23. R. C. Bray and P. F. Knowles, Electron spin resonance in enzyme chemistry: the mechanism of action of xanthine oxidase, Proc. Roy. Soc. (London) A302: 351 (1968).

24. S. Gutteridge and R. C. Bray, Studies by electron paramagnetic resonance on the nature and reactions of the molybdenum centre of xanthine oxidase, in: "Molybdenum-containing Enzymes," M. P. Coughlan, ed., Pergamon Press, Oxford (1979). In press.

25. V. Massey and D. Edmondson, On the mechanism of inactivation of xanthine oxidase by cyanide, J. Biol. Chem. 245: 6595 (1970).

26. F. M. Pick, M. A. McGartoll and R. C. Bray, Reaction of formaldehyde and of methanol with xanthine oxidase, Eur. J. Biochem. 18: 65 (1971).

27. S. J. Tanner and R. C. Bray, A new molybdenum electron-paramagnetic-resonance signal from treatment of functional xanthine oxidase with ethylene glycol, Biochem. Soc. Trans. 6: 232 (1979).

28. D. J. Lowe, M. J. Barber, R. T. Pawlik and R. C. Bray, A new non-functional form of milk xanthine oxidase containing stable quinquivalent molybdenum, Biochem. J. 155: 81 (1976).

FORMATE DEHYDROGENASE FROM CLOSTRIDIUM THERMOACETICUM

Lars G. Ljungdahl

Department of Biochemistry
University of Georgia
Athens, Georgia 30602, U.S.A.

INTRODUCTION

Clostridium thermoaceticum is an obligate anaerobic and thermophilic bacterium. It ferments various sugars to acetate, which is the only product, and from one mole of hexose almost three moles of acetate are formed.[1] The fermentation of glucose is via the Embden-Meyerhof glycolytic pathway to pyruvate.[2] Half of the pyruvate is oxidized by pyruvate-ferredoxin reductase to yield acetate, CO_2 and reduced ferredoxin.[3] Carbon dioxide serves as the acceptor of electrons generated during the fermentation of glucose and in the oxidation of pyruvate. The CO_2 is reduced to a methyl group, which with the carboxyl group of the other half of pyruvate formed in the glycolysis is converted to acetate.[4] The fermentation of glucose by C. thermoaceticum can be summarized as follows:

$$C_6H_{12}O_6 + H_2O \rightarrow CH_3COCOOH + CH_3COOH + CO_2 + 6H \qquad (1)$$

$$CH_3COCOOH + CO_2 + 6H \rightarrow 2CH_3COOH + H_2O \qquad (2)$$

$$\text{Sum:} \quad C_6H_{12}O_6 \rightarrow 3CH_3COOH \qquad (3)$$

The reactions leading to the formation of acetate from CO_2 and the carboxyl group of pyruvate have been discussed in detail elsewhere.[4-6] They involve reduction of CO_2 to formate, conversion of formate to formyltetrahydrofolate, transfer of the methyl group to a corrinoid-protein to form methylcorrinoid-protein, and finally, a reaction between the methyl group and the carboxyl group of pyruvate (transcarboxylation) to form acetate.

129

The first step in the synthesis of acetate from CO_2 in C. thermoaceticum is catalyzed by a soluble NADP-dependent formate dehydrogenase (Reaction 4).[7-9]

$$CO_2 + NADPH \rightleftharpoons HCOO^- + NADP^+ \tag{4}$$

C. thermoaceticum appears unique in that it is the only organism
known which contains a NADP-dependent formate dehydrogenase.

REQUIREMENT FOR SELENITE, TUNGSTATE AND MOLYBDATE FOR FORMATION
OF FORMATE DEHYDROGENASE IN C. THERMOACETICUM

In our earlier studies of formate dehydrogenase in C. thermoaceticum, we noticed that the levels of the enzyme varied among
batches of cells. We also noticed that the enzyme activity rapidly
increased in the beginning of the log phase of growth and that it
rapidly decreased thereafter to almost zero at the end of the log
phase.[10] These results suggested that the medium lacked a factor
or factors necessary for the formation of formate dehydrogenase.
In 1954, Pinsent discovered that the formate dehydrogenase activity
of Escherichia coli is greatly enhanced by including selenite and
molybdate in the growth medium.[11] It has now been shown by Enoch
and Lester[12] that the E. coli enzyme contains both selenium and
molybdenum and, in addition heme, non-heme iron and acid-labile
sulfide. In a manner similar to other molybdenum enzymes, the
formation of an active formate dehydrogenase in E. coli is inhibited
by tungstate, when present in the growth medium.[11,13] Noticing
Pinsent's results, we added selenite and molybdate to the growth
medium of C. thermoaceticum. The results are shown in Table 1.

Table I. Effects of additions of selenite, molybdate, tungstate
and vanadate in the growth medium on formate dehydrogenase activity
in Clostridium thermoaceticum.[14]

Additions to medium	Formate dehydrogenase activity in French pressure extracts Units/mg of protein
None	0.167
Na_2MoO_4 $10^{-4}M$	0.152
Na_2SeO_3 $10^{-6}M$	0.253
Na_2WO_4 $10^{-4}M$	0.262
Na_2MoO_4, Na_2SeO_3	0.572
Na_2WO_4, Na_2SeO_3	1.500
Na_2WO_4, Na_2SeO_3, Na_2MoO_4	2.000
$NaVO_3$, Na_2SeO_3	0.148

Depending on the time of cell harvest, we obtained an increase in the formate dehydrogenase activity between 10 to 1000-fold of that of the controls without selenite and molybdate. As found with E. coli, both selenite and molybdate had to be present in the medium of C. thermoaceticum. However, in contrast to the results with E. coli, we found that tungstate was not antagonistic toward molybdate but could replace molybdate in the growth medium. As a matter of fact, tungstate enhanced the formate dehydrogenase activity more than molybdate.[14] We have now repeatedly observed that the highest formate dehydrogenase activity is obtained when selenite, tungstate and molybdate are all together in the growth medium of C. thermoaceticum.

INCORPORATION OF SELENIUM, TUNGSTEN AND MOLYBDENUM IN FORMATE DEHYDROGENASE OF C. THERMOACETICUM

The nutritional requirements of selenite, tungstate and molybdate for formation of formate dehydrogenase in C. thermoaceticum indicate functional roles for selenium, tungsten and molybdenum in this enzyme. To establish the presence of these elements in this enzyme and to further establish its properties, we have attempted to purify it.[9] This purification has developed into a momentous task. The enzyme is extremely sensitive to oxygen and appears to disintegrate in dilute solutions. Frankly, we have not yet been able to obtain an enzyme preparation which meets the criteria of a purified enzyme. Nevertheless, 100-fold purification of the formate dehydrogenase in C. thermoaceticum has been obtained by working under strictly anaerobic conditions and with the use of a ferrous-thioglycolate complex as an oxygen scavenger in all solutions. The enzyme is conveniently assayed by following spectrophotometrically the reduction of NADP with formate. The purification, which includes heat treatment, ammonium sulfate fractionation, gel filtration and DEAE-cellulose chromatography yields an enzyme of about 50 units (μmol NADP reduced per min) per mg. Details of the assay and the purification procedures have been described.[15]

When C. thermoaceticum is grown on medium containing ^{75}Se-selenite and supplemented with tungstate or molybdate radioactive selenium is incorporated into several protein fractions.[14] This may indicate some non-specific incorporation of selenium into proteins. However, the formate dehydrogenase fraction purified as described above, when subjected to gel filtration emerged from the column as a single peak with protein, radioactive selenium and enzyme activity completely overlapping. The coincidence between the selenium content and the enzyme activity strongly supports the concept that formate dehydrogenase in C. thermoaceticum is a selenium protein. The number of selenium atoms in the enzyme has not been established.

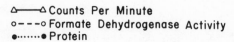

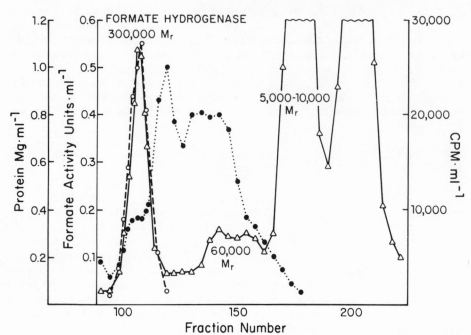

Fig. 1.   Fractionation of a cell-free extract of C. thermoaceticum
          on Ultrogel AcA 34 column. Cells were grown with 10^{-7}M
          ^{185}W-tungstate in the medium.

In experiments similar to that using ^{75}Se-selenite but with
^{185}W-tungstate, radioactive tungsten is incorporated into three
distinct fractions (Fig. 1).[16] From the elution pattern, the
molecular weights (M_r) of the tungsten-containing fractions have
been estimated to be 290,000, 60,000 and about 5,000 respectively.
The last fraction is most likely non-proteinaceous and it eluted
just ahead of salt and other low molecular weight material.

The formate dehydrogenase activity was present in the 290,000
M_r fraction and, as is shown in Fig. 1, the ^{185}W activity and
enzyme activity completely coincided. This fraction was further
purified by ammonium sulfate fractionation and DEAE cellulose
chromatography. The final product eluted from gel filtration
columns as a single protein containing ^{185}W and formate dehydrogenase activity. A M_r for formate dehydrogenase of about 300,000
has previously been obtained using sucrose gradient centrifugation.[9]

At the present time, we have no knowledge concerning the
function of the 60,000 M_r protein or the low M_r fraction. Known

molybdenum enzymes have molecular weights ranging from 120,000
and higher. Thus, it is unlikely that the protein represents a
known molybdenum-containing enzyme. However, a storage protein
for molybdenum has been found in Clostridium pasteurianum,[17] E.
coli[18] and in liver (K. V. Rajagopalan, personal communication).
The possibility exists that the C. thermoaceticum 60,000 M_r protein is a similar storage protein. It is interesting to note
that when cells are grown at a higher tungstate concentration
($10^{-5}M$) than that used in the experiment shown in Fig. 1 the 60,000
M_r tungsten peak becomes very prominent.[16] Leonhardt and Andreesen[19] found a similar tungsten protein in C. formicoaceticum, with
an M_r of 88,000. C. formicoaceticum synthesizes acetate from CO_2
via a pathway similar to that of C. thermoaceticum.[20] The 5,000
M_r fraction is, perhaps, a tungsten equivalent to the acid-stable
molybdenum factor associated with molybdenum enzymes.[21]

EFFECT OF MOLYBDATE ON [185]W-TUNGSTATE INCORPORATION INTO FORMATE
DEHYDROGENASE, 60,000 M_r FRACTION AND CELLS OF C. THERMOACETICUM

     Based on the requirement for tungsten or molybdenum and
selenium in the growth medium and on the incorporation of ^{75}Se and
[185]W into the active formate dehydrogenase, we have postulated that
this enzyme is a selenium-tungsten or a selenium-molybdenum protein.
The combination of selenium-tungsten yields a higher formate dehydrogenase activity in extracts than the combination of selenium-
molybdenum. However, the best activity is obtained when selenite,
tungstate and molybdate are all combined in the growth medium.
Therefore, we studied the effect of molybdate on the incorporation
of [185]W into the formate dehydrogenase of C. thermoaceticum.[6]

     C. thermoaceticum was grown in the presence of $10^{-5}M$ [185]W-
tungstate and with varying concentrations of molybdate (up to
$10^{-3}M$). The formate dehydrogenase and the 60,000 M_r protein were
separated as shown in Fig. 1. The incorporation of [185]W into
cells, formate dehydrogenase and the 60,000 M_r protein was determined. The following observations were made. Tungsten incorporation into the cell is markedly depressed in a competitive manner
by molybdate. Incorporation of tungsten into the 60,000 M_r protein
is similarly depressed. However, molybdate does not depress the
incorporation of tungsten into the formate dehydrogenase fraction.
On the contrary, an increase of tungsten in this fraction is seen
with increasing molybdate concentration. This increase corresponds
to the increase in the formate dehydrogenase activity, which is
always observed when molybdate is added together with selenite and
tungstate to the growth medium. Finally, the ratio of tungsten
incorporation over enzyme activity is not affected by molybdate.

     The fact that molybdate does not compete with incorporation
of tungstate into formate dehydrogenase of C. thermoaceticum and

that the ratio of ^{185}W incorporated over enzyme activity is
constant although the molybdate concentration was varied indicates
a need for tungsten in active formate dehydrogenase. That molybdate always increases formate dehydrogenase activity in the presence of tungstate can now also be explained. Shackelford (unpublished results) in my laboratory has performed tungsten and
molybdenum analyses on formate dehydrogenase preparation using
the method of Cardenes and Mortenson.[22] He found that these
preparations contained both tungsten and molybdenum, in a ratio of
1:2 (W:Mo). Thus, it is possible that C. thermoaceticum formate
dehydrogenase is a selenium-molybdenum-tungsten enzyme, with 1 or
2 atoms of tungsten per enzyme molecule.

EPR STUDIES OF FORMATE DEHYDROGENASE FROM C. THERMOACETICUM

    Recently, Dr. DerVartanian and I have started examining the
formate dehydrogenase from C. thermoaceticum using EPR. Unfortunately, these studies have been slow, partly because of the
difficult of obtaining active enzyme preparations. In enzyme
preparations with specific activities between 30 and 50 and in

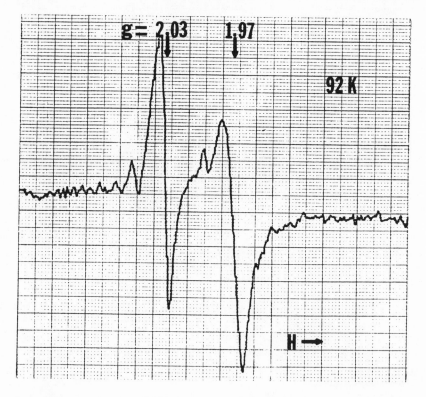

Fig. 2. EPR spectrum of formate dehydrogenase in the presence
         of formate.

the presence of formate (reduced enzyme), the major EPR species is of a g=1.97 signal attributed to Mo (V) (Figure 2). This signal disappears when the enzyme is oxidized. A correlation between enzyme activity and the strength of the EPR signal was observed.

A SUMMARY OF OTHER FORMATE DEHYDROGENASES CONTAINING TUNGSTEN, MOLYBDENUM AND SELENIUM

It has already been mentioned that E. coli formate dehydrogenase contains molybdenum, selenium, non-heme iron, acid-labile sulfide, and heme.[12] Recently, Scherer and Thauer[23] purified the formate dehydrogenase from Clostridium pasteurianum. This enzyme contains molybdenum, non-heme iron and acid-labile sulfide. Formate dehydrogenase activity is increased to 8 to 10-fold by including tungstate and selenite in the medium of Clostridium formicoaceticum[24] and Clostridium acidiurici.[25] Molybdate has no effect with C. acidiurici but may substitute for tungstate with C. formicoaceticum. The formate dehydrogenase from Clostridium cylindrosporum is perhaps a molybdenum-selenium enzyme, which is not affected by tungsten.[25] A molybdenum- and non-heme iron-containing formate dehydrogenase was recently purified from Vibrio succinogenes.[26]

ACKNOWLEDGEMENT   This work has been supported by Public Health Service grant AM-12913 and grant PCM-7726054 from National Science Foundation.

REFERENCES

1. F. Fontaine, W. H. Peterson, E. McCoy, M. J. Johnson, and G. Ritter, A new type of glucose fermentation by Clostridium thermoaceticum, J. Bacteriol., 43: 701 (1942).
2. H. G. Wood, Fermentation of 3,4-C[14]-and 1-C[14]-labeled glucose by Clostridium thermoaceticum, J. Biol. Chem., 199: 579 (1952).
3. S.-S. Yang, L. G. Ljungdahl, and J. LeGall, A four-iron, four-sulfide ferredoxin with high thermostability from Clostridium thermoaceticum, J. Bacteriol., 130: 1084 (1977).
4. M. Schulman, R. K. Ghambeer, L. G. Ljungdahl, and H. G. Wood, Total synthesis of acetate from CO_2. VII. Evidence with Clostridium thermoaceticum that the carboxyl of acetate is derived from the carboxyl of pyruvate by transcarboxylation and not by fixation of CO_2, J. Biol. Chem., 248: 6255 (1972).
5. L. G. Ljungdahl, and H. G. Wood, 1969, Total synthesis of acetate from CO_2 by heterotrophic bacteria, Ann. Rev. Microbiol., 23: 515 (1969).
6. L. G. Ljungdahl, and J. R. Andreesen, Reduction of CO_2 to acetate in homoacetate fermenting clostridia and the involvement of tungsten in formate dehydrogenase, in: "Microbial

Production and Utilization of Gases (H_2, CH_4, CO)," H. G. Schlegel, G. Gottschalk, and N. Pfennig, eds. E. Goltze KG, Gottingen (1976) pp. 163-172.

7. L.-F. Li, L. G. Ljungdahl, and H. G. Wood, Properties of nicotinamide adenine dinucleotide phosphate-dependent formate dehydrogenase from Clostridium thermoaceticum, J. Bacteriol., 92: 405 (1966).

8. R. K. Thauer, CO_2-reduction to formate by NADPH. The initial step in the total synthesis of acetate from CO_2 in Clostridium thermoaceticum, FEBS Lett., 27, 111 (1972).

9. J. R. Andreesen, and L. G. Ljungdahl, Nicotinamide adenine dinucleotide phosphate-dependent formate dehydrogenase from Clostridium thermoaceticum: purification and properties, J. Bacteriol., 120: 6 (1974).

10. J. R. Andreesen, A. Schaupp, C. Neurauter, A. Brown, and L. G. Ljungdahl, Fermentation of glucose, fructose, and xylose by Clostridium thermoaceticum: Effect of metals on growth yield, enzymes, and the synthesis of acetate from CO_2, J. Bacteriol., 114: 743 (1973).

11. J. Pinsent, The need for selenite and molybdate in the formation of formic dehydrogenase by members of the coli-aerogenes group of bacteria, Biochem. J., 57: 10 (1954).

12. H. B. Enoch, and R. L. Lester, The purification and properties of formate dehydrogenase and nitrate reductase from Escherichia coli, J. Biol. Chem., 250: 6693 (1975).

13. R. H. Scott, and J. A. DeMoss, Formation of the formate-nitrate electron transport pathway from inactive components in Escherichia coli, J. Bacteriol. 126: 478 (1976).

14. J. R. Andreesen, and L. G. Ljungdahl, Formate dehydrogenase of Clostridium thermoaceticum: Incorporation of selenium-75 and the effects of selenite, molybdate, and tungstate on the enzyme, J. Bacteriol., 116: 867 (1973).

15. L. G. Ljungdahl, and J. R. Andreesen, Formate dehydrogenase, a selenium-tungsten enzyme from Clostridium thermoaceticum, in: Methods in Enzymology," Vol. LIII, S. Fleischer, and L. Packer, eds., Academic Press, New York (1978) pp. 360-372.

16. L. G. Ljungdahl, and J. R. Andreesen, Tungsten, a component of active formate dehydrogenase from Clostridium thermoaceticum, FEBS Lett., 54: 279 (1975).

17. B. B. Elliott, and L. E. Mortenson, Regulation of molybdate transport by Clostridium pasteurianum. J. Bacteriol., 127: 770 (1976).

18. M. Dubourdieu, E. Andrade, and J. Puig, Molybdenum and chlorate resistant mutants in Escherichia coli K 12, Biochem. Biophys. Res. Commun., 70: 766 (1976).

19. U. Leonhardt, and J. R. Andreesen, Some properties of formate dehydrogenase accumulation and incorporation of [185]W-tungsten into proteins of Clostridium formicoaceticum. Arch. Microbiol., 115: 277 (1977).

20. W. E. O'Brien, and L. G. Ljungdahl, Fermentation of fructose and synthesis of acetate from carbon dioxide by Clostridium formicoaceticum. J. Bacteriol., 109: 626 (1972).

21. J. L. Johnson, H. P. Jones, and K. V. Rajagopalan, In vitro reconstitution of demolybdosulfite oxidase by molybdenum co-factor from rat liver and other sources, J. Biol. Chem., 252: 4994 (1977).

22. J. Cardenas, and L. E. Mortenson, Determination of molybdenum and tungsten in biological materials, Anal. Biochem., 60: 372 (1974).

23. P. A. Scherer, and R. K. Thauer, Purification and properties of reduced ferredoxin: CO_2 oxidoreductase from Clostridium pasteurianum, a molybdenum iron-sulfur protein, Eur. J. Biochem., 85: 125 (1978).

24. J. R. Anderson, E. El Ghazzawi, and G. Gottschalk, The effect of ferrous ions, tungstate and selenite on the level of formate dehydrogenase in Clostridium formicoaceticum, and formate synthesis from CO_2 during pyruvate fermentation, Arch. Microbiol., 96: 103 (1974).

25. R. Wagner, and J. R. Andreesen, Differentiation between Clostridium acidiurici and Clostridium cylindrosporum on the basis of specific metal requirements for formate dehydrogenase formation, Arch. Microbiol., 114: 219 (1977).

26. A. Kröger, E. Winkler, A. Innerhofer, H. Hackenberg, H. Schag-ger, The Formate Dehydrogenase Involved in Electron Transport from Formate to Fumarate in Vibrio succinogenes, Eur. J. Biochem., 94: 465 (1979).

STRUCTURAL CHEMISTRY OF MOLYBDENUM IN METALLOENZYMES AS ELUCIDATED BY EXAFS

Thomas D. Tullius, Steven D. Conradson
Jeremy M. Berg, and Keith O. Hodgson

Department of Chemistry
Stanford University
Stanford, California 94305, U.S.A.

INTRODUCTION

One of the most important perspectives in the study of metalloproteins is the relationship between chemical structure and function. A vast body of chemical research has shown structure and function to be irrevocably intertwined; it is upon this foundation that a good deal of the predictive power of chemistry rests. This paper describes a new structural tool, x-ray absorption spectroscopy (XAS),[1,2,3,4] and its application to the study of the coordination environment of molybdenum in biological systems.

Structure Determination and Biomacromolecules

The problem of obtaining structural information for molybdenum ions in biomolecules has indeed been a difficult one, and as a consequence two general approaches have arisen. The first of these employs measurements (either physical or in principle crystallographic) made on the molecule itself. The determination of a protein crystal structure[5] is a notoriously difficult task. Even when suitable single crystals of a molybdenum metalloprotein are grown, solution of the structure is likely to be a daunting and time-consuming task because of the large subunit size of molybdoproteins. Heavy-atom derivative crystals will probably be necessary to resolve the phase problem. The large number of atoms whose positions must be determined in a protein structure, and the limited number of data because of structural disorder of the crystal, greatly reduce the precision (typically <0.15-0.2Å) to which individual bond lengths can be measured.[6]

The lack of crystallographic results on molybdenum proteins has led to an even heavier reliance on other physical measurements to deduce the details of the structures of the metal sites. Analogies to known compounds are used to infer structure from physical data. Such reasoning sometimes proves inadequate, though, because proteins often bind metals in ways which have not been duplicated by synthetic complexes. The molybdenum sites of both xanthine oxidase[7] and nitrogenase[8] are examples which exhibit physical properties unlike known simple complexes of the metal itself. The unique properties of molybdenum-containing metalloproteins have stimulated much chemistry which seeks to model the metal sites with small inorganic complexes, in hopes of identifying the structural features which create these properties. This is the second general approach toward defining metalloprotein metal site structure, the so-called "analogue" approach, and is exemplified by the efforts of Holm and his collaborators in the iron-sulfur[9] and nitrogenase[10] areas. A simple comparison of physical properties of the analogue and the protein is suggestive of the fidelity of the model to its natural counterpart. The ultimate criterion must, however, be structural. Only when the coordination environment of the metal ion is similar in analogue and protein can questions be asked about the effect of the protein on the properties of the metal.

X-ray absorption spectroscopy can effectively integrate the two approaches described above, because XAS gives direct structural information about a metal's immediate coordination environment without concern for the rest of the molecule, be it a simple ligand or a protein. Metal-ligand bond lengths can be measured to an accuracy of a few hundreths of an angstrom,[3] nearly as good as by small-molecule crystallography, and almost an order of magnitude better than from the typical protein crystal structure.[6] XAS can, therefore, be used to determine the coordination environment of a metal ion in a structurally uncharacterized protein site, or it can provide metal-ligand distances to higher accuracy for a protein that has been previously studied by x-ray crystallography. In addition, a direct, accurate comparison of the environment the metal "sees" in both protein and analogue is possible.

X-Ray Absorption Spectroscopy - A Brief Illustration

Our methods for collection and analysis of x-ray absorption data have been described in much detail elsewhere.[1,3] Here, we will briefly review the relevant phenomena and illustrate the method through its application to an inorganic molybdenum complex.

At energies below an elemental absorption "edge", the x-ray absorption coefficient falls off monotonically. When the incident x-ray photons have sufficient energy to excite a core electron (for a K edge, a 1s electron), the absorption rises steeply. This core electron can be excited into higher-level bound states, thereby giving rise to fine structure superimposed on the edge itself. These transitions can be analyzed to give information about the electronic structure of the absorbing atom. As the energy of the incident x-ray photons increases the core electron escapes and propagates as a free photoelectron.

It is the interaction of the ejected photoelectron with neighboring atoms that gives x-ray absorption spectroscopy its power as a tool of molecular structure determination. The phenomenon which contains information about the atom type and distance (the metal) ligands is called the extended x-ray absorption fine structure (EXAFS). This is simply defined as the modulation of the x-ray absorption coefficient of a metal at energies above an absorption "edge". These sinusoidal modulations are the result of interference between the photoelectron wave leaving the metal, and the wave backscattered from nearby ligand atoms. As the wavelength of the outgoing photoelectron is varied by changing the energy of the incident x-ray photon, the interference shifts repeatedly from constructive to destructive. Thus, a sine wave is superimposed on the high-energy side of the classic x-ray absorption edge. The phase and frequency of the wave can be analyzed to give the type and distance of ligand atoms, and the amplitude of the wave is related to the number of ligands.

There are several ways of extracting the structural information contained in the EXAFS data. Since the frequency of the EXAFS wave is related to the metal-ligand distance, Fourier transformation of the data from k ($Å^{-1}$) space to distance (Å) space is a useful way of visualizing the metal-ligand distances involved. Because of the phase shift in the modulatory sine term, distances read directly from the transform are not the true metal-ligand distances. But if an empirical phase shift[1,3] for each metal-ligand pair of interest is evaluated from spectra of structurally characterized compounds, the Fourier transform can be used to give a first guess at the true distance in an unknown compound. In systems with complex first coordination spheres, there is often overlap of peaks in the transform due to each type of scattering atom, so that the interpretation of the transform in terms of actual distances is not straightforward. Also, there is no clear way of distinguishing atomic types of scatterers by a simple analysis of the Fourier transform. Because of these limitations, a more powerful method has been developed[1,3] to extract more fully the structural information contained in EXAFS data. This method involves fitting a sum of individual sine waves to the EXAFS data, each of which derives from one of the neighboring shells of atoms. Each shell of neighbors is described by a set of fixed parameters, which specify the type of atom in the shell, and variables, which determine the number of and distances to atoms in the shell.

The best way to illustrate this EXAFS analysis method is to present its application to the spectrum of a metal in a complex coordination environment. As an example, the dimeric complex of molybdenum, $Mo_2S_4[S_2CN(C_2H_5)_2]_2$ synthesized by W.E. Newton and collaborators,[11] will be used. It contains three distinct types of sulfur ligands as well as a Mo-Mo interaction. The crystal structure of this compound, by Huneke and Enemark[12], will be used as a measure of the success of the data analysis. The Mo EXAFS spectrum of the compound is complex, as illustrated in Figure 1. The presence of beat frequencies is especially prominent, providing clear evidence of more than one superimposed sine wave. A Fourier transform (Figure 2) confirms that there are at least two very different frequencies that make up the data, since there are two major peaks, separated by 0.55 Å. Application of phase shifts derived from model compound spectra[3] to this Fourier transform allows one to make an initial guess at the structure, which will guide attempts at curve fitting analysis. Assigning the shorter distance peak to Mo-S, and the longer one to Mo-Mo, gives 2.3 Å for Mo-S and 2.75 Å for Mo-Mo distances.

The first fit tried is with a single sulfur wave for an atom at a starting distance of around 2.3 Å. The least squares routine refines the shell of sulfur atoms to 2.354 Å (for reference, the average of the dithiocarbamate and the bridging sulfur distances in the crystal is 2.393 Å). The striking thing about this fit is that the curve fitting routine is able to pick out of the data a frequency corresponding so closely to the actual distance, even though this single wave obviously reproduces the data very poorly (Figure 3b). The Fourier transform shows evidence of a pronounced shoulder on the first peak, at around 1.6 Å (non-phase shifted). Application of the 0.46 Å Mo-S phase shift[3] gives a predicted Mo-S distance of 2.06 Å. If a single sulfur wave with a frequency corresponding to a distance of around 2.0 Å is used to fit the data, a shell of sulfur at 2.065 Å is found by the fitting routine. Again, the single wave is a poor model for the data, at least by eye (Figure 3a), but the fitting method has refined the Mo-S(terminal) distance to within 0.03 Å of the crystallographic value. When these two waves are combined, the fit begins to reproduce some of the beat frequencies in the data (Figure 3c), and gives distances similar to those found in the one-wave fits. When a Mo-Mo wave is added to the two Mo-S waves (Figure 3d), the three-wave fit finally reproduces most of the complex amplitude modulations of the data. The largest discrepancy between the crystal structure and the EXAFS three-wave fit is 0.031 Å, with the Mo-Mo distance only 0.009 Å from the crystallographic value.

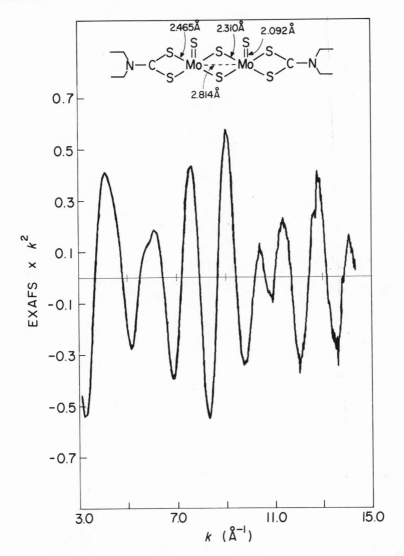

Fig. 1. Mo Model Compound EXAFS. The Mo coordination environment in this compound is complex, as shown by the M-L distances determined by x-ray crystallography. The EXAFS spectrum reflects this complexity, consisting of several superimposed sine waves. The interference of the waves causes amplitude modulations, and beat frequencies are clearly seen in the data. This spectrum is the result of one 20-min scan, taken under conditions of dedicated production of synchrotron radiation, at the Stanford Synchrotron Radiation Laboratory.

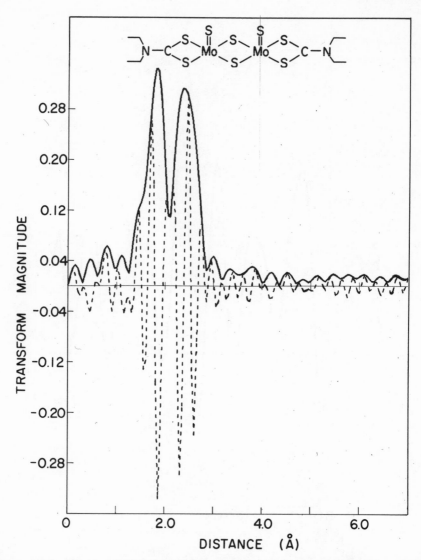

Fig. 2. Mo Model Compound Fourier Transform. Fourier transform
of the data of Figure 1. The two major peaks show the reason
for the beats in the EXAFS data - two Mo-L distances, separated
by 0.55 Å. The shorter distance peak is assigned to Mo-S, the
longer distance one to Mo-Mo. There is also some indication of
a short Mo-S bond, in the pronounced shoulder on the first
peak (around 1.6 Å not phase-shifted).

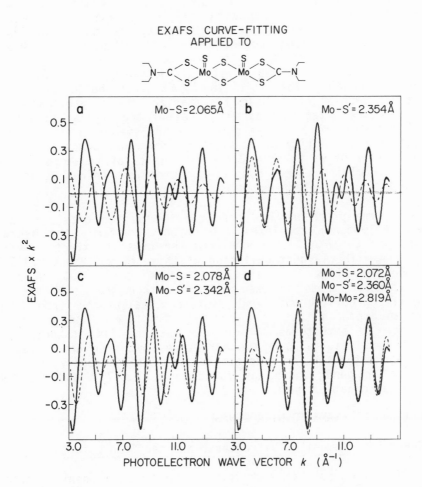

FIG. 3.   Curve Fitting of the Mo Model Compound EXAFS.   The EXAFS
data are shown as a solid line, the fit as a broken line.   The two
single sulfur wave fits (a and b) clearly do not describe the data
well, but the distances calculated from the fits are remarkably
close to the crystal structure values, showing the power of the
curve fitting method.   When two sulfur waves are combined (c), some
of the amplitude modulations appear.   If a Mo wave is added to the
two S waves (d), most of the beats are reproduced, especially at
high k (showing that the Mo contributes mostly to the high k data).
See the text for numerical results of these fits.

Including separate waves in the fit for the dithiocarbamate and bridging sulfurs, along with the Mo-Mo and Mo-S(terminal) waves, gives a composite wave which faithfully models the data, even all of the beat frequencies. This calculated wave is compared to the data in Figure 4. The average deviation of the EXAFS distances from those of the crystal structure is 0.010 Å. With appropriate rounding, the fit of the EXAFS gives the correct number of atoms in each shell. Table 1 summarizes the comparison of the EXAFS structure determination with that of the crystal structure. It is clear that our curve-fitting method, even with its assumptions of transferability of parameters from model compound spectra and additivity of waves from different shells of scatterers, does well in determining a very complex coordination sphere, using only two adjustable parameters per wave.

This same procedure has been carried out on data from some eighteen molybdenum complexes[3] of known structure; the average deviation of the EXAFS distance from the crystal structure value was 0.017 Å. The largest deviation of this set was about 0.035 Å. The number of atoms in a shell is determined to 20% accuracy. These extensive tests of the analysis method show that one can expect the precision of bond distances calculated from EXAFS spectra to approach that of a crystal structure. Application of the method to structures[3] which were at the time unknown also revealed good results once the structures had been determined by diffraction methods. For four complexes in this category, the mean difference in bond length was found to be 0.017 Å and the mean error in coordination number was 20%.[19]

APPLICATIONS OF EXAFS TO NITROGENASE

Molybdenum Environment in Nitrogenase and the FeMo Cofactor

Results from our EXAFS studies on the MoFe component of nitrogenase from Clostridium pasteuranium,[13] Azotobacter vinelandii[14] and the MoFe cofactor from Azotobacter vinelandii[14] have led to the following principal conclusions: (i) the molybdenum coordination sites in the native proteins and in FeMo-co are similar; (ii) the molybdenum atom is primarily ligated by sulfur and is separated by <3Å from another metal, most definitely not molybdenum and therefore iron, which is the only other metallic component of nitrogenase; and (iii) Mo=O interactions are absent in the resting (semi-reduced) state of the enzyme, and appear only when the enzyme is irreversibly aerobically deactivated.

Conclusion (ii), together with the known presence of inorganic sulfide and cysteinyl residues in nitrogenase MoFe proteins, implicates a Mo-Fe-S(R) polynuclear cluster structure with bridging sulfide and, possibly, terminal ligation by cysteinate.

TABLE 1

The   Accuracy   of   EXAFS   Structure   Determination   of $Mo_2S_4[S_2CN(C_2H_5)_2]_2$

	Crystallography	EXAFS	$\Delta R(\text{Å})$
Mo=S	2.092 Å (ave)	2.097 Å	0.005
Mo-S$_b$	2.310 Å	2.318 Å	0.008
Mo-S$_{dtc}$	2.465 Å	2.453 Å	0.012
Mo-Mo	2.814 Å	2.828 Å	0.014

ave. 0.010 Å

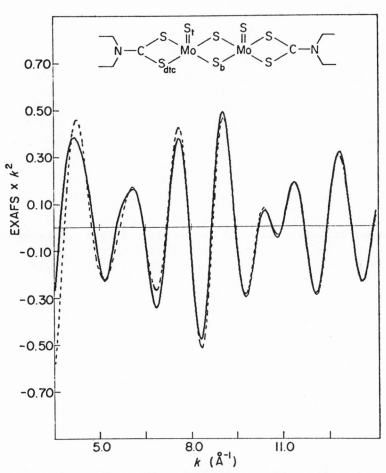

FIG. 4. Final Fit of Mo Model Compound EXAFS. This four-wave fit (Mo, S_t, S_b, S_{dtc}) gives a faithful description of the data, and the distances calculated from the fit are in excellent agreement with the crystal structure results (See Table 1 for a comparison of the EXAFS and crystallographic result).

Model Systems

Holm and his collaborators have undertaken a study of Mo-Fe-S clusters working under the hypothesis that such a cluster, when found naturally in a biomolecule, may be a thermodynamically favored, soluble reaction product and hence it, or some precursor, may spontaneously self-assemble from simple reactants. The first experiments[10] led to the preparation and structural characterization of the bridged double-cubane complex $[Mo_2Fe_6S_9(SEt)_8]^{3-}$ which contains two $MoFe_3S_4$ cubane-type cores linked by a $Mo(\mu-S)-(\mu_2-SEt)_2$ Mo bridging unit. Nearly simultaneously, Christou et al.[15,16] described the structure and several other properties of the related complex $[Mo_2Fe_6S_8(SPh)_9]^{3-}$. Subsequently, another product $[Mo_2Fe_6S_8(SEt)_9]^{3-}$ has been isolated and studied by us and Holm.[17]

The EXAFS spectra for $[Mo_2Fe_6S_9(SEt)_8]^{3-}$ ($\mathbf{1}$) and $[Mo_2Fe_6S_8(SEt)_9]^{3-}$ ($\mathbf{2}$) are shown in Figure 5. Even casual inspection of the data reveals certain common features, notably the large negative and positive peaks in the EXAFS at about 5 and 6.4 $\mathring{A}^{-1}$, respectively. At these values of k, the Mo-S(core) and Mo-Fe waves are approximately in phase. The major difference occurs in the beat region of the spectrum (k = 7.0-10.0 $\mathring{A}^{-1}$) where the symmetrically bridged complex shows a substantially greater positive excursion around k= 8.5 $\mathring{A}^{-1}$, which results from a more in-phase combination of waves from the three core sulfide and three bridge thiolate sulfur atoms. These results, which show visually the change in the EXAFS arising from the differences in bridge units $\mathbf{1}$ and $\mathbf{2}$ at essential parity of core structures, have been quantified by curve-fitting analysis. Fits to the data of both synthetic compounds and the MoFe protein of nitrogenase from C. pasteurianum have been carried out and the results are summarized in Table 2.

TABLE 2   Summary of EXAFS Curve Fitting Results[a]

	Nitrogenase			$[Mo_2Fe_6S_9(SEt)_8]^{3-}$			$[Mo_2Fe_6S_8(SEt)_9]^{3-}$		
	R ($\mathring{A}$)	No.	L	R ($\mathring{A}$)	No.	L	R ($\mathring{A}$)	No.	L
Mo-S	2.352	3.8		2.351	3.7		2.345	2.4	
Mo-Fe	2.690	2.5		2.730	2.6		2.744	2.4	
Mo-S'	2.490	1.1		2.551	2.2		2.548	2.2	

[a]Fits over the range of k = 4-14 $\mathring{A}^{-1}$ on Fourier filtered data. Numerical methods used in the fitting procedure are described elsewhere.[17] The function values, $x = [k(data-fit)^2/N]^{1/2}$, for the three-wave fits of nitrogenase, $[Mo_2Fe_6S_9(SEt)_8]^{3-}$, and $[Mo_2Fe_6S_8(SEt)_9]^{3-}$ were 0.51, 0.42, and 0.37, respectively.

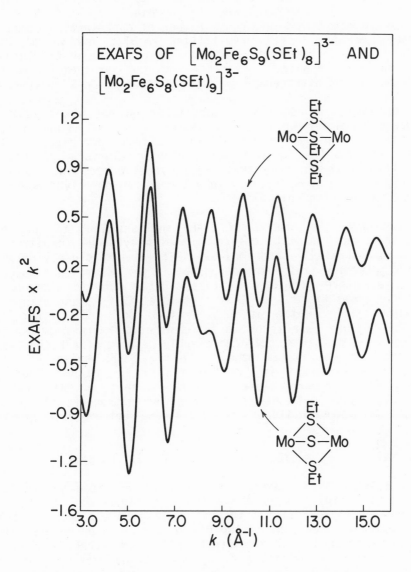

FIG. 5. A comparison of the Mo EXAFS spectra of [Mo₂Fe₆S₉(SEt)₈]³⁻ and [Mo₂Fe₆S₈ (SEt)₉]³⁻. The beat region (k ≈ 7-10 Å⁻¹) reflects the difference between the symmetric (1) and asymmetric (2) bridging units.

The mean difference in distances determined by EXAFS compared with those for the structures reported herein is 0.007 Å while the mean error in determination of the number of coordinating atoms in each shell is 16%. The curve-fitting results and visual comparisons of EXAFS data for synthetic compounds and nitrogenase provide the best available evidence for the involvement of the molybdenum atom of nitrogenase in a polynuclear Mo-Fe-S cluster.

Error limits in the EXAFS amplitudes do not yet permit an unambiguous characterization of the exact molecular nature of the molybdenum site in the enzyme. However, from comparison of the curve-fitting results for the two synthetic clusters and nitrogenase, it appears unlikely that the molybdenum atom in a Mo-Fe-S cluster is bridged symmetrically to another iron or molybdenum atom (e.g., as in bridging unit Å).

In summary, the results of EXAFS analyses lead to the conclusion that the immediate environment of molybdenum in nitrogenase, $(Mo_2Fe_6S_8(SEt)_9)^{3-}$, and $[Mo_2Fe_6S_9(SEt)_8]^{3-}$ are closely related and, thus, directly implicate molybdenum in the enzyme in a Mo-Fe-S cluster. The core structure of this cluster unit is such as to place 3-4 S atoms at $\sim$2.3 Å, 2-3 Fe atoms at $\sim$2.7 Å, and with much less certainty 1 S atom at $\sim$2.5 Å from Mo. The first two structural features are found in the synthetic $MoFe_3S_4$ cores, and the longer Mo-S distance could conceivably be part of a bridge unit to an iron or molybdenum atom. By the EXAFS criterion, the molybdenum environment in $[Mo_2Fe_6S_9(SEt)_8]^{3-}$ rather than in $[Mo_2Fe_6S_8(SEt)_9]^{3-}$ more closely resembles that in the enzyme. Further elucidation of this environment in nitrogenase by EXAFS requires improved amplitude parameters and better data on the enzyme.

EXAFS STUDIES OF THE MOLYBDENUM SITE OF XANTHINE OXIDASE

The molybdenum K absorption edges of oxidized and dithionite-reduced xanthine oxidase are shown in Figure 6. Unlike the smooth edge characteristic of the nitrogenase MoFe component, both xanthine oxidase edges exhibit two distinct inflection points. The presence and intensity of the first inflection at 20003 eV (assigned to a 1s-4d bound state transition) argues strongly for at least one, and more likely two, terminal oxo groups bound to molybdenum in the oxidized form of the protein. Based on the absorption edges of numerous other Mo complexes with biologically relevant ligands (N, O, or S), the position of the second inflection point at 20016 eV also suggests the presence of oxo groups. Upon reduction, the higher energy inflection is seen to move by about 3.1 eV to lower energy. This shift could result from either a formal two-electron reduction of the molybdenum (Mo(VI) to Mo (IV)) while maintaining the same ligation, or the loss of one oxo group, or both. For example, the edge of the ionic $MoCl_3$ is lower by about 3.9 eV than that of $MoCl_5$,[13] and the edge of $MoO_3(dien)$ is about 1.6 eV lower than that of MoO_4^{2-}. Only the Mo(IV) or the Mo(VI) oxidation states are consistent with the absence of a molybdenum EPR signal in the reduced enzyme. Considering the relatively low energy of the second inflection, the edge results are most consistent with reduced xanthine oxidase containing Mo(IV).

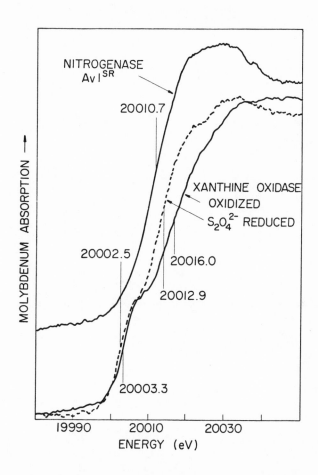

FIG 6. A comparison of the Mo K x-ray absorption edges for oxidized and dithionite-reduced xanthine oxidase with that of the semireduced MoFe component of <u>Azotobacter vinelandii</u> ($Av1^{Sr}$) nitrogenase. The energies of the inflection points as calculated numerically are shown. The edges of the xanthine oxidase show striking resemblance to those of molybdenum complexes containing one or two Mo=O groups. Upon air oxidation, the nitrogenase edge also takes on a shape and position clearly indicative of Mo=O groups.

To gain a more quantitative understanding of the structure of the molybdenum site, we have analyzed the Mo EXAFS of oxidized xanthine oxidase with our curve fitting techniques. The most obvious feature of the EXAFS data (Figure 7) is the presence of a "beat" in the amplitude envelope, which is direct evidence of at least two different Mo-L distances. The best two-wave fit determined 1.5 oxygen atoms at 1.71(2) Å, and 2.0 sulfur atoms at 2.54 (2) Å. Including another sulfur atom at 2.84(2) Å significantly improved the fit, especially in the beat region. This three-wave fit is compared to the Fourier filtered data in Figure 7.

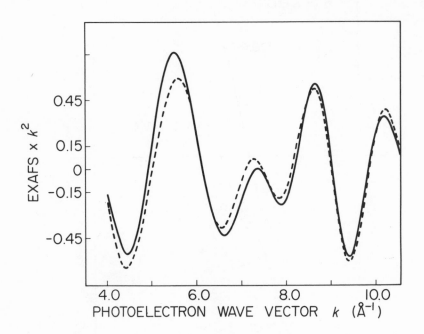

FIG. 7.   Fourier Filtered (k=4-10.5Å^{-1}, R=0.3-3.0 Å) oxidized xanthine oxidase data (——). The beat clearly indicates the presence of at least two different Mo-L distances. The best three-wave fit to the data (O,S,S' waves) (---) results in the distances and coordination numbers described in the text. Fit range, k=4-10.5 Å^{-1}, k^{6} weighting.

Fits using several other combinations of ligands and distances were tried, none of which gave reasonable results. For example, the oxo (short oxygen) wave was absolutely required. No other atom type but sulfur would satisfactorily reproduce the beat when combined with the oxo wave. The identity of the longer sulfur is also unique, and it could not be replaced by a sulfur at a shorter distance (see Table 3 for the results of one such fit). No really good evidence was found for the presence of terminal sulfur, alone or in combination with terminal oxygen. We illustrated earlier that our curve fitting method can determine accurate distances and coordination numbers for terminal sulfur, using as an example the dimeric molybdenum dithiocarbamate of Newton. Terminal sulfur (Mo=S) has been recently suggested by Bray[18] as the source of the "cyanolysable sulfur atom" of xanthine oxidase, based on interpretations of EPR spectra. This point maybe further elucidated by additional EXAFS studies.

The distances and coordination numbers derived from the best three-wave fit lead to a model for the oxidized state of xanthine oxidase which is in agreement with known molybdenum structural chemistry. The Mo-O distance of 1.71(2) Å is similar to that found for molybdenum complexes which have several terminal oxo groups,[19] and is not consistent with a bridging oxo group distance. The intermediate distance sulfurs are most likely thiolates, although the distances are slightly longer (ca. 0.1 Å) than observed for thiolates in other MoO_2L_4 complexes.[19] Although the EXAFS contains no information about the angular disposition of the ligands, this longer distance suggests that these sulfurs may be trans to the Mo=O groups (which are themselves normally cis in Mo(VI) complexes) due to the well known trans effect of Mo=O. The longer sulfur ligand could be a thioether, consistent with the structure reported for a molybdenum complex synthesized by Stiefel.[19] An alternative formulation for this sulfur could be as the second sulfur of a persulfide. Unfortunately, no examples of this type of coordination yet exist in molybdenum chemistry.

Persulfide ligation logically raises the question of the nature of the cyanide inactivation of xanthine oxidase. As our enzyme was not free of the "desulpho" form, the molybdenum site could reflect inhomogeneity due to the presence of both forms (assuming that the inactivation involves the primary coordination environment of molybdenum), which could account for our non-integral number of oxo groups. It is also possible that a fraction of a Mo=S group could be difficult to detect, given the quality and range of the data. It is not possible, given our present EXAFS data, to make any further conclusions regarding this question. However, the clear ability of EXAFS to distinguish the presence or absence of a single sulfur, even at 2.84 Å,[19] makes more accurate experiments quite feasible assuming enzyme homogeneity questions can be resolved.

Because of the preliminary nature of these experiments, the data were good only over a limited range. Another scattering atom of low atomic number at 2-2.5 Å would be difficult to observe in our present data. The amine nitrogen, which has been suggested as the ligand with the acidic proton which splits the Mo(V) epr signals[20], cannot be excluded on the basis of the presently available EXAFS data.

ACKNOWLEDGMENTS

We wish to acknowledge the close collaboration with Professor R.H. Holm and T.E. Wolff throughout the course of these studies. This research was supported by the National Science Foundation through Grant PCM 17105. Synchrotron Radiation Beam Time was provided by the Stanford Synchrotron Radiation Laboratory, supported by NSF Grant DMR-77-27489 in cooperation with the Stanford Linear Accelerator Center and the United States Department of Energy.

REFERENCES

1. S.P. Cramer and K.O. Hodgson, X-Ray Absorption Spectroscropy: A New Structural Method and Its Applications to Bioinorganic Chemistry, in: "Progress in Inorganic Chemistry," S.J. Lippard, ed., John Wiley and Sons, New York (1979).

2. S. Doniach, P. Eisenberger, and K.O. Hodgson, X-Ray Absorption Spectroscopy of Biological Molecules, in: "Synchrotron Radiation Research," H. Winick and S. Doniach, eds., Plenum Press, New York. In press.

3. S.P. Cramer, K.O. Hodgson, E.I. Stiefel, and W.E. Newton, A Systematic X-Ray Absorption Study of Molybdenum Complexes. The Accuracy of Structural Information from Extended X-Ray Absorption Fine Structure, J. Amer. Chem. Soc., 100:2748 (1978).

4. R.G. Shulman, P. Eisenberger and B.M. Kincaid, X-Ray Absorption Spectroscopy of Biological Molecules, Ann. Rev. Biophys. Bioeng., 7:559 (1978).

5. T.L. Blundell and L.N. Johnson, "Protein Crystallography", Academic Press, New York, N.Y. (1976).

6. J.L. Chambers and R.M. Stroud, The Accuracy of Refined Protein Structures: Comparison of Two Independently Refined Models of Bovine Trysin, Acta Cryst. In press.

7. R.C. Bray, Molybdenum Iron-Sulfur Flavin Hydroxylases and Related Enzymes, in:"The Enzymes", vol. XII, part B, P.D. Boyer, ed., Academic Press, New York, N.Y. (1975) p. 299.

8. W.H. Orme-Johnson and L.C. Davis, Current Topics and Problems in the Enzymology of Nitrogenase,in: "The Iron Sulfur Proteins", vol. 3, W. Lovenberg ed., Academic Press, New York, N.Y. (1977), p. 16.

9.   R.H. Holm, Identification of Active Sites in Iron-Sulfur
     Proteins, in: "Biological Aspects of Inorganic Chemistry",
     A.W. Addison, W.R. Cullen, D. Dolphin, B.R. James, ed.,
     Wiley-Interscience, New York, N.Y. (1977) p. 71.
10. T.E. Wolff, J.M. Berg, C. Warrick, K.O. Hodgson, R.H. Holm,
     and R.B. Frankel, The Molybdenum-Iron-Sulfur Cluster Complex
     $[Mo_2Fe_6S_9(SC_2H_5)_8]^{3-}$. A Synthetic Approach to the Molybdenum
     Site in Nitrogenase, J. Amer. Chem. Soc., 100:4630 (1978).
11. F.A. Schultz, V.R. Ott, D.S. Rolinson, D.C. Bravard, J.W.
     McDonald, and W.E. Newton, Synthesis and Electrochemistry of
     Oxo- and Sulfido-Bridged Molybdenum (V) Complexes with 1,
     1-Dithiolate Ligands, Inorg. Chem., 17:1758 (1978).
12. J.T. Huneke and J.H. Enemark, The Mo=S Bond Distance in
     Di-μ-sulfido-bis[sulfido (N,N-diethyl-dithiocarbamate)
     molybdenum (V)], Inorg. Chem., 17:3698 (1978).
13. S.P. Cramer, K.O. Hodgson, W.O. Gillum and L.E. Mortenson, The
     Molybdenum Site of Nitrogenase. Preliminary Structural
     Evidence from X-Ray Absorption Spectroscopy, J. Amer. Chem.
     Soc., 100:3398 (1978).
14. S.P. Cramer, W.O. Gillum, K.O. Hodgson, L.E. Mortenson, E.I.
     Stiefel, J.R. Chisnell, W.J. Brill, and V.K. Shah, The
     Molybdenum Site of Nitrogenase. 2. A Comparative Study of
     Mo-Fe Proteins and the Iron-Molybdenum Cofactor by X-Ray
     Absorption Spectroscopy, J. Amer. Chem. Soc., 100:3814. (1978)
15. G. Christou, C.D. Garner, F.E. Mabbs, A Molybdenum Derivative
     of a Four-Iron Ferredoxin Type Center, Inorg. Chim. Acta,
     29:189 (1978).
16. G. Christou, C.D. Garner, F.E. Mabbs and T.J. King, Crystal
     Structure of Tris(tetra-n-butylammonium) Tri-μ-benzenethiolato-
     bis{tri-μ-sulfido-[μ3-sulfido-tris-(benzenethiolatoiron)]-
     molybdenum}$(Bu_4N)_3[{(PhSFe)_3MoS_4}_2(SPh)_3]$; an Fe_3MoS_4 Cubic
     Cluster Dimer, J.C.S. Chem. Comm., 17:740 (1978).
17. T.E. Wolff, J.M. Berg, K.O. Hodgson, R.B. Frankel and R.H.
     Holm, Synthetic Approaches to the Molybdenum Site in
     Nitrogenase. Preparation and Structural Properties of the
     Molybdenum-Iron-Sulfur "Double-Cubane" Cluster Complexes
     $[Mo_2Fe_6S_8(SC_2H_5)_9)]^{3-}$ and $[Mo_2Fe_6S_9(SC_2H_5)_8]^{3-}$, J. Amer. Chem.
     Soc., 101:4140 (1979).
18. D.J. Lowe and R.C. Bray, Magnetic Coupling of the Molybdenum
     and Iron-Sulphur Centers in Xanthine Oxidase and Xanthine
     Dehydrogenases, Biochem. J., 169:471 (1978).
19. J.M. Berg, K.O. Hodgson, S.P. Cramer, J.L. Corbin, A. Elsberry,
     N. Pariyadath, and E.I. Stiefel, Structural Results Relevant
     to the Molybdenum Sites of Xanthine Oxidase and Sulfite Oxidase.
     Crystal Structures of $MoO_2L,L=(SCH_2CH_2)_2NCH_2CH_2X$ with $X=SCH_3$,
     $N(CH_3)_2$, J. Amer. Chem. Soc., 101:2774 (1979).
20. E.I. Stiefel, The Coordination and Bioinorganic Chemistry of
     Molybdenum, in: "Progress in Inorganic Chemistry," Vol.22, S.J.
     Lippard, ed., John Wiley and Sons, New York (1977) p.1.

EXAFS AND OTHER STUDIES OF SULFITE OXIDASE MOLYBDENUM

S. P. Cramer

Exxon Research and Engineering Company
P. O. Box 45
Linden, New Jersey 07036, USA

H. B. Gray, N. S. Scott

Department of Chemistry
California Institute of Technology
Pasadena, California 91109, USA

M. Barber, K. V. Rajagopalan

Department of Biochemistry
Duke University
Durham, North Carolina 27710, USA

INTRODUCTION

Sulfite oxidase is a molybdenum- and heme-containing enzyme which catalyzes the oxidation of sulfite to sulfate, with cytochrome c as the ultimate physiological electron acceptor.[1] Sulfite is oxidized at the molybdenum site, which is, in turn, reoxidized by the b_5-like heme.[2] The enzyme thereby couples the two-electron oxidation of sulfite with two one-electron reductions of cyctochrome c, and the catalytic cycle shown on page 158.

The enzyme from beef[3], chicken[4] or rat[5] liver has a molecular weight of about 110,000 and exists as a dimer of roughly 55,000 dalton subunits, each containing a molybdenum and heme prosthetic group.

Sulfite oxidase from rat liver can be proteolytically cleaved to yield molybdenum and heme "domains", each of which preserves

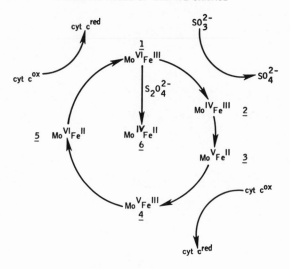

the essential features of the particular metal site.[6] The chicken
enzyme can also be cleaved, but in this case only the heme domain
remains intact.[7] We have recently discovered that the beef enzyme
can also be cleaved into intact molybdenum and heme domains, and
the results of such a cleavage experiment are presented in Figure
1.

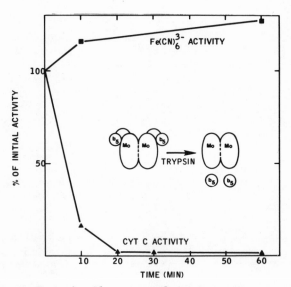

Figure 1   Tryptic Cleavage of Beef Liver Sulfite Oxidase

SULFITE OXIDASE REDOX POTENTIALS

As a complement to kinetic and structural[8] studies of sulfite oxidase, the redox potentials of both heme and molybdenum sites have been investigated under a variety of conditions. The heme potentials are similar to those obtained from other b_5-like cytochromes,[9-12] while the molybdenum potentials are between those previously reported for xanthine oxidase[13] and nitrate reductase.[14] Figure 2 illustrates a typical redox titration of sulfite oxidase, in which the heme is monitored in a thin-layer spectroelectrochemical cell. Under all conditions examined, the heme displayed simple Nernstian behavior. The potentials obtained are given in Table 1.

The Mo potentials of sulfite oxidase were obtained from EPR-monitored titrations as shown in Figure 3. The potentials shift to lower values when the pH is raised, as expected for a site with a dissociable proton. The molybdenum potentials are compared with those of other proteins in Table 2.

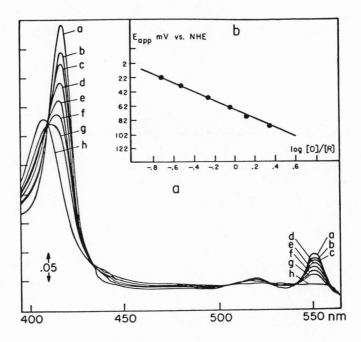

Figure 2   Typical Spectroelectrochemical Titration of Sulfite Oxidase

Table 1 – Cytochrome b_5-Like Potentials

Protein	State	Conditions	E_o (mV)	Reference
	beef, intact	100 mM Tris pH 9	55	
oxidase sulfite	beef, cleaved	100 mM Tris pH 9	66	
	beef, intact	100 mM KP_i pH 7	61	this work
	beef, cleaved	100 mM KP_i pH 7	57	
	chicken, intact	100 mM KP_i pH 7	84	
cytochrome b_5	intact	50 mM KP_i pH 7	24	10
	cleaved	50 mM KP_i pH 7	20	
lactate dehydro-enase	intact	100 mM P_i pH 7	-28	11
	cleaved	100 mM P_i pH 7	0	

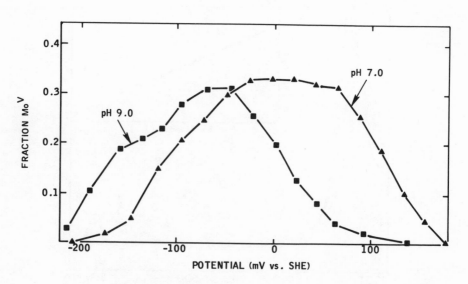

Figure 3 – EPR monitored titration of sulfite oxidase
molybdenum. (Recent recalibration of the
integrated EPR indicates a maximum of 50%
Mo^V.)

Table 2 - Redox Potentials of Molybdenum Sites in Proteins

Protein	Conditions	Redox Couple	E_O (mV)	Ref.
nitrate reductase	100 mM MOPS pH 7.14	$Mo^{VI} - Mo^{V}$ $Mo^{V} - Mo^{IV}$	+220 +180	14
sulfite oxidase	5 mM Tris-Cl pH 9	$Mo^{VI} - Mo^{V}$ $Mo^{V} - Mo^{IV}$	+117 -110	this work
	5 mM Tris-Cl pH 7	$Mo^{VI} - Mo^{V}$ $Mo^{V} - Mo^{IV}$	+38 -163	
xanthine dehydrog- enase	50 mM P_i pH 8.2	$Mo^{VI} - Mo^{V}$ $Mo^{V} - Mo^{IV}$	-350 -362	16
milk xanthine oxidase	55 mM P_i pH 8.2	$Mo^{VI} - Mo^{V}$ $Mo^{V} - Mo^{IV}$	-355 -355	13
"desulfo" milk x.o.	55 mM P_i pH 8.2	$Mo^{VI} - Mo^{V}$ $Mo^{V} - Mo^{IV}$	-440 -480	13
Mo-(2Fe-2S) protein	100 mM Tris- Cl pH 8.5	$Mo^{VI} - Mo^{V}$ $Mo^{V} - Mo^{IV}$	-415 -530	15

X-RAY ABSORPTION SPECTROSCOPY OF SULFITE OXIDASE MOLYBDENUM

The technique of x-ray absorption spectroscopy has recently become a powerful tool for the study of metalloproteins. In the past, the only significant structural probe of molybdenum in enzymes was EPR spectroscopy. The high g-values observed for Mo(V) in sulfite oxidase pointed to the presence of sulfur ligands, and the proton superhyperfine splitting demonstrated the proximity of a hydrogen atom to the molybdenum. However, a complete structural picture is not available from EPR, and the Mo(IV) and Mo(VI) states are inaccessible to this technique. Since a detailed explanation of the x-ray absorption technique is available in the preceding paper, we will concentrate on the results only.

Absorption Edges

The molybdenum absorption edges of sulfite oxidase in the oxidized (as-isolated) and dithionite-reduced states are shown in Figure 4. The important feature in these edges is the shoulder on the low energy side with an inflection point at about 20003 eV. It has been observed that molybdenum compounds with Mo=O or Mo=S bonds have edges similar to these sulfite oxidase edges. From the present data, we can immediately say that in both states the molybdenum of sulfite oxidase has one or two Mo=O or Mo=S bonds.

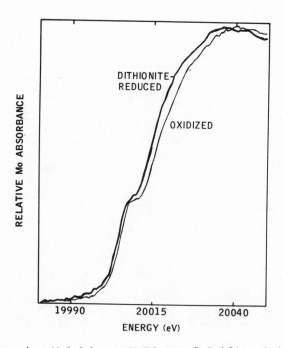

Figure 4 - Molybdenum K-Edges of Sulfite Oxidase

EXAFS

    The EXAFS Fourier transforms for the molybdenum atoms in
sulfite oxidase in the oxidized, sulfite-reduced, and dithionite-
reduced forms are illustrated in Figure 5a.   These transforms
are a very rough approximation to a radial distribution function,
but the peaks are shifted from the true distances by 0.2 to 0.5 Å.
The sulfite oxidase transforms bear a strong resemblance to those
of other molybdenum compounds with a mixture of oxo and sulfur
ligands (Figure 5b), and it is reasonable to assign the first
major peak in the sulfite oxidase transform to Mo=0, and the
second peak to Mo-S.

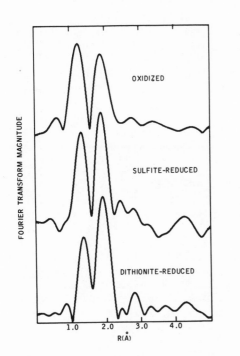
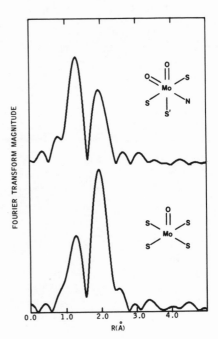

Figure 5a) Sulfite Oxidase Mo EXAFS Fourier Transforms
Figure 5b) (Top) MoO_2 $((SCH_2CH_2)_2NCH_2CH_2SCH_3)$, (Bottom)
$MoO(S_2CN(C_2H_5)_2)_2$

A more detailed analysis of the EXAFS spectra can be made by curve-fitting the data with empirical phase shift and amplitude functions. Beyond about 50 eV above th absorption edge, the EXAFS modulations can be described by the following formula:

$$\chi\,(k) = \sum_{s} \frac{N_s \left| f_s\,(\pi,k) \right| e^{-2\sigma^2_{as} k^2} \sin\,(2kR_{as} + \alpha_{as}(k))}{kR^2_{as}}, \qquad (1)$$

where $f_s\,(\pi,k)$ is the backscattering amplitude of the neighboring atom, R_{as} is the mean distance between adsorber (a) and scatterer (s), σ^2_{as} is the mean square deviation of that distance, N_s is the number of equivalent scatterers, and α_{as} is the total phase shift for propagation from a to s and back to a. Examination of equation (1) shows that in order to obtain distance information from EXAFS, knowledge of the total phase shift $\alpha(k)$ is required, while calculation of scatterer numbers requires knowing the scattering amplitude and Debye-Waller factor as well. Although both $\alpha_{as}(k)$ and $\left| f_s(\pi,k) \right|$ can now be calculated from first principles, phase shifts and total amplitudes can also be obtained experimentally from the spectra of suitable model compounds.

We have analyzed the sulfite oxidase EXAFS by minimizing the difference (f) between χ_{obs} and χ_{calc} in a non-linear least squares routine:

$$f = (\sum_{n} k^6 (\chi_{calc} - \chi_{obs})^2)/n, \qquad (2)$$

where

$$\chi_{calc} = \sum_{s} \frac{N_s c_0 e^{-c_1 k^2} \sin\,(2kR_{as} + a_0 + a_1 k + a_2 k^2)}{R^2_{as} k^{c_2}} \qquad (3)$$

For each different elemental type of scatterer in the fits, a different set of phase shift (a_0, a_1, a_2) and amplitude (c_0, c_1, c_2) parameters is required. Mo-S, Mo-O, and Mo-N parameters were obtained from the EXAFS of $Mo(S_2C_6H_4)_3$, MoO_4^{2-}, and $Mo(NCS)_6^{3-}$ model compounds. The results of these fits are shown in Figure 6 and Table 3.

Table 3 - Best Fits for Sulfite Oxidase and Models

Sample	Mo-0	Mo-S	Mo-S'	Mo-(0 or N)
S.O. (oxidized)	1.71(1.8)	2.42(2.0)	2.84(.7)	
(sulfite-reduced)	1.72(1.9)	2.36(2.9)	2.90(.7)	2.01(1.0)
(dithionite-reduced)	1.69(1.0)	2.38(3.3)	2.85(.9)	2.06(1.0)
$MoO_2[(SCH_2CH_2)_2NCH_2CH_2$ $SCH_3^2]$	1.69(2.1) [1.69(2)]	2.40(1.7) 2.40(2)	2.80(.5) 2.79(1)	
$MoO[S_2CN(C_2H_5)_2]_2$	1.66(1.2) [1.66(1)]	2.43(3.7) 2.41(4)]		

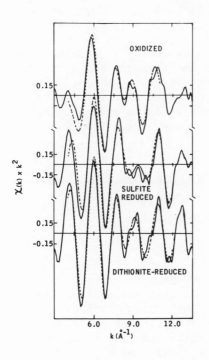 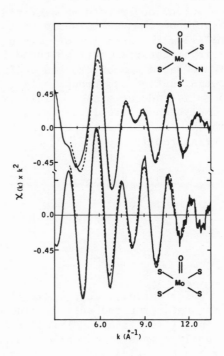

Figure 6a) EXAFS (——) and fits (-----) for sulfite oxidase
Mo in various states

6b) Model compound EXAFS (——) and fits (-----)

For all three oxidation states, the major features of the
EXAFS may be accounted for by including a short-length Mo-O wave
and a medium-length Mo-S wave in the fits.   However, a further
improvement in the quality of the fits, is obtained through introduction of a third wave corresponding to a long Mo-S bond.   For the
EXAFS of the two reduced forms, but not the oxidized enzyme, a
fourth wave corresponding to either oxygen or nitrogen at about
2.05 Å was also found to improve the fits.

Two facts from oxomolybdenum chemistry allow us to propose
a fairly detailed structure for the molybdenum site.   First,
oxomolybdenum groups are almost invariably cis to each other.
Second, bonds trans to oxo groups are lengthened relative to
comparable bonds in a cis geometry.   With this in mind, a reasonable structure for the oxidized, Mo(V) state of sulfite oxidase is I, while dithionite-reduced, Mo(V) sulfite oxidase
probably has a structure like II.   Ligand Y in structure I might
be a sulfur donor with an elogated Mo-S bond, or alternatively
it could be a water molecule or some other loosely bound ligand.
Similarly, the identity of ligand X in structure II is unclear.
The important results from these fits are that: 1) sulfite
oxidase has a mix of oxo and sulfur donor ligands; and 2) it
appears that an oxo group is lost by molybdenum between oxidized
and fully reduced states.   This is consistent with the idea that
the molybdenum site is acting as an oxo-transfer reagent.   Presumably, the molybdenum site picks up a new oxygen from water
on being reoxidized to Mo(VI).

I   II

Unfortunately, we can say little at this point about the
Mo(V) state from the sulfite-reduced EXAFS data.   Although we
originally felt that addition of excess sulfite should result
in nearly complete conversion to the Mo(V) state, it appears
from the titration data that a mix of Mo(VI), Mo(V), and Mo(IV)
states will always be encountered.   In the future, we will ti-
trate the enzyme to a known potential, quantitate the Mo(V)
present by EPR, and attempt to subtract out the contributions
to the EXAFS of the Mo(VI) and Mo(IV) calculated to be present.

REFERENCES

1. H. J. Cohen, S. Betcher-Lange, D. Kessler, and K. V.
    Rajagopalan, Hepatic Sulfite Oxidase-Congruency
    in Mitochoncria of Prosthetic Groups and Activity,
    J. Biol. Chem. 247, 7759 (1972).
2. J. L. Johnson and K. V. Rajagopalan, Electron Paramagnetic
    Resonance of the Tungsten Derivative of Rat Liver
    Sulfite Oxidase, J. Biol. Chem. 251, 5505 (1976).
3. H. J. Cohen, and I. Fridovich, Hepatic Sulfite Oxidase-
    Purification and Properties, J. Biol. Chem. 246, 359
    (1970).
4. D. L. Kessler and K. V. Rajagopalan, Purification and
    Properties of Sulfite Oxidase from Chicken Liver,
    J. Biol. Chem. 247, 6566 (1972).
5. D. L. Kessler, J. L. Johnson, H. J. Cohen, and K. V.
    Rajagopalan, Visualization of Hepatic Sulfite Oxidase
    in Crude Tissue Preparations by Electron Paramagnetic
    Resonance Spectroscopy, Biochem. Biophys. Acta. 334,
    86 (1974).
6. J. L. Johnson and K. V. Rajagopalan, Tryptic Cleavage of
    Rat Liver Sulfite Oxidase, J. Biol. Chem. 252, 2017
    (1977).
7. B. Guiard and F. Lederer, The "b5-like" Domain from
    Chicken Liver Sulfite Oxidase, Eur. J. Biochem. 74,
    181 (1977).
8. S. P. Cramer, H. B. Gray, and K. V. Rajagopalan, The
    Molybdenum Site of Sulfite Oxidase-Structural Evidence
    from X-Ray Absorption Spectroscopy, J. Am. Chem. Soc.
    101, 2772 (1979).
9. S. Velick and P. Strittmatter, The Oxidation-Reduction
    Stoichiometry and Potential of Microsomal Cytochrome,
    J. Biol. Chem. 221, 265 (1956).
10. H. Weber, W. Weiss, and H. Stauginder, Bestimmung des
    Standardredoxpotentials von Cytochrom b5, Hoppe-Seyler's
    Z. Physiol. Chem. 352, 109 (1971).
11. F. Labeyrie, O. Groudinsky, Y. Jacquot-Armand, and
    L. Maslin, Proprietes d'un Noygau Cytochromique b2
    Resultant d'une Proteolyse de la L-Lactate: Cytochrome
    C Oxydoreductase de la Levure, Biochem. Biophys. Acta.
    128, 492 (1966).
12. B. Chance, D. F. Wilson, P. L. Dutton, and M. Erecinska,
    Energy Coupling Mechanisms in Mitochondria, Proc. Nat.
    Acad. Sci. 66, 1175 (1970).
13. R. Cammack, M. J. Barber, and R. C. Bray, Oxidation-Reduction
    Potentials of Molybdenum, Flavin, and Iron-Sulfur Centres
    in Milk Xanthine Oxidase, Biochem. J. 157, 469 (1976).

14. S. P. Vincent, Oxidation-Reduction Potentials of Molybdenum
     and Iron-Sulfur Centres in Nitrate Reductase from
     Escherichia Coli, Biochem. J. 177, 757 (1979).
15. J. J. G. Moura, A. V. Xavier, R. Cammack, D. O. Hall,
     M. Bruschi, and J. LeGall, Oxidation-Reduction
     Studies of the Mo-(2Fe-2S) Protein from Desulfovibrio
     Gigas, Biochem. J. 173, 419 (1978).
16. M. J. Barber, R. C. Bray, R. Cammack, and M. P. Coughlan,
     Oxidation-Reduction Potentials of Turkey Liver Xanthine
     Dehydrogenase, Biochem. J. 163, 279 (1977).

IRON–MOLYBDENUM COFACTOR AND

MO–FE–S CUBANE CLUSTERS

METABOLISM OF MOLYBDENUM BY NITROGEN-FIXING BACTERIA

Winston J. Brill and Vinod K. Shah

Department of Bacteriology and Center
  for Studies of Nitrogen Fixation
University of Wisconsin
Madison, Wisconsin 53706, U.S.A.

INTRODUCTION

The role of molybdenum in nitrogen fixation has been studied
for many decades but only now is it being understood at the
molecular level. This paper focuses on the use of mutant strains
to understand molybdenum metabolism and its role in nitrogenase.

BIOCHEMISTRY OF MOLYBDENUM METABOLISM

In 1930, Bortels[1] discovered that molybdenum stimulated growth
of Azotobacter when the bacterium was grown on N_2. The explanation
of this phenomenon was obvious after Bulen and LeComte separated
the two fractions of nitrogenase and showed that one of them contained molybdenum. Nitrogenase has been purified from many
bacteria[3] and all nitrogenases contain two proteins, component I
(Mo-Fe protein) and component II (Fe protein). Molybdenum is
always found in component I.

When A. vinelandii is grown in a molybdenum-deficient medium,
it is unable to fix N_2. This observation is not surprising since
molybdenum is an integral part of component I. However, molybdenum-
starved cells do not synthesize component I, not even in an inactive form.[4] Inactive component I can readily be detected
serologically or by polyacrylamide gel electrophoresis. This same
dependence of nitrogenase synthesis on molybdenum also holds true
with Klebsiella pneumoniae.[5] Thus, these N_2-fixing bacteria have
regulatory mechanisms which prevent an inactive, molybdenum-
deficient component I of nitrogenase from being synthesized when

the cells are starved for molybdenum. Such a regulatory system
saves the cell energy and metabolites. There are examples in nature
of molybdenum-deficient soils and this regulation might have a very
important role.[6,7]

In the case of A. vinelandii, the addition of tungstate to a
molybdenum-deficient medium allowed an inactive nitrogenase to be
formed.[4] When molybdate is added to the tungsten-grown cells,
nitrogenase is activated and this in vivo activation does not
require the synthesis of new protein. Activation by molybdate,
however, did not occur in vitro in extracts of tungsten-grown cells.

There was good evidence that molybdoenzymes contain their
molybdenum in some form of cofactor.[8,9] Acid treatment of a molybdoenzyme inactivated the enzyme but did not destroy the cofactor.
When acid-treated component I was added to extracts of tungsten-
grown cells activation of inactive nitrogenase occurred.[10] It is
not yet known how this activation proceeds. Exchange of tungsten
by molybdenum may be all that is necessary or that the tungsten-
analog of the molybdenum cofactor is replaced by the molybdenum-
containing cofactor. Another possibility is that the tungsten-grown
cells do not have tungsten in nitrogenase at all, but rather that
tungsten inhibits a reaction required for the synthesis of the
molybdenum cofactor.

Nitrogenaseless mutants (Nif[-]) of A. vinelandii[11] and K.
pneumoniae[12,13] have been isolated and characterized. Some of the
mutants made an active component II and an inactive component I.
Such mutants might be a result of a mutation in the structural genes
for the two subunits of component I or they might have a lesion
in a gene required specifically for synthesis of the molybdenum
cofactor. The latter possibility was tested by adding acid-treated
component I (a source of the cofactor) to extracts of the mutant
strains. Several mutants had an inactive component I that could
be activated in vitro upon addition of acid-treated component I.[10,12,13]

It is very easy to purify large amounts of pure component I.[14]
Extracts of the molybdenum cofactor-negative mutants could be used
as a simple assay for the cofactor. Thus, it seemed a simple
matter to purify the molybdenum cofactor from nitrogenase component
I. However, acid treatment did not solubilize the cofactor.
The activity was associated with the denatured protein. Many
techniques were attempted in order to solubilize the cofactor, but
all conventional methods normally used in biochemistry failed. One
technique was successful--extraction of the cofactor from acid-
precipitated component I with N-methylformamide.[15] Component I was
precipitated with citrate and the precipitate washed with dimethylformamide. The washed precipitate contained the molybdenum, but
lost about half of the iron in component I. N-methylformamide was

then used to extract the cofactor. The extracted material contained all of the molybdenum in component I as well as the remaining iron.

We did not expect to find iron as part of the molybdenum cofactor since it had previously been shown that the molybdenum cofactor from nitrogenase could replace the molybdenum cofactor from
nitrate reductase in Neurospora crassa (the Nit-1 mutant).[8,16,17]
Since nitrate reductase does not contain non-heme iron, it would be
expected that the nitrogenase molybdenum cofactor also not contain
non-heme iron. Further purification of the component I-extracted
cofactor was, therefore, necessary to determine if the extracted
iron was actually part of the cofactor. When the N-methylformamide-
extracted material was applied to a Sephadex G-100 column (run in
N-methylformamide), the fractions containing both molybdenum and
activity also had iron.[15] Electrophoresis of the cofactor also
showed co-migration of molybdenum and iron with activity. The
cofactor contains 8 iron and 6 acid-labile sulfide atoms per atom
of molybdenum. Since this discovery, the cofactor from nitrogenase
has been called the iron-molybdenum cofactor (FeMo-co). FeMo-co
has a molecular weight less than 5,000 daltons.

FeMo-co has been isolated in a similar manner from component I
purified from a variety of different N_2-fixing organisms.[15] All
of these cofactors had the same activity when they were assayed by
activation of a FeMo-co-negative mutant of A. vinelandii. This
result was especially interesting in the case of Clostridium
pasteurianum, which is known to produce a component I that is unable
to fix N_2 in vitro with component II from A. vinelandii.[18]
Because the FeMo-co in these two bacteria are similar, the inability
to achieve cross complementation must reside in the structural
proteins of component I.

One of the properties of FeMo-co that initially hampered
purification is its extreme sensitivity to oxygen. However, as long
as FeMo-co is kept anaerobically in N-methylformamide, it will remain
stable for days even at room temperature.[15] Because component I is
oxygen labile, we wondered whether a FeMo-co-negative mutant might
have an apo-protein that is stable to oxygen. In other words, is
FeMo-co the sole oxygen-sensitive site of component I? However, the
inactive component I from such a mutant could not be activated by
FeMo-co when the extract was exposed to oxygen prior to the activation assay. When apo-component I from the mutant strain UW45 of A.
vinelandii, which can be activated by FeMo-co in crude extracts, was
purified, it could still be activated by FeMo-co. This observation
demonstrated that activation did not require another enzyme.

FeMo-co did not[19] activate Nit-1, the nitrate reductase mutant
of N. crassa, in direct contrast to previous work which showed that
purified component I contained a factor able to activate Nit-1 in

<u>vitro</u>.[8,16,17] We were unable to repeat those observations, even with crystallized component I that had been acid treated.[19] It turned out that the component I used previously was contaminated with another type of molybdenum cofactor that could activate Nit-1. This second cofactor, called Mo-co, is common to xanthine oxidase and nitrate reductase, but is inactive with nitrogenase. FeMo-co and Mo-co can be separated from each other on a Sephadex G-100 column.[19]

Because molybdenum is such an unusual metal to be found in an enzyme, it was expected to be a part of the active site of nitrogenase. The evidence pointing to FeMo-co as the active site is now quite good. Examination of electron paramagnetic resonance (EPR) spectra of Nif⁻ mutants showed that the unique signal at g=3.65 is caused by the active site of component I.[11] The site responsible for the signal undergoes oxidation and reduction during N_2 fixation.[20] FeMo-co, in the presence of thiophenol, also has this signal.[21] Further support of FeMo-co as the active site comes from the observation that FeMo-co itself catalyzes acetylene reduction to ethylene in the presence of a strong reducing agent like sodium borohydride.[22] The reduction of acetylene to ethylene is a non-physiological reaction that is catalyzed by all nitrogenases. Both nitrogenase- and FeMo-co-catalyzed acetylene reduction are inhibited by carbon monoxide.[22]

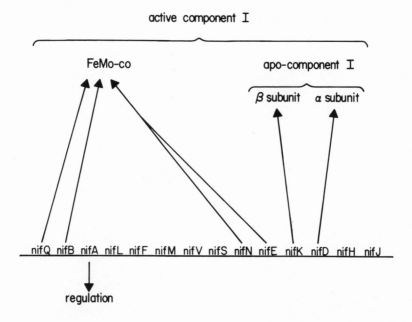

Fig. 1. Genes coding for active component I in <u>Klebsiella</u> <u>pneumoniae</u>.

GENETICS OF MOLYBDENUM METABOLISM

A detailed genetic study of Nif⁻ mutants was performed with K. pneumoniae.[23] Fourteen nif genes were shown to be required specifically for N_2 fixation (Figure 1). This result is surprising because most enzymatic reactions require one or two genes. The number of nif genes points to the tremendous complexity of nitrogenase. The protein products of most of the genes have been identified and the roles of most of the proteins have now been determined.[13] Four of the nif genes code for proteins necessary for synthesis of an active FeMo-co and two nif genes are required for the structure of the two subunits of component I. At present, mutants with lesions in the FeMo-co genes are being used to determine the sequence of reactions involved in synthesizing FeMo-co. The protein products of nifN and nifE interact as a complex.[13] All fourteen nif genes are regulated through a regulatory gene, nifA.

In the presence of excess ammonium, a cell has no need for nitrogenase; therefore, ammonium-grown cells have mechanisms for switching off the synthesis of nitrogenase. FeMo-co synthesis also is prevented when ammonium is included in the growth medium.[13] Interestingly, Mo-co synthesis occurs in ammonium-containing medium.[19] Mutants that are insensitive to ammonium repression have been isolated and these mutants produce FeMo-co whether or not ammonium is in the medium.[24,25]

CONCLUSION

Many questions have yet to be answered. What is the structure of FeMo-co? How do electrons, protons, and N_2 interact on FeMo-co? How is FeMo-co synthesized and placed onto nitrogenase? These questions are expected to be answered by integrated work from genetic, biochemical, and chemical approaches to the problems.

REFERENCES

1. H. Bortels, Molybdan als Katalysator bei der biologischen Stickstoffbindung, Archiv. Mikrobiol. 1:333 (1930).
2. W. A. Bulen and J. R. LeComte, The nitrogenase system for Azotobacter: two enzyme requirement for N_2 reduction, ATP-dependent hydrogen evolution and ATP hydrolysis, Proc. Natl. Acad. Sci. U.S.A. 56:979 (1966).
3. R. R. Eady and J. R. Postgate, Nitrogenase, Nature 249:805 (1974).
4. H. H. Nagatani and W. J. Brill, Nitrogenase V. The effect of Mo, W and V on the synthesis of nitrogenase components in Azotobacter vinelandii, Biochim. Biophys. Acta 362:160 (1974).

5. W. J. Brill, A. L. Steiner, and V. K. Shah, Effect of molyb-
    denum starvation and tungsten on the synthesis of nitro-
    genase components in Klebsiella pneumoniae, J. Bacteriol.
    118:986 (1974).

6. H. J. Evans, E. R. Purvis, and F. E. Bear, Effects of soil
    reaction on availability of molybdenum, Soil Sci. 71:117
    (1951).

7. A. J. Anderson, Molybdenum in relation to pasture improvement
    in south Australia, J. Council Sci. Indus. Res. 19:1 (1946).

8. A. Nason, K.-Y. Lee, S.-S. Pan, P. A. Ketchum, A. Lamberti,
    and J. DeVries, In vitro formation of assimilatory reduced
    nicotinamide adenine dinucleotide phosphate: nitrate
    reductase from a Neurospora mutant and a component of
    molybdenum-enzymes, Proc. Natl. Acad. Sci. U.S.A. 68:3242
    (1971).

9. P. A. Ketchum, H. Y. Cambier, W. A. Frazier III, C. W.
    Madansky, and A. Nason, In vitro assembly of Neurospora
    assimilatory nitrate reductase from protein subunits of a
    Neurospora mutant and the xanthine oxidizing or aldehyde
    oxidase systems of higher animals. Proc. Natl. Acad. Sci.
    U.S.A. 66:1016 (1970).

10. H. H. Nagatani, V. K. Shah, and W. J. Brill. Activation of
    inactive nitrogenase by acid-treated component I, J. Bac-
    teriol. 120:697 (1974).

11. V. K. Shah, L. C. Davis, J. K. Gordon, W. H. Orme-Johnson,
    and W. J. Brill, Nitrogenase III. Nitrogenaseless mutants
    of Azotobacter vinelandii. Activities, cross-reactions
    and EPR spectra, Biochim. Biophys. Acta 292:246 (1973).

12. R. T. St. John, H. M. Johnston, C. Seidman, D. Garfinkel,
    J. K. Gordon, V. K. Shah, and W. J. Brill, Biochemistry and
    genetics of Klebsiella pneumoniae mutant strains unable to
    fix N₂, J. Bacteriol. 121:759 (1975).

13. G. P. Roberts, T. MacNeil, D. MacNeil, and W. J. Brill,
    Regulation and characterization of protein products coded
    by the nif (nitrogen fixation) genes of Klebsiella
    pneumoniae, J. Bacteriol. 136:267 (1978).

14. V. K. Shah and W. J. Brill, Nitrogenase IV. Simple method of
    purification to homogeneity of nitrogenase components from
    Azotobacter vinelandii, Biochim. Biophys. Acta 305:445
    (1973).

15. V. K. Shah and W. J. Brill, Isolation of an iron-molybdenum
    cofactor from nitrogenase, Proc. Natl. Acad. Sci. U.S.A.
    74:3249 (1977).

16. C. E. McKenna, N. P. L'vov, V. L. Ganelin, N. S. Sergeev, and
    V. L. Kretovich, Existence of low-molecular factor, common
    to different molybdenum-containing enzymes, Dokl. Akad.
    Nauk. SSSR 217:228 (1974).

17. W. G. Zumft, Die Abtrennung niedermolekularer componenten aus
    dem Molybdan-Eisen-Protein der Nitrogenase von Clostridium
    pasteurianum, Ber. Deutsch. Bot. Ges. Bd. 87:135 (1974).

18. D. W. Emerich and R. H. Burris, Interactions of heterologous nitrogenase components that generate catalytically inactive complexes, Proc. Natl. Acad. Sci. U.S.A. 73:4369 (1976).

19. P. T. Pienkos, V. K. Shah and W. J. Brill, Cofactors from molybdoenzymes and in vitro reconstitution of nitrogenase and nitrate reductase activities, Proc. Natl. Acad. Sci. U.S.A. 74:5468 (1977).

20. W. H. Orme-Johnson, W. D. Hamilton, T. Ljones, M. Y. Tso, R. H. Burris, V. K. Shah and W. J. Brill, Electron paramagnetic resonance (EPR) of nitrogenase and nitrogenase components from Clostridium pasteurianum W5 and Azotobacter vinelandii, Proc. Natl. Acad. Sci. 69:3142 (1972).

21. J. Rawlings, V. K. Shah, J. R. Chisnell, W. J. Brill, R. Zimmermann, E. Münck and W. H. Orme-Johnson, Novel metal cluster in the iron-molybdenum cofactor of nitrogenase: spectroscopic evidence, J. Biol. Chem. 253:1001 (1978).

22. V. K. Shah, J. R. Chisnell, and W. J. Brill, Acetylene reduction by the iron-molybdenum cofactor from nitrogenase, Biochem. Biophys. Res. Commun. 81:232 (1978).

23. T. MacNeil, D. MacNeil, G. P. Roberts, M. A. Supiano, and W. J. Brill, Fine-structure mapping and complementation analysis of nif (nitrogen fixation) genes in Klebsiella pneumoniae, J. Bacteriol. 136:253 (1978).

24. J. K. Gordon and W. J. Brill, Mutants that produce nitrogenase in the presence of ammonia, Proc. Natl. Acad. Sci. 69:3501 (1972).

25. D. MacNeil and W. J. Brill. 6-Cyanopurine, a color indicator useful for isolating mutations in the nif (nitrogen fixation) genes of Klebsiella pneumoniae, J. Bacteriol. 136: 247 (1978).

STUDIES ON THE IRON-MOLYBDENUM COFACTOR FROM THE NITROGENASE

Mo-Fe PROTEIN OF KLEBSIELLA PNEUMONIAE

Barry E. Smith

Agricultural Research Council
Unit of Nitrogen Fixation
University of Sussex, Brighton, BN1 9RQ Sussex, U.K.

INTRODUCTION

Nitrogenase,[1] the enzyme which reduces nitrogen to ammonia, consists of two metalloproteins, the Fe protein and the Mo-Fe protein. Mechanistic studies have shown that the Fe protein acts as a MgATP-activated reductant for the the Mo-Fe protein which contains the substrate-binding and -reducing site. The Mo-Fe protein has a molecular weight near 220,000 and consists of two each of two types of subunit with molecular weights of 50,000 and 60,000 respectively. The most active preparations contain two molybdenum atoms and 24-33 iron atoms with approximately equivalent numbers of acid-labile sulfur atoms per molecule.[1] The iron atoms are in at least three distinct environments.[2] Circumstantial evidence indicates that the molybdenum is at the enzyme active site.[3]

Early genetical results and biochemical studies on the nit-1 mutant of Neurospora crassa, which is deficient in nitrate reductase activity, indicated that all molybdenum enzymes might have a common molybdenum-containing cofactor.[4] This cofactor could apparently be liberated from these molybdenum-containing proteins by treatment with acid (which destroyed the enzyme activity) followed by neutralization. Interpretation of these studies was complicated by the extreme instability of the cofactor and the consequent failure to purify it.

A considerable advance was reported orally by Shah and Brill at the 2nd International Symposium of Nitrogen Fixation, 1976 and subsequently published.[5] In their method the cofactor from the nitrogenase Mo-Fe protein was extracted into N-methylformamide (NMF)

179

from the acid-treated protein after neutralization and washing of
the precipitate with N,N-dimethylformamide (DMF). The cofactor
prepared in this manner, although extremely oxygen sensitive, was
relatively stable anaerobically and could activate crude extracts
from a nif B⁻ mutant. Extracts of nif B⁻ mutants contain active
Fe protein but inactive Mo-Fe protein. Preparations of the
cofactor from a variety of nitrogenases all exhibited similar
activating abilities.

Although a molybdenum-containing cofactor from xanthine oxidase, prepared in a similar way, could activate extracts of the
nit-l mutant of N.crassa, the nitrogenase cofactor could not.
Furthermore, cofactors from the two sources were chromatographically
distinct. Thus it was concluded that at least two classes of
molybdenum cofactors exist, one associated with xanthine oxidase
and nitrate reductase and the other with the Mo-Fe protein of
nitrogenase.[6]

The cofactor isolated from nitrogenase was reported to contain
molybdenum, iron and acid-labile sulfur in the ratios 1:8:6 and was[5]
called FeMo-co to distinguish it from the other class of cofactors.
FeMo-co exhibited an epr* spectrum similar in form, though with
slightly distorted symmetry and broader lines, to that observed
from the Mo-Fe protein.[7] In addition, the molybdenum EXAFS* spectra
of the Mo-Fe proteins and FeMo-co were essentially indistinguishable.[8] These data indicated that the molybdenum and iron environments in FeMo-co and the Mo-Fe protein were very similar and that
the cofactor was extracted without major structural alterations.
Analysis of the EXAFS spectra indicated a molybdenum environment
consisting of 3-4 S atoms at 2.36A°, 1-2 S atoms at 2.5A° and 1-2
Fe atoms at 2.73A°.[8] No molecular weight data on FeMo-co have
been published and no evidence on the nature of the ligands to the
metals other than acid-labile sulfur has been presented, although
it has been suggested that amino acids are present.[9]
In this paper, I shall describe the preparation of FeMo-co
from Klebsiella pneumoniae by an acid-treatment similar to the
method of Shah and Brill and by a solvent precipitation method and
shall compare the properties of both preparations. I shall present
evidence related to the molecular weight of the cofactor and discuss
the nature of the ligands other sulfide.

MATERIALS AND METHODS

Klebsiella pneumoniae M5A1 was grown and harvested and the
nitrogenase proteins purified as described previously.[10,11]
Klebsiella pneumoniae strain Kp5058, a nif B⁻ strain,[12] was grown
as KpM5A1 but with the addition of 0.1 g l⁻¹ aspartic acid

---

* epr = electron paramagnetic resonance
EXAFS = Extended X-ray absorption fine structure spectroscopy

dissolved in sodium hydroxide, 0.025 g l^{-1} histidine, 0.4 g l^{-1} yeast extract, 0.2 g l^{-1} sodium chloride and 0.2 g l^{-1} bacto-tryptone to the medium. This medium was slightly repressive and cells were not harvested until the optical density was between 35-40.

Metal analyses,[10] amino acid analyses[10] and epr spectroscopy[13] were carried out as described earlier.

Assays

FeMo-co was assayed by incubating aliquots of the NMF solution with crude extract of Kp5058. Incubations were for 30 minutes at 30°C in a shaking water bath, after which time, a sample of the incubation mixture was added to a standard reaction mixture.[10] Additional Kp2 was added and the reaction mixture assayed for acetylene reduction at 30°C for 10 min. The assay was terminated by addition of 0.1 ml of 30% trichloroacetic acid. Ethylene production was determined as described earlier.[10] Kp5058 crude extract,when assayed with additional Kp2, exhibited a small amount of activity, which was subtracted when determining the ethylene produced due to activation by FeMo-co.

Preparation of FeMo-co

Two methods of preparation of FeMo-co were used. The first (Method 2, Scheme 1) was developed following the description given by Shah at Salamanca in 1976 but differs somewhat from his published method[5] (Method 1, Scheme 1). All methods of preparation of FeMo-co require rigorous anaerobic precautions and so all operations were carried out under a positive pressure of N_2 in glass centrifuge tubes sealed with Suba seals.

In Method 2 (Scheme 1), ice-cold 1 M citric acid (previously sparged with N_2) was added to ice-cold Kp1* until the pH of the solution reached 2.0. After 2-3 minutes incubation, the pH was adjusted to 5.0 by addition of 0.5 M Na_2HPO_4, pH 9.0, 1mM $Na_2S_2O_4$. The resultant precipitate was centrifuged at about 1500 g and the colorless supernatant discarded. The greyish precipitate was then washed twice with 2 ml DMF containing 1% 200 mM aqueous $Na_2S_2O_4$ and then extracted twice with 2 ml NMF containing 5 mM Na_2HPO_4, pH 8.0, (by addition of 0.5 M aqueous Na_2HPO_4) and 2 mM $Na_2S_2O_4$ (by addition of 200 mM aqueous $Na_2S_2O_4$). After centrifugation, the NMF solution of FeMo-co was yellow-brown and the precipitate was white.

In Method 3 (Scheme 1), ice-cold Kp1 solution was slowly treated with an equal volume of dimethyl sulfoxide (DMSO)

---

* Kp1 is the Mo-Fe protein from K.pneumoniae

containing 1% 200 mM $Na_2S_2O_4$. The brown precipitate was
centrifuged and the colorless supernatant discarded. The
precipitate was then treated as in Method 2 except that the
extraction with NMF was slower, each extraction requiring about
1 hr.

Chemicals

Biochemicals were obtained from Sigma Chemical Co., Poole,
Dorset; other chemicals from BDH Chemicals Ltd., Poole, Dorset, U.K.;
Sephadex G100 from Pharmacia Ltd., Hounslow, Middlesex, U.K.; and
N-methylformamide from Aldrich Chemical Co., Gillingham, Dorset,
U.K. The N-methylformamide was distilled in vacuo before use.

Scheme 1. Schematic representation of methods of preparing FeMo-co

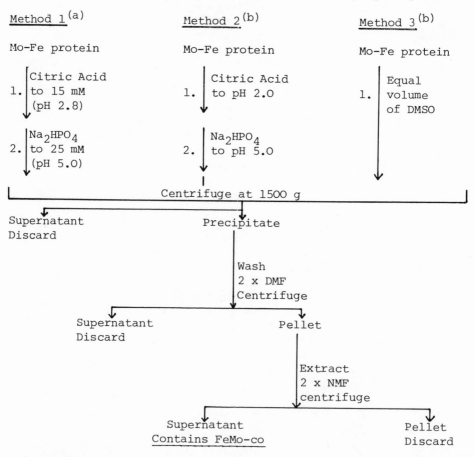

(a) Ref. 5.
(b) Current work

ISOLATION AND PROPERTIES OF FeMo-co

The differences between the method of Shah and Brill[5] (Method 1, Scheme 1) and the acid treatment used in this work (Method 2, Scheme 1) are relatively slight, viz. the pH of the acid treatment and, consequently, the ionic strength of the solution after neutralisation and from which the protein was precipitated. The conditions of Method 2 were originally chosen because Kp1 was not fully inactivated unless treated at pH 2.0. However, Fig. 1 shows the results of an experiment in which Kp1 solutions were given different acid treatments before adjustment to pH 5 followed by extraction of FeMo-co. The data clearly show that under the conditions of the experiment the optimal pH was 2.0-2.2.

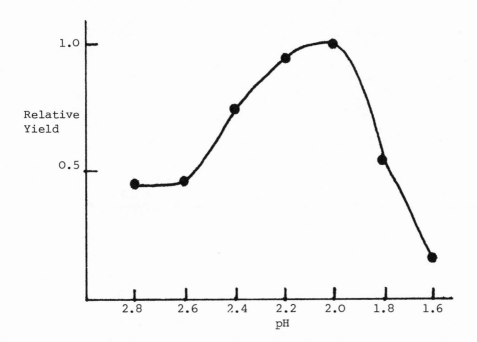

Fig. 1. Effect of pH of acid treatment on the relative yield of FeMo-co from Kp1. To 8 mg Kp1 in 1 ml 25 mM Tris pH 8.7 + 0.1 mg/ml Dithiothreitol + 1 mM $Na_2S_2O_4$ was added to 1 M Citric acid at ice temperature until the pH reached that shown. The mixture was incubated on ice for 2 min and then 0.5 M Na_2HPO_4 pH 9 + 2 mM $Na_2S_2O_4$ added to adjust the pH to 5.0. Further treatment and assay is described in Scheme 1 and the Methods section.

Also extraction of FeMo-co from the precipitate formed after acidification to pH 2.8 followed by neutralization appeared slower than
that formed after the pH 2.0 treatment. This observation would
explain the observed pH optimum if treatment at pH values below 2.0
damaged FeMo-co. Another feature of Methods 1 and 2 which did not
accord with Shah and Brill's description,[5] was that the DMF washes
always extracted some FeMo-co, even after purification of the
solvent by vacuum distillation. In fact, DMF extracted more FeMo-co
after the pH 2.8 than after the pH 2.0 treatment. With the latter
treatment, about 10% of the total units extracted were found in DMF
washes. Rigorous adherence to Method 1 proved less reliable in my
hands than Method 2 and, thus, was not studied further.

Method 3 (Scheme 1) was developed during an investigation of
the conditions necessary for FeMo-co extraction since these might
give information on the nature of the bonding between FeMo-co and
the protein. In particular, it was important to establish whether
acidification of the protein was a prerequisite of FeMo-co extraction
If the citric acid and Na_2HPO_4 were premixed before addition to the
protein, then FeMo-co could be extracted from the resultant
precipitate although the pH did not drop below pH 5.0. However,
this technique did not always result in complete precipitation of
the protein and so other methods were investigated. Some success
was achieved with $(NH_4)_2SO_4$-precipitated Kp1 but extraction was
hindered by the residual $(NH_4)_2SO_4$. DMSO precipitation yielded a
precipitate from which it was relatively easy, although a little
slow, to extract FeMo-co into NMF, although no FeMo-co was extracted
into the DMF washes. The yields of FeMo-co from this method were
at least the equivalent of those obtained with Method 2 and both
methods gave FeMo-co with similar activating ability per molybdenum
atom (see below). Nevertheless, chromatography on Sephadex G100 in
a 1 x 60 cm column in NMF (containing Na_2HPO_4 pH 8, 2 mM $Na_2S_2O_4$
and 2% H_2O) demonstrated that FeMo-co2 (from Method 2) and FeMo-co3
(from Method 3) were not identical (Table 1). Before discussing the
significance of this finding, it is pertinent to consider some other
properties of these FeMo-co preparations.

Metal content

FeMo-co2 and FeMo-co3 were each subjected to G100 chromatography
in NMF (containing 2% H_2O, 5 mM Na_2HPO_4, pH 8.0 and 2 mM $Na_2S_2O_4$) in
order to remove impurities. Each was then assayed for activity and
subjected to metal analysis. Typical results for the ratios Mo:Fe:
activity for FeMo-co2 were 1:10:280 and for FeMo-co3 were 1:8:315.
Some excess iron in the original extracts was removed on the G100
columns without loss of activity. However, some may still remain so
methods 2 and 3 probably yield FeMo-co with a Mo:Fe ratio which is
not significantly different from the ratio of 1:8 reported by Shah
and Brill.[5]

Amino acid content

    After acid-hydrolysis, analysis of a sample of the FeMo-co2
extract before chromatography indicated the presence of all the
amino acids present in Kpl. However, after G100 chromatography,
neither FeMo-co2 nor FeMo-co3 preparations contained significant
amounts of amino acids. For example, analysis of FeMo-co3 after
alkaline hydrolysis showed traces of most amino acids, the most
concentrated of which was glycine but at only, 0.17 molecules per
molybdenum atom. In a separate experiment, samples of FeMo-co2 were
acid-hydrolysed for 24 hrs in the presence and absence of 50 µg
bovine serum albumin (BSA). The amino acid analyses of these,
compared with a control hydrolysis of 50 µg BSA alone, showed that
the presence of FeMo-co in the hydrolysis did not significantly
enhance the rate of degradation of any of the amino acids. Thus,
it was concluded that amino acids are not a constituent part of
active FeMo-co prepared by Methods 2 and 3.

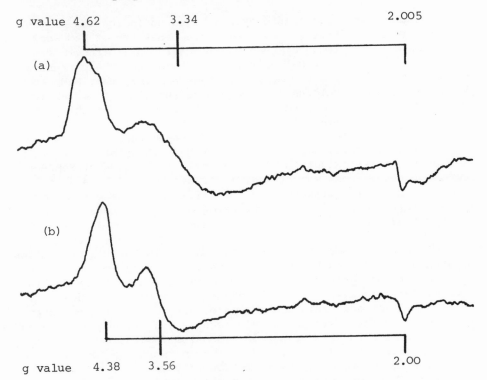

Fig. 2. Effect of thiophenol on the electron paramagnetic resonance
spectrum of FeMo-co. (a) FeMo-co as prepared by Method 3,
Scheme 1. (b) as (a) but with the addition of 10 mM thiophenol; the mixture was frozen after 5 min incubation.
Spectra were recorded at 8°K and a microwave frequency of
9.3 GHz. Microwave power was 150 mW.

Electron paramagnetic resonance (e.p.r.) spectroscopy

FeMo-co1 exhibits an e.p.r. spectrum that is similar in form to
that of the $Na_2S_2O_4$-reduced Mo-Fe protein but has much broader lines
and altered g values indicating a slightly distorted symmetry.[7] The
e.p.r. spectrum was sharpened and rendered much more like the Mo-Fe
protein spectrum by the addition of 10 mM thiophenol indicating some
ligand replacement on FeMo-co by thiophenol. The differing elution
volumes of FeMo-co2 and FeMo-co3 on G100 chromatography in NMF could
be associated with the presence of different ligands and, thus, thiophenol might not have the same effect on the two forms of cofactor.
Fig. 2 shows the e.p.r. spectra of FeMo-co3 in the absence and
presence of 10 mM thiophenol. As with FeMo-co1, a marked sharpening
of the g_1 and g_2 features was observed indicating that the two forms
of the cofactor react similarly with thiophenol.

Sephadex G100 chromatography in NMF

The above structural information gives no indication of the
cause(s) of the differing chromatographic behaviors of FeMo-co2 and
FeMo-co3. Our knowledge of FeMo-co structure is limited to its
molybdenum, iron and acid-labile sulfur ratios and the current work
shows the absence of a peptide. It seems possible that FeMo-co
consists of: (a) metal and acid-labile sulfur and some metabolic
product which helps to bind FeMo-co to the protein; or (b) just the
metals and acid-labile sulfur extracted as an intact cluster. In
this latter case, any of the materials used in the extraction could
act as metal ligands, e.g., the solvent NMF, phosphate, water,
dithionite or its oxidation products, or, in the case of FeMo-co3,
DMSO. In either case (a) or (b), it is necessary to postulate
modification of the structure by the different treatments in order
to explain the differing G100 chromatographic behaviors of FeMo-co2
and FeMo-co3.

A literature search revealed no information on the properties
of Sephadex G100 in NMF. However some information on the chromatography of inorganic compounds and complexes on Sephadex G25 in
aqueous media is available.[14] Sephadex, a cross-linked dextran, is
mainly used for gel exclusion chromatography, i.e., it separates
substances on the basis of their molecular weight. Aqueous Sephadex
G100 has a fractionation range of 4,000-150,000 for peptides and
globular proteins. However, the charge on inorganic ions is often
important in determining their elution volumes.[14] Thus, the
difference between FeMo-co2 and FeMo-co3 could be one of charge.
Accordingly, an investigation into the behavior of some inorganic
compounds and ions during anaerobic G100 chromatography in NMF
(containing 2% H_2O, 5 mM Na_2HPO_4 plus the oxidation products of 2 mM
$Na_2S_2O_4$) was undertaken. The results are summarized in Table 1.

Table 1. Elution volumes (V_e) on Sephadex G100 in N-methylformamide[a]

Compound or ion	Molecular weight	V_e ml
$[WH_4(dppe)_2]$[b]	974	44
$[WBr_2(N_2H_2)(dppe)_2]^+$[c]	1170	59.4
$[Mo(NNCOOCH_3)\{S_2CN(CH_3)_2\}_3]$	543	46.7
$[Mo\{NN(CH_3)_2\}\{S_2CN(CH_3)_2\}_3]^+$[c]	514	50
$[Mo_2Fe_6S_8(SPh)_9]^{3-}$[d]	1766	40.3
$[Fe_4S_4(SBu^t)_4]^{2-}$[d]	708	33.6
$[Fe_4S_4(SPh)_4]^{2-}$[d]	788	46
$[Mo_2Fe_6S_8(SCH_2CH_2OH)_9]^{3-}$[d]	1477	26.2
FeMo-co2	?	32.5±1
FeMo-co3	?	24±2

[a] N-methylformamide containing 5 mM Na_2HPO_4 pH 8.0 + 2 mM $Na_2S_2O_4$ + 2% H_2O. Chromatography was on a 60 x 1 cm glass column under an N_2 atmosphere. [b] dppe = $Ph_2PCH_2CH_2PPh_2$. [c] Cations were loaded as the BPh_4^- salts. [d] Anions were loaded as the $(CH_3)_4N^+$ salts.

Comparison of the neutral complexes $[WH_4(dppe)_2^*]$ and $[Mo(NNCOOCH_3)\{S_2CN(CH_3)_2\}_3]$ with the similar cations $[WBr_2(N_2H_2)-(dppe)_2]^+$ and $[Mo\{NN(CH_3)_2\}\{S_2CN(CH_3)_2\}_3]^+$ respectively demonstrated that positively-charged complexes tend to bind to the gel. Both cations were eluted in volumes greater than the bed volume of the column. Furthermore, the elution profiles of both cations were skewed whereas the neutral complexes had symmetrical elution profiles. The anion $Mo_2Fe_6S_8(SPh)_9$ $^{3-}$ has a symmetrical elution profile and an elution volume of 40.3 ml, significantly more than either form of FeMo-co. However, the smaller anion $[Fe_4S_4(SBu^t)_4]^{2-}$ had an elution volume similar to that of FeMo-co2. This anomaly was resolved by comparison with $[Fe_4S_4(SPh)_4]^{2-}$ which had an elution volume of 46 ml. Clearly, ligands containing phenyl groups cause complexes to bind to the gel and distort the pattern of the elution volumes expected on the basis of molecular weight alone. Chromatography of $[Mo_2Fe_6S_8(SCH_2CH_2OH)_9]^{3-}$ confirmed this observation since it had an elution volume of 26.2 ml, much less than that of the thiophenol analogue, and very similar to that of FeMo-co3.

---

*dppe = $Ph_2PCH_2CH_2PPh_2$

Consideration of the results of Table 1 leads to the following conclusions:-

(1)   Sephadex G100 chromatography in NMF can discriminate between inorganic complexes of molecular weight 500-1500.

(2)   FeMo-co2 and FeMo-co3 apparently need have molecular weights no larger than 7-800 and about 1500 respectively.

(3)   Changing the charge on a complex can markedly affect its elution volume and apparent molecular weight.

(4)   The ligands of the complex may also interact with the gel to increase the elution volume and thus decrease the apparent molecular weight.

One further experiment was carried out to investigate the relationship between FeMo-co2 and FeMo-co3. Because FeMo-co2 was exposed to citric acid during its preparation whereas FeMo-co3 was not and citrate is a good ligand for iron, FeMo-co2 might be the citrate analogue of FeMo-co3. In order to test this hypothesis, 0.1 ml aqueous 1 M sodium citrate, pH 8.5, was added to 1 ml NMF solution of FeMo-co3 and the mixture incubated at room temperature for 15 minutes. Much of the citrate precipitated but, when the mixture was chromatographed on Sephadex G100 in NMF as normal, the elution volume of FeMo-co3 had risen from 24 ml to 29.5 ml. This value is still less than the elution volume of FeMo-co2 but does indicate that the citrate had reacted and had had the hypothesized effect. It is possible that only partial reaction had taken place and that more than one citrate ligand is present in FeMo-co2.

SUMMARY OF CONCLUSIONS

The two methods of preparation of FeMo-co described above gave rise to different species, FeMo-co2 and FeMo-co3, which nevertheless had similar activating abilities and similar ratios of molybdenum to iron. Neither form contained significant amounts of amino acids.

Sephadex G100 chromatography in NMF indicated that the molecular weight of FeMo-co3 is probably about 1500. The molybdenum, iron and acid-labile sulfur contents account for about half of this molecular weight the remainder presumably consisting of other ligands to the metal atoms. These ligands might be specific metabolic products involved in the attachment of FeMo-co to the protein but the reaction with thiophenol suggests that at least some of them are replaceable. Thus, it seems probable that at least some of the ligands of FeMo-co3 are derived during its extraction, i.e., that they are NMF, DMSO, phosphate, water or dithionite ion or its decomposition products.

The apparent molecular weight of FeMo-co2 is only about the sum of its constituent metal and sulfur atoms. It is unlikely that there are not other ligands to the metal atoms and so it is probable that the ligands of FeMo-co2 interact with the Sephadex G100 thus

increasing its elution volume. Reaction of FeMo-co3 with citrate
had a similar effect and, thus, it is probable that FeMo-co2 is a
citrate analogue of FeMo-co3.

The presence of a relatively small ligand that is not an amino
acid yet is derived from the Mo-Fe protein is not excluded by these
studies. Such a ligand would presumably be essential for activity
and thus should be common to FeMo-co2 and FeMo-co3.

ACKNOWLEDGEMENTS

I thank my colleagues Drs. R. Burt, J. Dilworth and R.L.
Richards for providing the inorganic complexes used in this work;
Dr D.J. Lowe for running the e.p.r. spectra; Janice Hojka for skilled
technical assistance; and K. Baker and Helen Kent for the growth of
the bacteria.

REFERENCES

1.   R.R. Eady and B.E. Smith, Physicochemical properties of
         nitrogenase and its components, in "Dinitrogen Fixation",
         Section 2, R.W. Hardy and R.C. Burns, eds., Wiley Inter-
         science, New York (1979) p.399.
2.   B.E. Smith and G. Lang, Mössbauer Spectroscopy of the
         nitrogenase proteins from Klebsiella pneumoniae.
         Structural assignments and mechanistic conclusions,
         Biochem.J. 137:169 (1974).
3.   B.E. Smith, The structure and function of nitrogenase: A review
         of the evidence for the role of molybdenum, J.Less-Common
         Metals 54:465 (1977).
4.   For review, see R.C. Bray, Molybdenum Iron-sulfur Flavin
         hydroxylases and related enzymes, in "The Enzymes"
         Vol.XIII, P.D. Boyer, ed., Academic Press, New York,
         San Francisco, London, (1975) p.299.
5.   V.K. Shah and W.J. Brill, Isolation of an iron-molybdenum
         cofactor from nitrogenase, Proc.Natl.Acad.Sci.U.S. 74:3249
         (1977).
6.   P.T. Pienkos, V.K. Shah and W.J. Brill, Molybdenum cofactors
         from molybdoenzymes and in vitro reconstitution of
         nitrogenase and nitrate reductase, Proc.Natl.Acad.Sci.U.S.
         74:5468 (1977).
7.   J. Rawlings, V.K. Shah, J.R. Chisnell, W.J. Brill, R. Zimmermann,
         E. Munck and W.H. Orme-Johnson, Novel metal cluster in the
         iron-molybdenum cofactor of nitrogenase, J.Biol.Chem. 253:
         1001 (1978).
8.   S.P. Cramer, W.O. Gillum, K.O. Hodgson, L.E. Mortenson,
         E.I. Stiefel, J.R. Chisnell, W.J. Brill and V.K. Shah,
         The molybdenum site of nitrogenase. 2. A comparative
         study of Mo-Fe proteins and the Iron-molybdenum cofactor
         by X-ray absorption spectroscopy, J.Amer.Chem.Soc. 100:

3814 (1978).

9. V.K. Shah, Paper presented at the Steenbock-Kettering International Symposium on Nitrogen Fixation, Madison, Wisconsin, June 12-16, 1978.

10. R.R. Eady, B.E. Smith, K.A. Cook and J.R. Postgate, Nitrogenase of Klebsiella pneumoniae. Purification and properties of the component proteins, Biochem.J. 128:655 (1972).

11. B.E. Smith, R.N.F. Thorneley, M.G. Yates, R.R. Eady and J.R. Postgate, Structure and function of nitrogenase from Klebsiella pneumoniae and Azotobacter chroococcum, in "Proceedings of the 1st International Symposium on Nitrogen Fixation", Vol.1, W.E. Newton and C.J. Nyman, eds, Washington State University Press, Pullman (1976) p.150.

12. R.A. Dixon, C.K. Kennedy, A. Kondorosi, V. Krishnapillai and M.J. Merrick, Complementation analysis of Klebsiella pneumoniae mutants defective in nitrogen fixation, Mol. Gen.Genetics 157:189 (1977).

13. B.E. Smith, D.J. Lowe and R.C. Bray, Studies by electron paramagnetic resonance on the catalytic mechanism of nitrogenase of Klebsiella pneumoniae, Biochem.J. 135:331 (1973).

14. N. Yoza, Gel chromatography of inorganic compounds, J.Chromatog. 86:325 (1973).

CHEMICAL PROPERTIES OF THE Fe-Mo COFACTOR FROM NITROGENASE

William E. Newton, Barbara K. Burgess and
Edward I. Stiefel

Charles F. Kettering Research Laboratory
Yellow Springs, Ohio 45387, U.S.A.

INTRODUCTION

The existence of a molybdenum cofactor common to molybdenum-containing enzymes was first postulated by Pateman et al.[1] who discovered a common genetic determinant for nitrate reductase and xanthine dehydrogenase in Aspergillus nidulans. The cofactor concept was further developed by Nason and co-workers who studied a mutant of Neurospora crassa (nit-1), which was blocked in the terminal portion of its nitrate reductase electron transport chain (where molybdenum is located), and which also lacked xanthine dehydrogenase activity.[2,3] A key discovery in the field was that crude extracts prepared from the nit-1 mutant could be restored to activity by treatment with an acid hydrolysis product obtainable from any molybdenum-containing enzyme,[4,5] including the molybdenum-iron protein of nitrogenase.[6] Additional studies showed that although molybdate could enhance the activity produced by the nit-1 extract plus the acid-treated products of the molybdoenzymes, neither molybdate nor any of the simple molybdenum complexes tested could activate the nit-1 extracts alone. Furthermore, when $^{99}MoO_4^{2-}$ was used, it became incorporated into the newly activated nitrate reductase, but only in the presence of the acid-treated products of the molybdoenzymes. These data led to the hypothesis of a labile molybdenum complex being one of the products of all acid-treated molybdoenzymes.[7]

In 1974, Nagatani et al.[8] demonstrated that the UW 45 mutant of A. vinelandii had an inactive molybdenum-iron protein which could be activated in vitro by adding acid-treated products (themselves inactive) of purified molybdenum-iron proteins from nitrogenases obtained from a number of N_2-fixing organisms. The complementation property of this mutant could be simulated by growing A. vinelandii cells on tungstate where the inactive molybdenum-iron protein produced could be similarly activated in vitro. The availability of this nitrogenase-based cofactor assay system in addition to the nitrate reductase-based assay has greatly stimulated research into the purification and properties of these molybdenum cofactors. Mutants analogous to nit-1 and UW 45,[9] which appear to lack a molybdenum cofactor, have now been isolated from a number of different sources including E. coli,[10,11] Klebsiella pneumoniae nif-B,[12] and Rhizobium japonicum.[13] However, the mutants of nitrogen-fixing organisms lacked only the nitrogenase activity while retaining full activity for nitrate reduction.

Several reports have appeared on the isolation of the cofactor from different nitrogenase sources. Kretovich and coworkers[14] claimed the isolation of a low molecular weight (<1000) molybdenum-containing peptide from A. vinelandii nitrogenase but did not show its ability to activate any of the cofactor-deficient proteins. Zumft[15,16] reported the isolation of two different molybdenum-containing species from C. pasteurianum molybdenum-iron protein, which were suggested to be thiomolybdate derivatives. The most recent report by Brill and Shah described the isolation of a small molybdenum-containing entity from the molybdenum-iron protein of A. vinelandii, which is capable of activating the A. vinelandii UW 45 extracts. This cofactor is reported to contain molybdenum, iron, and labile sulfide in a 1:8:6 atomic ratio and is quite stable in certain organic solvents,[17] but less so in water and very susceptible to O_2. This discovery of non-heme iron associated with the molybdenum in the cofactor from nitrogenase was surprising because fungal nitrate reductase does not contain non-heme iron. Other studies based on attempted reconstitution of the crude extracts of nit-1, UW 45 and demolybdosulfite oxidase have shown that, at least, two molybdenum-containing cofactors exist.[18-20] It appears that one cofactor, called Mo-co, is common to all known molybdoenzymes except nitrogenase, which has its own peculiar molybdenum-containing prosthetic group called FeMo-co.

Insight into the composition of FeMo-co has been gained using a combination of Mössbauer and electron paramagnetic resonance (EPR) spectrscopy[21-23] and from molybdenum X-ray absorption spectroscopy.[24,25] This last technique indicates that the molybdenum environments in the molybdenum-iron protein from A. vinelandii, C. pasteurianum, and FeMo-co from A. vinelandii are extremely similar and quite probably identical. A similar conclusion can be drawn from the Mössbauer and EPR studies about the environments of the iron atoms. These studies also showed that the FeMo-co is the magnetic center of the enzyme responsible for the g = 4.3, 3.65, 2.01 EPR signal. Thus, removal of FeMo-co from the protein apparently does not destroy the integrity of the cofactor. The probable location of molybdenum at the active site of nitrogenase[26,27] suggests that FeMo-co may possess some of the chemical and catalytic properties of the enzyme. Thus, the remainder of the enzyme system, consisting (insofar as prosthetic groups are concerned) of Fe-S clusters, may simply be required to deliver electrons to FeMo-co at the appropriate potential.

This paper deals with our recent work on FeMo-co from A. vinelandii nitrogenase and describes our efforts to determine its composition, and both physical and chemical properties.

ISOLATION AND PURIFICATION OF FeMo-co

Our FeMo-co samples are prepared using the procedure of Shah and Brill.[17] All reagents are kept at 0-4°C, except Na_2HPO_4, degassed and are 1.2 mM in $Na_2S_2O_4$. To each of eight degassed clinical centrifuge tubes containing 30 mg of purified Azotobacter molybdenum-iron protein in 2 ml of 0.025 M Tris (pH 7.4, 0.25 M NaCl) is added water (4 ml), then 0.1 M citric acid (1 ml), mixed, and allowed to stand for 165 seconds. 0.2 M Na_2HPO_4 (1ml) is then added, mixed and allowed to stand for 25 minutes. Centrifugation at slow speed on a clinical centrifuge for 10 minutes allows the pellet, which contains denatured protein and FeMo-co, to be collected and the supernatant discarded. Freshly distilled N,N-dimethylformamide (DMF; 4 ml) is added to the pellet, the mixture vortexed for 30 seconds, and centrifuged at medium speed for 10 minutes. The DMF supernatant is discarded. DMF (4 ml) is again added and the last step repeated. Freshly distilled N-methylformamide (NMF; 2.4 mM in Na_2HPO_4; 2 ml) is then added, the mixture vortexed for 5 minutes and centrifuged at top speed of the clinical centrifuge for 10 minutes. The supernatant is removed and saved and a second identical extraction with NMF performed. The combined NMF extracts (∿24 ml) are then concentrated at room temperature under high vacuum to 4 ml in 1-2 hours.

FeMo-co at this stage of purity contains a small but measurable quantity of protein, whose amino acid composition closely parallels that of the Mo-Fe protein. We believe this material to be contaminating denatured Mo-Fe protein, because anaerobic Sephadex G-100 chromatography in NMF or, much more simply, centrifugation and filtration of the supernatant through a sintered glass frit results in an amino acid-free preparation of similar specific activity. Preparations treated in this way routinely contain 7 iron atoms per molybdenum and, when used to reconstitute crude extracts of the mutant Azotobacter vinelandii UW 45,[17] have activities of ∿300 nmol C_2H_4 produced/min/nmol Mo.

As there is no report concerning the charge on protein-free FeMo-co, we applied anaerobic polyacrylamide gel electrophoresis in the usual buffer to an NMF solution. We observed its migration into the gel and toward the anode, indicating it to be anionic in nature. Its migration was very rapid compared to the Mo-Fe protein, consistent with its (presumed) much smaller size. The anionic nature of FeMo-co gains further support from our preliminary studies involving ion exchange resins. FeMo-co in NMF solution (1 ml) was applied in parallel to 0.1g of Bio-Rad AG1-X2 in its acetate form or Dowex 50-X8 in its sodium form, both in NMF (2 ml). The mixtures were shaken for 5 minutes, allowed to stand for 5 minutes, and centrifuged. An aliquot (0.25 ml) of each supernatant was transferred to separate EPR tubes, sodium dithionite (30 μl of 0.1 M solution in 0.025 M Tris pH 7.4) added to each, then frozen and their EPR spectra obtained at 14°K. Only the supernatant from the cation exchange resin, Dowex 50-X8, exhibited the appropriate EPR (g ∿4.7, 3.4, 2.0) signal. No signal appeared in the spectrum of the NMF supernatant from the anion exchange resin, Bio-Rad AG1-X2. Consistent with this observation was the brown color of FeMo-co assumed by this resin, together with the colorless supernatant. Although preliminary attempts to remove this color (presumably FeMo-co) from the resin with $(CH_3)_4NCl$ (1M in 80% NMF/water) have failed, this binding indicates FeMo-co to be anionic.

REDOX PROPERTIES AND AVAILABLE OXIDATION STATES

After the concentration and filtration steps of our isolation procedure (see above), FeMo-co solutions suitable for freezing and storage and containing several hundreds of nanomoles of molybdenum per milliliter are obtained. For our various studies of its properties, we often dilute this stock solution with NMF or DMF. In early experiments, we frequently found no EPR signal from spectral analysis of an aliquot of these diluted solutions or sometimes a very weak g ∿4.3 signal, typical of high-spin ferric iron, was observed. We found that the disappearance of the g ∿4.7, 3.4, 2.0 EPR signal (that is observed in the presence of excess sodium dithionite and hereafter is called the EPR-active, reduced or "act.-red." signal or state) correlated approximately with the care with which we degassed the diluent, i.e., with the presence of trace amounts of O_2. These observations indicate that our concentration procedure also caused most (if not all) of the excess dithionite to precipitate and to be removed in the pellet. Further studies have shown that this EPR-silent state of FeMo-co can be obtained in a controlled manner using methylene blue as oxidant. This EPR-silent form of FeMo-co (hereafter called the EPR-silent, oxidized or "sil.-ox." state) was susceptible to reduction by aqueous dithionite or thiophenol to reform FeMo-co(act.-red.). With thiophenol, this EPR signal was in the "thiol" form as reported previously.[21] Thus, this is a second redox state of FeMo-co, which is presumably analogous to the "sil.-ox" state observed with the intact molybdenum-iron protein.[22,28-31]

To test this trace O_2 theory further, 10 μl of O_2 (400 nmoles) was added to an NMF solution of FeMo-co(act.-red.) containing 50 nmoles of molybdenum under argon. After mixing, allowing to stand for 3 minutes and freezing, its EPR spectrum at 14°K showed a new intense signal at g = 4.38 and another weaker signal at g ∿2. Rereduction with excess aqueous dithionite eliminated the g ∿4.4 signal completely and the g ∿2 signal by ∿85%, but only partially restored (∿25%) the act.-red. signal, indicating significant denaturation of the chromophore, which is consistent with its known oxygen sensitivity.[17] The g ∿4.4 signal might then be indicative of the presence of Fe^{3+} produced by oxidative destruction of FeMo-co (but see below).

As oxidation of this FeMo-co chromophore while incorporated within the intact Mo-Fe protein of nitrogenase[22,23,28] occurs with appropriate redox-active dyes without denaturation, methylene blue oxidation was attempted on isolated FeMo-co. On addition of an excess of aqueous methylene blue, again a sharp signal at g = 4.38 and a weaker signal at g ∿2.0 appeared, but now of only ∿35% of the intensity of the signals in the O_2-oxidized spectrum. Addition of excess aqueous dithionite regenerated the act.-red. state in ∿95% yield based on a comparison of EPR spectra. These redox experiments can best be interpreted in terms of another available redox state of FeMo-co, an EPR-active, oxidized or "act.-ox." state, which can be reached via oxidation with either dyes or O_2, and is not characteristic of denaturation of FeMo-co. The latter reagent can, however, react further to cause FeMo-co to lose its integrity. The existence and viability of this third redox state for FeMo-co gains support from its activity in the UW 45 reconstitution assay (see below). It also is consistent with the redox experiments carried out by Watt (these Proceedings) on the intact Mo-Fe protein or nitrogenase. These three redox states (see figure) are related as in equation (1), where n and n' are the as yet unknown numbers of electrons (e^-) required to interconvert the appropriate redox states.

$$\text{FeMo-co(act.-ox.)} \xrightleftharpoons[-n'e^-]{+n'e^-} \text{FeMo-co(sil.-ox.)}$$

$$\xrightleftharpoons[-n\ e^-]{+n\ e^-} \text{FeMo-co(act.-red.)} \qquad (1)$$

PHOTOREDUCTION REACTIONS

Our discovery of the two previously unknown oxidized states of FeMo-co led us to search for other redox states more reduced than the EPR-active, reduced (act.-red.) state. This highly reduced

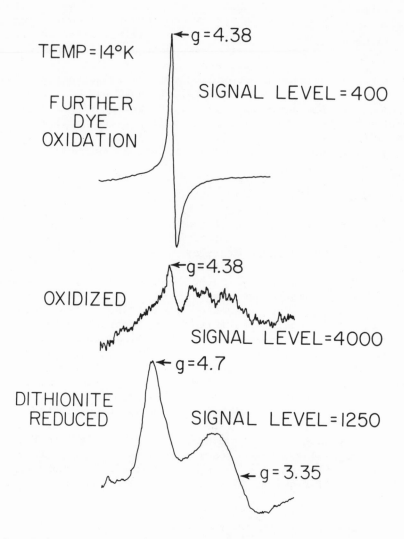

Fig. 1.   EPR spectra of the three redox states of the FeMo-co from
A. vinelandii nitrogenase in frozen NMF solution at 14°K.

state would likely correspond to the proposed substrate-reducing,
EPR-silent form of the Mo-Fe protein in the intact, fixing nitrogenase system.[29-31]   A recent report[21] implied that this "sil.-red."
state had been achieved using the highly reducing system consisting
of photo-activated 5-deazaflavin and EDTA (as the electron donor)
but only under a carbon monoxide atmosphere. We attempted to reproduce the conditions described for this photoreduction, but were
unable to achieve a final concentration of 10 mM EDTA in an NMF

solution of FeMo-co. This concentration could only be reached by adding aliquots of an aqueous EDTA solution. However, in our hands, precipitation occurred of some, as yet unidentified, material within 15 minutes of the addition of the EDTA solution to FeMo-co in NMF. Other donors were then sought. Dimethylaminoethanol (DMAE), a commonly used donor in photoredox systems,[32,33] was chosen as the donor as it is miscible with NMF and was found to be very effective in the 5-deazaflavin-mediated photo-reduction of both methyl violo-gen in NMF and the Azotobacter flavoprotein in aqueous solution.

On illuminating FeMo-co at 28°C with a 300-watt lamp for 30 minutes with or without DMAE, with or without 5-deazaflavin, and with or without dithionite under a CO, N_2 or argon atmosphere, to our surprise, no discernible change could be detected in the original act.-red. EPR spectrum. Similar experiments were then attempted with EDTA, even though precipitation occurred in the EPR tube during illumination, because the 5-deazaflavin/EDTA/light system did, nevertheless, effectively and rapidly reduce methyl viologen in NMF. The act.-red. signal of the control (containing no additions) persisted after illumination, but was lost completely in the presence of 5-deazaflavin, EDTA, and dithionite under 100% CO. However, control experiments with all additions, but under argon and with no 5-deazaflavin present, also did not exhibit an EPR signal. These results show that EDTA alone interacts with the EPR chromo-phore. We also found that the act.-red. signal did not reappear on thawing and replacement of CO by argon and could not be regen-erated by addition of aqueous dithionite to this same sample. Thus, in our hands, this system does not operate as a photoredox system as previously reported.[21] The strongly reducing conditions generated by 5-deazaflavin and light apparently have nothing to do with the loss of the act.-red. EPR signal in the presence of EDTA; CO is not required for this signal loss; and this reaction is not reversed by evacuation. The nature of the interaction of EDTA with FeMo-co is not presently understood. It cannot, however, be a destructive perturbation as FeMo-co treated in this way is fully active in the UW 45 reconstitution assay (see below).

EPR SPECTRUM AND RECONSTITUTION ACTIVITY

As the three redox states of FeMo-co identified above are readily interconvertible and all assume the act.-red. state in the presence of excess aqueous dithionite, it was expected that all three would show comparable activities on a molybdenum basis when used to reconstitute the C_2H_2-reducing activity of crude extracts of A. vinelandii UW 45. As shown in the Table, this expectation was well founded. Aliquots of the reaction mixtures from the photo-reduction assays were tested for UW 45 reconstitution activity to give somewhat surprising results. Although the mixture con-taining dimethylaminoethanol had an act.-red. EPR signal, it did

Table. Attempted Correlation of EPR Signal with UW 45
        Reconstituion Activity.

EPR Signal	Activity[a]	How Treated
Act.-red.	211	As isolated
Act.-red.	269	Concentrated
Act.-ox.	252	1:1 dilution DMF[b]
EPR-silent	253	EDTA present
Act.-red	0	DMAE/DAZ present[b]

[a] Nmole C_2H_4 evolved per min. per nmole Mo on
    reconstitution of crude UW 45 extracts.
[b] DMF is N,N-dimethylformanide; DMAE is dimethyl-
    aminoethanol; DAZ is 5-deazaflavin.

not reduce C_2H_2 in combination with the UW 45 extracts. In contrast,
the mixture containing EDTA, which was EPR-silent, when recombined
with the UW 45 extracts reduced C_2H_2 just as efficiently as the
control.

The UW 45 activity assay has the advantage of indicating whether
or not the added sample has genuine biological activity, but it
suffers from the drawback that added reagents might prevent the
reconstitution of the UW 45 nitrogenase even if the added FeMo-co
were structurally intact. Dimethylaminoethanol appears to be just
such a case because control reactions containing 5-deazaflavin that
have been illuminated show no loss of activity. Thus, the observation of an act.-red. EPR signal is not necessarily sufficient to
establish biochemical viability. The lack of an act.-red. EPR
signal, even in the presence of added excess dithionite, also cannot be taken as an indication of loss of UW 45 reconstitution
activity as shown by our EDTA experiments. Therefore, no single
spectral criteron, as yet, exists for biochemical activity for
FeMo-co.

CONCLUSIONS

We have demonstrated the existence of three interconvertible
redox states of FeMo-co based on EPR spectroscopy, all of which
restore the C_2H_2-reducing capacity of A. vinelandii UW 45 extracts.
FeMo-co behaves as an anion towards ion exchange resins and on
polyacrylamide gel electrophoresis. It reacts with EDTA to produce
an EPR-silent entity which is still capable of activating UW 45
extracts, indicating, together with other data, that there is no
direct correlation between the EPR spectrum and biological activity.
No evidence presently exists for a fully reduced, EPR-silent form
of FeMo-co which might be expected to reduce substrate.

ACKNOWLEDGEMENTS

    We thank Prof. J.A. Howard of University of Minnesota for the
amino acid analyses, Prof. W.J. Brill for kindly providing us with
the A. vinelandii UW 45 sample, Dr. J.L. Corbin for synthesizing
5-deazaflavin, Dr. J.W. McDonald for the design and construction
of the concentration apparatus, Dr. G.D. Watt for much valuable
advice, and D. Jacobs and S. Lough for skilled technical assistance.
We are indebted to the ASRA Division of the National Science
Foundation for support of this research through grant PFR-77-27269.
This manuscript constitutes Contribution No. 666 from the Charles
F. Kettering Research Laboratory.

REFERENCES

1. J.A. Pateman, D.J. Cove, B.M. Rever, and D.B. Roberts, A
    common cofactor for nitrate reductase and xanthine
    dehydrogenase which also regulates the synthesis of
    nitrate reductase, Nature (London) 201:58 (1964).
2. S.-S. Pan, R.H. Erickson, K.-Y. Lee, and A. Nason, Molybdenum-
    containing component shared by the molybdenum enzymes
    as indicated by the in vitro assembly of assimilatory
    nitrate reductase using the Neurospora mutant nit-1, in
    "Proceedings of First International Symposium on Nitrogen
    Fixation," W.E. Newton and C.J. Nyman, eds., Washington
    State University Press, Pullman (1976) Vol. 1, p. 293.
3. G.J. Sorger, Nitrate reductase electron transport systems in
    mutant and wild-type strains of Neurospora, Biochim.
    Biophys. Acta 118:484 (1966).
4. A. Nason, A.D. Antoine, P.A. Ketchum, W.A. Frazier, III, and
    D.K. Lee, Formation of assimilatory nitrate reductase by
    in vitro inter-cistronic complementation in Neurospora
    crassa, Proc. Nat. Acad. Sci. USA 65:133 (1970).
5. P.A. Ketchum, H.Y. Cambier, W.A. Frazier, III, C.H. Madansky,
    and A. Nason, In vitro assembly of Neurospora assimilatory
    nitrate reductase from protein subunits of a Neurospora
    mutant and the xanthine oxidizing or aldehyde oxidase
    systems of higher animals, Proc. Nat. Acad. Sci. USA
    66:1016 (1970).
6. A. Nason, K.-Y. Lee, S.-S. Pan, P.A. Ketchum, A. Lamberti, and
    J. DeVries, In vitro formation of assimilatory reduced
    nicotinamide adenine dinucleotide phosphate: nitrate
    reductase from a Neurospora mutant and a component of
    molybdenum-enzymes, Proc. Nat. Acad. Sci. USA 68:3242
    (1971).
7. K.-Y. Lee, S.-S. Pan, R.H. Erickson, and A. Nason, Involvement
    of molybdenum and iron in the in vitro assembly of assim-
    ilatory nitrate reductase utilizing Neurospora mutant
    nit-1, J. Biol. Chem. 249:3941 (1974).

8. H.H. Nagatani, V.K. Shah, and W.J. Brill, Activation of inactive
    nitrogenase by acid-treated component I, J. Bacteriol.
    120:697 (1974).

9. P.E. Bishop and W.J. Brill, Genetic analysis of Azotobacter
    vinelandii mutant strains unable to fix nitrogen, J.
    Bacteriol. 130:954 (1977).

10. J. Ruiz-Herrara, M.K. Showe, and J.A. DeMoss, Nitrate reductase
    complex of Escherichia coli K12: Isolation and char-
    acterization of mutants unable to reduce nitrate, J.
    Bacteriol. 97:129 (1969).

11. J.H. Glaser and J.A. DeMoss, Phenotypic restoration by molybdate
    of nitrate reductase activity in chlD mutants of
    Escherichia coli, J. Bacteriol. 198:854 (1971).

12. C. Kennedy and J.R. Postgate, Expression of Klebsiella
    pneumoniae nitrogen fixation genes in nitrate reductase
    mutants of Escherichia coli, J. Gen. Micro. 98:551 (1977).

13. I.R. Kennedy, J. Rigand, and J.C. Trinchant, Nitrate reductase
    from bacteriods of Rhizobium japonicum: Enzyme char-
    acteristics and possible interaction with nitrogen fixation,
    Biochim. Biophys. Acta 397:24 (1975).

14. V.L. Ganelin, N.P. L'vov, N.S. Sergeev, G.L. Shaposhnikov, and
    V.L. Kretovich, Isolation and properties of a molybdenum-
    containing peptide from component I of the nitrogen-fixing
    complex of Azotobacter vinelandii, Doklady Akad. Nauk SSSR
    206:1236 (1972); C.E. McKenna, N.P. L'vov, V.L. Ganelin,
    N.S. Sergeev, and V.L. Kretovich, Existence of a low-
    molecular-weight factor common to various molybdenum-
    containing enzymes, Doklady Akad. Nauk SSSR 217:228 (1974).

15. W.G. Zumft, Die abtrennung nieder molekularer komponenten aus
    dem molybdan-eisen protein der nitrogenase von Clostridium
    pasteurianum, Ber. Deutsch. Bot. Ges. 87:135 (1974).

16. W.G. Zumft, Isolation of thiomolybdate compounds from the
    molybdenum-iron protein of clostridial nitrogenase, Eur.
    J. Biochem. 91:345 (1978).

17. V.K. Shah and W.J. Brill, Isolation of an iron-molybdenum
    cofactor from nitrogenase, Proc. Nat. Acad. Sci. USA
    74:3249 (1977).

18. P.T. Pienkos, V.K. Shah, and W.J. Brill, Molybdenum cofactors
    from molybdoenzymes and in vitro reconstitution of nitrogen-
    ase and nitrate reductase, Proc. Nat. Acad. Sci. USA
    74:5468 (1977).

19. J.L. Johnson, H.P. Jones, and K.V. Rajagopalan, In vitro
    reconstitution of demolybdosulfite oxidase by a molybdenum
    cofactor from rat liver and other sources, J. Biol. Chem.
    252:4994 (1977).

20. B.K. Burgess, D. Jacobs, and E.I. Stiefel, Azotobacter vinelandii
    nitrogenase: Large-scale, anaerobic, high-yield, high-
    activity preparation using modified Schlenk techniques,
    submitted for publication.

21. J. Rawlings, V.K. Shah, J.R. Chiswell, W.J. Brill, R.
        Zimmermann, E. Münck, and W.H. Orme-Johnson, Novel metal
        cluster in the iron-molybdenum cofactor of nitrogenase,
        J. Biol. Chem. 253:1001 (1978).
22. R.Z. Zimmermann, E. Münck, W.J. Brill, V.K. Shah, M.T. Henzl,
        J. Rawlings, and W.H. Orme-Johnson, Nitrogenase X:
        Mössbauer and EPR studies on reversibly oxidized MoFe
        protein from Azotobacter vinelandii OP. Nature of the
        iron centers, Biochim. Biophys. Acta 537:185 (1978).
23. B.H. Huynh, E. Münck, and W.H. Orme-Johnson, Nitrogenase XI:
        Mössbauer studies on the cofactor centers of the MoFe
        protein from Azotobacter vinelandii OP, Biochim. Biophys.
        Acta 576:192 (1979).
24. S.P. Cramer, K.O. Hodgson, E.I. Stiefel and W.E. Newton, A
        systematic X-ray absorption study of molybdenum complexes.
        The accuracy of structural information from extended X-ray
        absorption fine structure, J. Am. Chem. Soc. 100:2748
        (1978).
25. S.P. Cramer, W.O. Gillum, K.O. Hodgson, L.E. Mortenson, E.I.
        Stiefel, J.R. Chisnell, W.J. Brill, and V.K. Shah, The
        molybdenum site of nitrogenase. 2. A comparative study
        of Mo-Fe proteins and the iron-molybdenum cofactor by
        X-ray absorption spectroscopy, J. Am. Chem. Soc. 100:3814
        (1978).
26. B.E. Smith, The structure and function of nitrogenase: a review
        of the evidence for the role of molybdenum, J. Less-
        Common Metals 54:465 (1977).
27. V.K. Shah, L.C. Davis, J.K. Gordon, W.H. Orme-Johnson, and
        W.J. Brill, Nitrogenase. III. Nitrogenaseless mutants
        of Azotobacter vinelandii: Activities, cross-reactions
        and EPR spectra, Biochim. Biophys. Acta 292:246 (1973).
28. G. Watt, Paper presented at the Steenbock-Kettering Interna-
        tional Symposium on Nitrogen Fixation, Madison, Wisconsin,
        June 1978.
29. W.H. Orme-Johnson, W.D. Hamilton, T. Ljones, M.-Y. W. Tso,
        R.H. Burris, V.K. Shah, and W.J. Brill, Electron para-
        magnetic resonance of nitrogenase and nitrogenase
        components from Clostridium pasteurianum W5 and Azotobacter
        vinelandii OP, Proc. Nat. Acad. Sci. USA 69:3142 (1972).
30. L.E. Mortenson, W.G. Zumft, and G. Palmer, Electron paramagnetic
        resonance studies on nitrogenase. III. Function of magne-
        sium adenosine 5'-triphosphate and adenosine 5'-diphosphate
        in catalysis by nitrogenase, Biochim. Biophys. Acta 292:
        422 (1973).
31. B.E. Smith, D.J. Lowe, and R.C. Bray, Studies by electron
        paramagnetic resonance on the catalytic mechanism of
        nitrogenase of Klebsiella pneumoniae, Biochem. J.
        135:331 (1973).

32. W.R. Frisell, C.W. Chung, and C.G. MacKenzie, Catalysis of
     oxidation of nitrogen compounds by flavin coenzymes in
     the presence of light, J. Biol. Chem. 234:1297 (1959).
33. A.I. Krasna, Proflavin catalyzed photoproduction of hydrogen
     from organic compounds, Photochem. Photobiol. 29:267
     (1979).

SYNTHESIS AND CHARACTERIZATION OF IRON-MOLYBDENUM-SULFUR CLUSTERS

C. David Garner, George Christou, Stephen R. Acott,
Frank E. Mabbs, Richard M. Miller, Trevor J. King[1a],
Michael G.B. Drew[1b], Charles E. Johnson[1c], James D.Rush[1c]
and (the late) G.W. Kenner[1d]

Chemistry Department, Manchester University,
Manchester M13 9PL, England

INTRODUCTION

The structure and function of the nitrogenase enzymes is a topic which continues to attract considerable attention. A significant contribution to the understanding of the role of molybdenum in these enzymes has been the measurement and interpretation of the extended X-ray absorption fine structure (EXAFS) associated with the molybdenum K-edge of the molybdenum iron proteins from Azotobacter vinelandii and Clostridium pasteurianum and the FeMo-cofactor of the former protein.[2,3] The environment of the molybdenum is very similar in each of these three systems and is suggested to consist of three or four sulfur atoms at a distance of 2.35Å, one or two sulfur atoms at 2.49Å, and two or three iron atoms at 2.72Å. A structurally attractive possibility is that the molybdenum is present in an {MoFe$_3$S$_4$} cubane-like cluster. We have, therefore, directed our efforts to the synthesis and characterization of compounds containing such clusters.

SYNTHESIS

The convenient synthesis of [Fe$_4$S$_4$(SR)$_4$]$^{2-}$ complexes by the anaerobic reaction between iron(III) chloride, the sodium salt of the thiol, and elemental sulfur in methanol solution[4] suggested that the corresponding reaction with [MoS$_4$]$^{2-}$, instead of elemental sulfur, should be investigated. The reaction scheme (1)-(3), under anaerobic condition, leads to the formation of [Fe$_6$Mo$_2$S$_8$(SR)$_9$]$^{3-}$ complexes (for R = Ph,[5-6] p - Cl-C$_6$H$_4$,[6] or Et [6]). The compounds are obtained in good (> 60%) yield as black needles or

microcrystalline powders which, after recrystallisation from a
suitable solvent, e.g., MeOH, MeCN/MeOH, may be converted into
well-formed needle-like crystals. This type of complex is the only
one isolated so far in our studies, despite syntheses having been
attempted (for R = Ph) over a range of thiol:iron stoichiometries.
However, Holm et al.[7] have prepared $[NEt_4]_3[Fe_6Mo_2S_9(SEt)_8]$ from a
reaction sequence similar to (1)-(3).

$$6FeCl_3 + 24NaSR \xrightarrow{MeOH} \frac{6}{n}[Fe(SR)_3]_n + 18NaCl + 6NaSR \qquad (1)$$

$$6/n[Fe(SR)_3]_n + 2[NH_4]_2[MoS_4] \rightarrow [NH_4]_3[Fe_6Mo_2S_8(SR)_9]$$
$$+ 4RSSR + [NH_4]SR \qquad (2)$$

$$[NH_4]_3[Fe_6Mo_2S_8(SR)_9] + (excess) [NR'_4]Br \rightarrow$$
$$[NR'_4]_3[Fe_6Mo_2S_8(SR)_9] + 3[NH_4]Br \qquad (3)$$

Exchange of the alkylthiolato- groups of $[Fe_4S_4(SR)_4]^{2-}$
complexes is readily accomplished[8] and this has been established for
the $[Fe_6Mo_2S_8(SR)_9]^{3-}$ (R = alkyl) species also. Treatment of the
R = Et complex with an excess of 2-hydroxyethanethiol in MeCN at
ambient temperatures, and removal of liberated ethanethiol in vacuo,
effects the quantitative conversion to the $[Fe_6Mo_2S_8(SCH_2CH_2OH)_9]^{3-}$
complex,[9] indicating that both the bridging and terminal thiolato-
groups are susceptible to exchange.

CRYSTAL STRUCTURE

The unit cell parameters of those compounds containing
$\{Fe_3MoS_4\}$ cubane-like clusters reported thus far are detailed in
Table 1. The three $[NEt_4]^+$ salts are isomorphous and crystallize
in the hexagonal space group $P6_3/m$ with the cluster dimer centred
on a site of C_{3h} ($\overline{6}$) symmetry, the Mo---Mo vector being parallel
to the crystallographic c axis. The shorter Mo---Mo separation in
$[Fe_6Mo_2S_9(SEt)_8]^{3-}$ as compared to $[Fe_6Mo_2S_8(SEt)_9]^{3-}$ (vide infra)
is apparent in the relative length of the unit cell c dimension.
The Fe-Mo-S framework of the anion in $[NBu^n_4]_3[Fe_6Mo_2S_8(SPh)_9]$ also
closely approximates to C_{3h} symmetry.

The atomic arrangement of the central portion of $[Fe_6Mo_2S_8-$
$(SEt)_9]^{3-}$ is illustrated in Figure 1. Each $[Fe_6Mo_2S_8(SR)_9]^{3-}$
anion consists of two $\{Fe_3MoS_4\}$ cubane-like clusters linked across
their molybdenum centers by three μ_2-thiolato- groups.
$[Fe_6Mo_2S_9(SEt)_8]^{3-}$, however, involves one μ_2- sulphido- and two μ_2-
ethanethiolato- groups, disordered over the crystallographic C_3
axis of the anion, bridging the two cubes. The interatomic
dimensions of the anions are summarized in Table 2. The anions
with the the three μ_2-thiolato- groups have an Mo---Mo separation

Table 1. Unit cell parameters of $[NR_4]_3[Fe_6Mo_2S_{8+n}(SR)_{9-n}]$ $(n = 0$ or $1)$ complexes

Compound	System	Space Group	Parameters (Å, degrees)	Z
$[NBu^n_4]_3[Fe_6Mo_2S_8(SPh)_9]^a$	monoclinic	$\underline{Cc}$	$\underline{a} = 16.452(2)$, $\underline{b} = 27.469(4)$; $\underline{c} = 26.754(4)$, $\beta = 94.81(2)$	4
$[NEt_4]_3[Fe_6Mo_2S_8(SCH_2CH_2OH)_9]\cdot Me_2CO^b$	hexagonal	$\underline{P6_3/m}$	$\underline{a} = 17.001(7)$, $\underline{c} = 16.381(9)$	2
$[NEt_4]_3[Fe_6Mo_2S_8(SEt)_9]^c$	hexagonal	$\underline{P6_3/m}$	$\underline{a} = 17.261(4)$, $\underline{c} = 16.346(4)$	2
$[NEt_4]_3[Fe_6Mo_2S_9(SEt)_8]^d$	hexagonal	$\underline{P6_3/m}$	$\underline{a} = 17.230(6)$, $\underline{c} = 15.999(4)$	2

$\underline{a}$ Ref. 5; $\underline{b}$ Ref. 9; $\underline{c}$ Ref. 6; $\underline{d}$ Ref. 7.

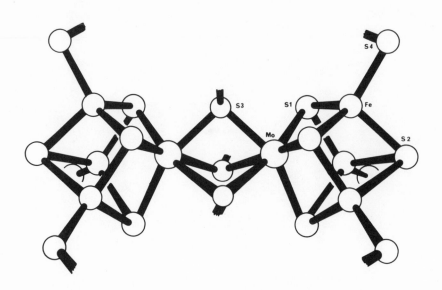

Figure 1.   Atomic arrangement of the central portion of
             $[Fe_6Mo_2S_8(SEt)_9]^{3-}$

of $\underline{ca}$. 3.66Å, whereas the Mo-(S)(SEt)$_2$-Mo arrangement involves the
molybdenum atoms 3.306(3)Å apart. Other than this difference, the
values of the corresponding dimensions of these units are not
significantly different within the accuracy of the determinations.
A comparison of these values with those obtained for $[Fe_4S_4(SR)_4]^{2-}$
complexes, Table 3, shows that substitution of molybdenum for an
iron atom results in only small perturbations in the structure.
These changes include slightly reduced Fe---Fe contacts (by $\underline{ca}$.
0.04Å) and Fe-(μ_3-)S (by $\underline{ca}$. 0.025Å) and Fe-S(R) (by $\underline{ca}$. 0.03Å)
bonds.

        The outstanding conclusion which emerges from the structural
data is that the three Mo-(μ_3-)S bonds of length 2.345Å, the three
Mo---Fe contacts of 2.73Å, and the three Mo-S(R) bonds of length
2.57Å, compare very closely with the structural predictions derived
from the EXAFS data.[2,3] This conclusion has been augmented by the
demonstration[7] that the EXAFS associated with the molybdenum K-edge
absorption of $[NEt_4]_3[Fe_6Mo_2S_9(SEt)_8]$ is remarkably similar to that
observed for the nitrogenases. We suggest that the difference
between the Mo-S(R) bond lengths of these dimeric units and those
in the enzymes may be due to the latter involving cysteinyl sulphur
atoms, which do not bridge across two metal clusters.

Table 2. Dimensions[a] of $[Fe_6Mo_2S_8(SR)_9]^{3-}$ complexes and $[Fe_6Mo_2S_9(SEt)_8]^{3-}$

	$[Fe_6Mo_2S_8(SR)_9]^{3-}$			$[Fe_6Mo_2S_9(SEt)_8]^{3-}$
	Ph	CH_2CH_2OH	Et	
Distance (Å)				
Fe - S(1)	2 25(2)	2.28(2)	2.259(4)	2.264(6)
Fe - S(2)	2.26(2)	2.25(2)	2.276(4)	2.282(6)
Fe - S(4)	2.25(3)	2.25(1)	2.228(4)	2.242(6)
Mo - S(1)	2.34(2)	2.39(2)	2.347(3)	2.340(6)
Mo - S(3)	2.58(2)	2.55(2)	2.568(3)	(2.55)
Fe ... Fe	2.69(2)	2.72(1)	2.699(3)	
Mo ... Fe	2.71(2)	2.75(1)	2.726(2)	2.730(2)
S(1) ... S(1)	3.65(3)	3.72(2)	3.668(5)	
S(1) ... S(2)	3.55(3)	3.55(2)	3.577(5)	
Mo ... Mo	3.685(3)	3.64(1)	3.662(2)	3.306(3)
Angle (degrees)				
S(1)-Fe-S(1)	108.7(7)	109.4(6)	108.5(1)	
S(1)-Fe-S(2)	104(2)	103.4(6)	104.1(1)	
S(1)-Fe-S(4)	116(3)	116.0(7)	116.0(4)	
S(2)-Fe-S(4)	107(3)	107.0(7)	107.0(2)	
S(1)-Mo-S(1)	102.7(5)	102.3(5)	102.7(1)	
S(1)-Mo-S(3)	86.2(1.5) 93.4(1.3)	89.9(8)	88.3(1) 90.8(1)	
S(3)-Mo-S(3)	74.6(6)	74.6(7)	74.8(1)	
Mo-S(1)-Fe	72.5(7)	72.1(8)	72.5(1)	(72.7)
Fe-S(1)-Fe	73.4(9)	73.2(8)	73.4(1)	
Fe-S(2)-Fe	73.2(1.1)	74.4(6)	72.7(2)	

a For specification of atoms see Figure 1.

Table 3. Comparison of some corresponding interatomic dimensions[a] of $(Fe_6Mo_2S_8(SEt)_9)^{3-}$ and $(Fe_4S_4(SCH_2Ph)_4)^{2-}$

	[b]$(Fe_6Mo_2S_8(SEt)_9)^{3-}$	[c]$(Fe_4S_4(SCH_2Ph)_4)^{2-}$
Distance (Å)		
Mo-S(1)	2.347	
Fe-S(1)	2.259	2.286
Fe-S(2)	2.276	
Mo-S(3)	2.568	
Fe-S(4)	2.228	2.251
Mo---Fe	2.726	
Fe---Fe	2.699	2.747
S(1)---S(1)	3.668	
S(1)---S(2)	3.577	3.616
Angle (degrees)		
S(1)-Mo-S(1)	102.7	
S(1)-Fe-S(1)	108.5	104.1
S(1)-Fe-S(2)	104.1	
S(1)-Fe-S(4)	116.0	
S(2)-Fe-S(4)	107.0	114.4
Mo-S(1)-Fe	72.5	
Fe-S(1)-Fe	73.4	73.8
Fe-S(2)-Fe	72.7	

a   For specification of atoms see Figure 1; the corresponding values for $(Fe_4S_4(SCH_2Ph)_4)^{2-}$ are given; b Ref.6; c Ref.10.

REDOX PROPERTIES

Electrochemical studies have established that the $[Fe_6Mo_2S_8(SR)_9]^{3-}$ complexes undergo an irreversible one-electron oxidation and two reversible one-electron reductions. These processes are illustrated in Figures 2 and 3 for $[NEt_4]_3[Fe_6Mo_2S_8(SEt)_9]$, the oxidation of which occurs at -0.46v and the reduction at -1.56 and -1.76v (in DMSO vs Ag/Ag^+ (0.1 mol ℓ^{-1})). These latter data compare with the two one-electron reductions of $[NBu^n_4]_2[Fe_4S_4(SEt)_4]$ which occur at -1.59 and -2.16v under the same conditions.

The oxidation product $[Fe_6Mo_2S_8(SEt)_9]^{2-}$ contains the collection of metal atoms in the same formal oxidation state as $[NEt_4]_3[Fe_6Mo_2S_9(SEt)_8]$. Given the crystallographic symmetry of the latter unit and its spectroscopic properties,[7] we suggest that oxidation removes an electron from the dimeric entity as a whole rather than from one $\{Fe_3MoS_4\}$ center. Similarly, the reductions are considered to lead to the addition of one and two electrons to the dimeric entity. These suggestions imply that the two $\{Fe_3MoS_4\}$ centers are strongly coupled over the $Mo-(\mu_2-SR)_3-Mo$ bridging network.

SPECTROSCOPIC PROPERTIES

Mössbauer spectra

The ^{57}Fe Mössbauer spectra of $[NBu^n_4]_3[Fe_6Mo_2S_8(SPh)_9]$ at 4, 77, and 195K are shown in Figure 4. These spectra indicate that the iron atoms are nearly, but not precisely, electronically equivalent and that the inequivalance becomes more pronounced at lower temperatures. The spectra have been interpreted in terms of three different iron sites and the average values of the isomer shifts and quadrupole splittings from the curve fitting procedures are given in Table 4. The values[7] of the ^{57}Fe isomer shift and quadrupole splitting for $[NEt_4]_3[Fe_6Mo_2S_9(SEt)_8]$ at 77K have been given as 0.27(2) and 0.95(3) mm s^{-1}, respectively, however, the reference for the former value was not given. The corresponding values for the iron-molybdenum cofactor from nitrogenase are 0.37 and 0.75 mm s^{-1} at 90K.[12]

A comparison of the isomer shift of the complexes listed in Table 4 with those obtained for iron atoms in the redox active centers of the iron-sulfur proteins and their synthetic analogues implies that the iron atoms of the $[Fe_6Mo_2S_8(SR)_9]^{3-}$ complexes have an average oxidation state of ca. + 2.5. This result suggests that these clusters involve molybdenum(III)-(IV). The ^{57}Fe Mössbauer spectrum of a sample of $[NBu^n_4]_3[Fe_6Mo_2S_8(SPh)_9]$, reduced by treatment with an excess (> 2 equiv.) of sodium acenaphthalenide, may be interpreted on the basis of average δ and ΔE_Q values of 0.47(2) and 1.2(2) mm s^{-1} (at 77K relative to

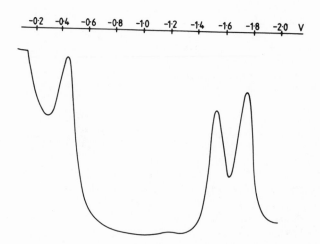

Figure 2.   Differential pulse polarogram of $[NEt_4]_3[Fe_6Mo_2S_8(SEt)_9]$
            (5×10^{-3} mol ℓ^{-1}) in $KClO_4$ (0.1 mol ℓ^{-1}) in DMSO vs.
            Ag/Ag^+ (0.1 mol ℓ^{-1})

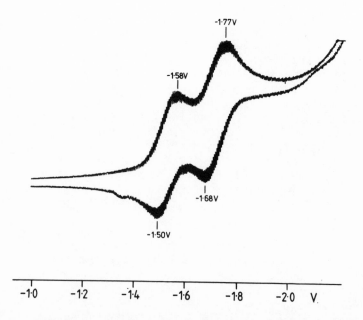

Figure 3.   Staircase cyclic voltammogram of reductions of
            $[NEt_4]_3[Fe_6Mo_2S_8(SEt)_9]$($5 \times 10^{-3}$ mol ℓ^{-1}) in $KClO_4$ (0.1
            mol ℓ^{-1}) in $KClO_4$ 0.1 mol ℓ^{-1}) in DMSO vs Ag/Ag^+ (0.1
            mol ℓ^{-1}) at 20°C (16 mV s^{-1}).

Table 4. ^{57}Fe Mössbauer data for $[Fe_6Mo_2S_8(SR)_9]^{3-}$ complexes.[a]

Compound	Temp (K)	δ (mm s^{-1})	ΔE_Q (mm s^{-1})
$[NBu^n_4]_3[Fe_6Mo_2S_8(SPh)_9]$	293	0.30(3)	0.96(5)
	77	0.41(1)	1.13(2)
	4.2	0.42(1)	1.25(2)
$[NEt_4]_3[Fe_6Mo_2S_8(SEt)_9]$	77	0.40(3)	1.12(2)
$[NEt_4]_3[Fe_6Mo_2S_8(SCH_2CH_2OH)_9] \cdot Me_2CO$	77	0.41(2)	1.20(2)

a relative to iron metal at room temperature; cf. $[NBu^n_4]_2[Fe_4S_4(SPh)_4]$

δ = 0.44(1) ΔE_Q = 0.86(2) mm s^{-1}.

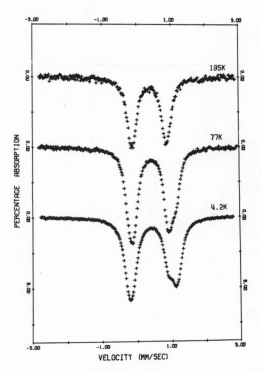

Figure 4.   Zero-field ^{57}Fe Mössbauer spectrum of
              $[NBu^n{}_4]_3[Fe_6Mo_2S_8(SPh)_9]$ at 4, 77 and 195K.

iron metal at room temperature). If reduction to $[Fe_6Mo_2S_8(SPh)_9]^{5-}$
has been achieved, the small shift in the ^{57}Fe isomer shift implies
that it is the molybdenum, rather than the iron, which experiences
the major change in electron density upon electron addition.

Electron paramagnetic resonance spectra

The $[Fe_6Mo_2S_8(SR)_9]^{3-}$ complexes exhibit no significant e.p.r.
activity, even when cooled <10^{0}K. Each $\{Fe_3MoS_4\}$ center in these
complexes has a total of 19 electrons which may be considered to
be accommodated in valence orbitals associated primarily with the
metal atoms. Thus, one explanation for this lack of e.p.r. activity
could be strong coupling between the two clusters, as suggested
above, leading to an overall S = 0 ground state. Reduction of
$[NBu^n{}_4]_3[Fe_6Mo_2S_8(SPh)_9]$ with an excess (> 2 equiv.)of sodium
acenaphthalenide in MeCN solution, followed by rapid freezing,
produces a species which is e.p.r. active, exhibiting a sharp
feature at g = 2.0 and broad features at apparent g-values of 5.1
and 4.3, none of which display any [95,97]Mo hyperfine splittings.[12]
The iron-molybdenum cofactor from nitrogenase is considered to have
an S = $^3/2$ ground state and exhibits g-values at 4.6, 3.3 and 2.0.[11]

[NEt$_4$]$_3$[Fe$_6$Mo$_2$S$_9$(SEt)$_8$] has a strong e.p.r. spectrum with apparent g-values near ~10, 4.0, 1.9 and 1.2 (4.2K, DMF).[7]

^{1}H-n.m.r. spectra

The ^{1}H-n.m.r. spectrum of [NEt$_4$]$_3$[Fe$_6$Mo$_2$S$_8$(SEt)$_9$] in CD$_3$CN at ca. 20°C exhibits signals due to the cations, plus two others at 53 and 17 ppm downfield of Me$_4$Si in an intensity ratio of 2:1. These compare with the signals at 54.2 downfield and 3.1 ppm up-field of Me$_4$Si in a ca. 3:1 intensity ratio observed[7] for the [Fe$_6$Mo$_2$S$_9$(SEt)$_8$]$^{3-}$ ion. The data are consistent with the assignment of the former signal of each pair to the Fe-SCH$_2$ (or CH$_3$) protons and the latter to the Mo-SCH$_2$ (or CH$_3$) protons. Therefore, consistent with the tentative conclusions based on the Mossbauer data, these ^{1}H-n.m.r. data could be taken to indicate that it is the electronic structure about the molybdenum, as opposed to the iron, centers that experience the greatest change in electronic structure for a redox change of these dimers. However, the change in the atomic arrangement of the bridging region may also contribute to the different ^{1}H-n.m.r. properties of the above complexes.

The ^{1}H-n.m.r. spectrum of [NBu$^n_4$]$_3$[Fe$_6$Mo$_2$S$_8$(SPh)$_9$] in (CD$_3$)$_2$SO at 35°C shows, in addition to signals characteristic of the cation, resonances at 13.5 and 5.7 ppm downfield and 3.6 ppm upfield of Me$_4$Si. These are assigned, respectively, to the meta, ortho, and para hydrogens of the phenyl rings. The temperature dependence of these resonance positions is the opposite of that typical[13] of [Fe$_4$S$_4$(SR)$_4$]$^{2-}$ centers. As the temperature is increased from 35 to 90°C, the separations between these peaks decrease, so that at this higher temperature, the resonances occur at 12.7 and 6.0 ppm downfield and 2.0 ppm upfield of Me$_4$Si.[5]

Optical absorption spectra

The optical absorption spectrum of a [Fe$_6$Mo$_2$S$_8$(SR)$_9$]$^{3-}$ complex shows an intense band in the ultra-violet 'tailing' into the visible region.   [NEt$_4$]$_3$[Fe$_6$Mo$_2$S$_8$(SCH$_2$CH$_2$OH)$_9$].Me$_2$CO exhibits two absorption maxima which in water occur at λ_{max} 277 nm (ε = 67.0 x 10^3 ℓ mol^{-1} cm^{-1}) and 371 nm (ε = 40.1 x 10^3 ℓ mol^{-1} cm^{-1}).[4] The positions of these two features are not particularly sensitive to: (i) the nature of the thiol because [NEt$_4$]$_3$[Fe$_6$Mo$_2$S$_8$(SEt)$_9$] in (CH$_3$)$_2$SO has λ_{max} of 280 nm (ε = 56.0 x 10^3 mol^{-1} cm^{-1}) and 390 nm (ε = 35.5 x 10^3 mol^{-1} cm^{-1}),[7] or (ii) to the oxidation state of the complex because [NEt$_4$]$_3$[Fe$_6$Mo$_2$S$_9$(SEt)$_8$] has λ_{max} of 274 nm (ε = 58.1 x 10^3 ℓ mol^{-1} cm^{-1}) and 395 nm (ε = 38 x 10^3 ℓ mol^{-1} cm^{-1}).[7]

CONCLUSIONS

The existence of {Fe$_3$MoS$_4$} cubane-like cluster dimers is now

established and the environment of the molybdenum within these
entities is remarkably similar to that suggested[2,3] for this element
in the nitrogenase enzymes and their iron-molybdenum cofactor.
Physical measurements for these systems have allowed some
description of the electronic characteristics to be presented. The
chemical reactivity established has so far been limited to simple
redox and substitution reactions.   However, $[NEt_4]_3[Fe_6Mo_2S_8(SCH_2$
$CH_2OH)_9]$ has been shown to have the ability to replace ferredoxins
in a hydrogen-evolving system using components from Clostridium
pasteurianum and dithionite.[14] Further developments, in particular
the synthesis of monomeric Fe_3MoS_4 centers, are necessary before
the authenticity of these moieties as analogues of the molybdenum
center(s) in the nitrogenases can be assessed in detail.

REFERENCES

1a.   Chemistry Department, Nottingham University, Nottingham NG7
      2RD; b Chemistry Department, The University, Whiteknights
      Reading RG6 2AD; c Oliver Lodge Laboratory, University of
      Liverpool, Liverpool L69 3BX; d Robert Robinson Laboratories,
      University of Liverpool, Liverpool L69 3BX.
2.   S.P. Cramer, K.O. Hodgson, W.O. Gillum and L.E. Mortenson,
      'The Molybdenum Site in Nitrogenase. Preliminary
      Structural Evidence from X-Ray Absorption Spectroscopy'.
      J. Amer. Chem. Soc.,100:3398 (1978).
3.   S.P. Cramer, W.O. Gillum, K.O. Hodgson, L.E. Mortenson, E.I.
      Stiefel, J.R. Chisnell, W.J. Brill and V.K. Shah. 'The
      Molybdenum Site of Nitrogenase. 2. A Comparative Study
      of Mo-Fe Proteins and the Iron-Molybdenum Cofactor by
      X-Ray Absorption Spectroscopy'. J. Amer. Chem. Soc.,
      100:3814 (1978).
4.   G. Christou and C.D. Garner, 'A Convenient Synthesis of Tetra-
      kis[thiolato-μ_3-sulphido-iron](2-) Clusters'. J. Chem.
      Soc. Dalton Trans.,1093 (1979).
5.   G. Christou, C.D. Garner, F.E. Mabbs and T.J. King, 'Crystal
      Structure of Tris(tetra-n-butylammonium)Tri-μ-benzene-
      thiolato-bis{tri-{-sulphido-[μ_3-sulphido-tris(benzene-
      thiolatoiron)]molybdenum},$[Bu_4N]_3[\{(PhSFe)_3MoS_4\}_2(SPh)_3]$;
      an Fe_3MoS_4 Cubic Cluster Dimer', J. Chem. Soc. Chem.
      Comm.,740 (1978).
6.   S.R. Acott, G. Christou, F.E. Mabbs, T.J. King and C.D. Garner,
      'Isolation and Crystal Structure of $[Et_4N]_3[Fe_6Mo_2S_8(SEt)_9]$'
      Inorg. Chim. Acta.,35:L337 (1979).
7.   T.E. Wolff, J.N. Berg, C. Warrick, K.O. Hodgson, R.H. Holm
      and R.B. Frankel, 'The Molybdenum-Iron-Sulfur Cluster
      Complex $[Mo_2Fe_8S_9(SC_2H_5)_8]^{3-}$. A Synthetic Approach to
      the Molybdenum Site in Nitrogenase', J. Amer. Chem. Soc.,
      100:4630 (1978).

8.   L. Que, Jr., M.A. Bobrik, J.A. Ibers and R.H. Holm, 'Synthetic
      Analogs of the Active Sites of Iron-Sulfur Proteins. VII.
      Ligand Substitution Reactions of the Tetranuclear Clusters
      $[Fe_4S_4(SR)_4]^{2-}$ and the Structure of $[(CH_3)_4N]_2[Fe_4(SC_6H_5)_4]$
      J. Amer. Chem. Soc.,96:4168 (1974).
9.   G. Christou, C.D. Garner, F.E. Mabbs and M.G.B. Drew, 'Thiol
      Exchange Reactions of Iron-Molybdenum-Sulphur Clusters;
      Preparation and X-Ray Crystal Structure of $[Et_4N]_3[Fe_6Mo_2$
      $S_8(SCH_2CH_2OH)_9]$, a Water Soluble Iron-Molybdenum-Sulphur
      Cluster", J. Chem. Soc. Chem. Comm.,91 (1979).
10.   B.A. Averill, T. Herskovitz, R.H. Holm, and J.A. Ibers,
      'Synthetic Analogs of the Active Sites of Iron-Sulfur
      Proteins. II. Synthesis and Structure of Tetra[mercapto-
      μ_3-sulfido-iron] Clusters, $[Fe_4S_4(SR)_4]^{2-}$', J. Amer. Chem
      Soc.,95:3523 (1973).
11.   J. Rawlings, V.K. Shah, J.R. Chisnell, W.J. Brill, R.
      Zimmermann, E. Münck and W.H. Orme-Johnson, 'Novel Metal
      Cluster in the Iron-Molybdenum Cofactor of Nitrogenase'.
      J. Biol. Chem.,253:1001 (1978).
12.   R. Cammack, G. Christou and C.D. Garner, Unpublished results.
13.   R.H. Holm, W.D. Phillips, B.A. Averill, J.J. Mayerle, and
      and T. Herkovitz, 'Synthetic Analogs of the Active Sites
      of Iron-Sulfur Proteins. V. Proton Resonance Properties
      of the Tetranuclear Clusters $[Fe_4S_4(SR)_4]^{2-}$. Evidence
      for Dominant Contact Interactions.', J. Amer. Chem. Soc.,
      96:2109 (1974).
14.   M.W.W. Adams, K.K. Rao, D.O. Hall, G. Christou and C.D.
      Garner, 'Biological Activity of Synthetic Molybdenum-
      Iron-Sulphur, Iron-Sulphur and Iron-Selenium Analogues of
      Ferredoxin-Type-Centers' in press.

CHEMICAL AND PHYSICAL STUDIES OF POLYNUCLEAR Mo-Fe-S AND

MONONUCLEAR Mo(III) COMPLEXES

B. A. Averill,* H. C. Silvis,* R. H. Tieckelmann* and
W. H. Orme-Johnson[†]

*Department of Chemistry, Michigan State University,
East Lansing, MI 48824; [†]Department of Biochemistry,
University of Wisconsin, Madison, WI 53706

INTRODUCTION

Molybdenum is an essential component of a number of enzymes
that catalyze biological oxidation-reduction reactions.[1] Of particular interest to us are the reductive enzymes, nitrogenase and nitrate reductase. The reactions catalyzed by both of these enzymes
can be viewed as involving a coupled transfer of pairs of protons
and reducing equivalents to substrate.[2]

One of the major physical techniques used to examine these
enzymes has been electron paramagnetic resonance (EPR) spectroscopy,
since both Mo(V) and Mo(III) are expected to be paramagnetic and
thus, potentially observable by EPR. Although Mo(V) signals from a
number of enzymes, most notably xanthine oxidase, have been extensively studied by EPR, no signals have been reported that are definitely attributable to Mo(III) in nitrate reductase or in nitrogenase. This is due at least in part to conflicting reports of the
EPR properties of simple Mo(III) complexes. Thus, the axial signals
observed near g = 2 for bacterial nitrate reductase[3] at 80-100 K are
quite similar to those reported for a series of complexes of Mo(III)
with sulfur-containing ligands.[4] In contrast, the rhombic signal
(g = 4.3, 3.7, 2.0) of the Mo-Fe protein of nitrogenase,[5] observable
only at $\leq$ 20 K, and a similar signal observed in reduced samples of
nitrate reductase from Neurospora crassa[5] are analogous to that reported for molybdenum(III) acetylacetonate, Mo(acac)$_3$.[6] The molybdenum-containing oxidases examined to date all show narrow resonances
analogous to those of mononuclear Mo(V) complexes, with <g> $\sim$ 2.0
and <A> (^{95}Mo) of 30-70 gauss.[1] In contrast, no ^{95}Mo hyperfine was
observed for Mo(acac)$_3$[6] or for the Mo-Fe protein of nitrogenase

217

containing ^{95}Mo.[5,7] In order to ascertain the general properties
of such signals (g-values, linewidth, temperature dependence, and
^{95}Mo hyperfine coupling constants), the EPR spectra of a series of
Mo(III) complexes have been examined.

Despite the conceptual similarity of the reactions catalyzed
by nitrate reductase and nitrogenase, it is clear that there are
fundamental differences in the metallocomponents of the enzymes.
Nitrate reductase possesses a low molecular weight molybdenum-containing cofactor (Mo-co) common to other, oxidative molybdoenzymes,[8]
while the iron-molybdenum cofactor[9] (FeMo-co) isolated from nitrogenase is clearly a novel inorganic cluster.[10,11] The spin-coupled
nature of this cluster[12] suggests strongly that an EPR spectrum due
to simple mononuclear Mo(III) is not to be expected. Extended X-
ray absorption fine structure (EXAFS) spectroscopy results have been
interpreted as showing that the molybdenum has as nearest neighbors
4-6 sulfur atoms and 3 iron atoms at $\sim$ 2.7 Å, suggesting a cage
structure.[13] Inasmuch as only one such structural unit, containing
an MoFe$_3$S$_4$ cluster, has been reported to date,[13,14] we are investigating the synthesis and chemical properties of other compounds of
this type. In particular, a recent report[15] that acid hydrolysis
of the Mo-Fe protein of nitrogenase produces thiomolybdate (MoS$_4^{2-}$)
makes compounds containing (Fe-S)$_n$ and MoS$_4^{2-}$ units especially pertinent.

EPR SPECTRA OF Mo(III) COMPLEXES

Preparation of Samples

Because of the extreme oxygen sensitivity of Mo(III) complexes
and the small amount of isotopically enriched molybdenum available,
the complexes to be examined by EPR were generated in situ from an
air-stable Mo(III) source, taking advantage of the high solubility
of neutral metal tris-chelates in organic solvents. The procedure
involved reaction of [NH$_4$]$_3$[MoCl$_6$] with an aqueous or aqueous/N,N-
dimethylacetamide solution of the ligand, extraction of the neutral
complex with toluene, and transfer of the supernatant to an EPR tube.
Starting materials and ionic byproducts are thereby left in the
aqueous phases. Small and variable amounts of coextracted Mo(V)
complexes (probably MoOL$_3$) were observed in the g = 2 region, but
were not examined in detail. EPR spectra were recorded on a Varian
109 spectrometer operating at X-band with 100 KHz modulation; sample
temperature was varied over the range 5-80 K using boiloff gas from
liquid helium as the coolant.

Results and Discussion

The EPR properties of the compounds studied are summarized in

the Table. Five of the complexes ($[MoCl_6]^{3-}$, $[MoCl_5(H_2O)]^{2-}$, Mo-
(acac)$_3$, Mo(ox)$_3$, and $Mo[S_2P(OEt)_2]_3$) display similar very broad,
axial or slightly rhombic spectra centered at g = 4 and 2. These
signals are highly temperature dependent, disappearing above 30 K,
and are not saturated at relatively high microwave power (3 mW).
The similarity of the spectra of the three tris-chelate complexes to
those of $[MoCl_6]^{3-}$ and the previously examined[6] Mo(acac)$_3$ indicates
that they are all due to Mo(III) in an approximately octahedral environment. Although the tris(diethyldithiophosphate) complex of
Mo(III) has apparently not been reported previously, the analogous
difluorodithiophosphate complex is known to be monomeric, presumably
octahedral.[16] The observed spectra are typical of an S = 3/2 system
subjected to a relatively large zero-field splitting and an axial or
rhombic ligand field.[7,17] For such a system with axial symmetry,
values of $g_\perp$ = 4 and $g_{||}$ = 2 are predicted; small rhombic distortions
will split the g = 4 line in an approximately symmetrical fashion.

The Mo(acac)$_3$ system has been studied previously in an Al(acac)$_3$
matrix by Jarrett[6a] and Schoffman.[6b] The former reported values of
$g_\perp$ = 3.94 and $g_{||}$ = 1.94, while the latter obtained values of 4.3,
3.5, and 1.9. These compare favorably with the present work (apparent g values of 4.37, 3.59, and 2.00), considering the width of
the signals and the difference in samples (crystals vs. dilute
glasses). Parallel experiments in which the Mo(acac)$_3$ (final con-

Table. EPR Spectra Parameters of Complexes Examined

Complex	Color	Apparent g-Values
Mo(acac)$_3$	red	4.37, 3.59, 2.00
$Mo[S_2P(OEt)_2]_3$	orange-red	4.58, 2.67, 1.92
$[MoCl_6]^{3-a}$	salmon	5.37, b, 1.83
$[MoCl_5(H_2O)]^{2-c}$	orange-red	5.22, b, 1.88
Mo(ox)$_3$ ^{d}	green	5.51, b, e
$[MoCl_6]^{3-}$ + o-H$_2$NC$_6$H$_4$SH	brown	2.00
$[MoCl_6]^{3-}$ + oxSH^f	brown	2.29, 2.17, 1.97

a. Solvent: 12 N HCl. b. Exact position not well-defined due to
width of lines and pronounced rhombicity of spectrum. c. Solvent:
6 N HCl. d. ox = mono-anion of 8-hydroxyquinoline. e. Position
obscured by Mo(V) impurity. f. αSH = 8-mercaptoquinoline.

centration $\leq$ 1 mM) was diluted with 0.5 $\underline{M}$ Al(acac)$_3$ demonstrated
that the observed linewidth ($\sim$ 250 G at g = 4.37) was not due to
magnetic interactions caused by aggregation during freezing. Although no reduction in linewidth was observed, the spectra with ^{95}Mo
were also obtained in the presence of 0.5 $\underline{M}$ Al(acac)$_3$. That neither
Jarrett nor Schoffman were able to observe the hyperfine interaction
due to natural abundance (15.7%) ^{95}Mo is thus not surprising, since
the A (^{95}Mo) observed in Mo(V) compounds is typically 70 G at most.
In order to ascertain whether Mo(III) complexes show comparable
hyperfine, we prepared Mo(acac)$_3$ enriched in ^{95}Mo (97%). Figure 1
compares the EPR spectra of Mo(acac)$_3$ containing natural abundance
and enriched ^{95}Mo. It is obvious that there is a significant broadening upon incorporation of ^{95}Mo. A rough simulation, obtained by
recording the $g_\perp$ portion of the natural abundance spectrum six times
at positions corresponding to an assumed hyperfine of 60 G and sum-
ming on a minicomputer, is compared to the g = 4 region of the natural abundance spectrum in the inset of Figure 1. In view of the
natural linewidth, the approximations made in this approach (purely
axial symmetry, coincident g and A tensors), and the neglect of the
contribution of the naturally occurring ^{95}Mo to the observed linewidth, the fit is extremely good, indicating a value of $A_\perp$ (^{95}Mo) =
60 $\pm$ 5 G. The presence of residual amounts of Mo(V) impurities in
the g = 2 region made a similar procedure impossible, although there
is clearly some broadening of the $g_{||}$ resonance as well. Similar
broadening was observed when ^{95}Mo was incorporated into Mo(ox)$_3$ and
Mo[S$_2$P(OEt)$_2$]$_3$, but due to the quality of the spectra and the lack
of a suitable isostructural diamagnetic diluent, no attempts at spectral simulation were made.

Although the $\underline{N},\underline{N}$-diethyldithiocarbamate (dtc) ligand might be
expected to give a complex analogous to those obtained with the
ligands discussed above, no evidence for an S = 3/2 species was observed in the low temperature EPR spectrum. Instead, a mixture of
two complexes, whose relative concentrations varied from preparation
to preparation, was produced with signals near g = 2. These signals
are very similar to those variously assigned to a dimeric Mo(III)
species,[4] [Mo(dtc)$_3$]$_2$, or to a monomeric eight-coordinate Mo(V)
species,[18] [Mo(dtc)$_4$]$^+$. Similarly, 2-aminothiophenol does not give
an EPR spectrum attributable to a monomeric tris complex of Mo(III);
only an apparently isotropic signal at $g \sim$ 2.00 is observed. This
spectrum is similar to that reported recently for the [Mo(NHC$_6$H$_4$S)$_3$]$^-$
complex,[19] and suggests that Mo(NH$_2$C$_6$H$_4$S)$_3$, if produced initially,
is being oxidized (possibly by trace amounts of disulfide in the
ligand).

Reaction of 8-mercaptoquinoline (oxSH) with [NH$_4$]$_3$[MoCl$_6$] gives
a yellowish-brown toluene solution, together with a brown solid which
remains at the interface of the two layers. This observation agrees
with the findings of Lindoy et al.,[20] who reported the formation of
a brown polymer in refluxing ethanol. Examination of the super-

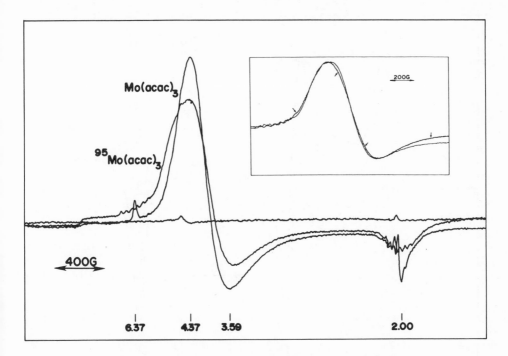

Figure 1. EPR spectra of natural abundance and [95]Mo-enriched Mo-(acac)$_3$, compared to a baseline obtained with the cavity under the same conditions. Inset: Comparison of [95]Mo-enriched spectrum and spectral simulation assuming $A_\perp$ ([95]Mo) = 60 G in $g_\perp$ region. Arrows indicate the simulated spectrum. Conditions of EPR spectroscopy: microwave frequency, 9.2 kHz; microwave power, 3 mW; modulation amplitude, 10 G; magnetic field sweep rate, 1000 G min^{-1}; time constant 0.3 sec; sample temperature, 13 K; instrument gain, 1250. In the inset, the amplitude and microwave power were adjusted to match the simulation and the [95]Mo(acac)$_3$ spectrum as closely as possible. The small feature at g = 6.37 is present in variable amounts in all preparations; its position and the observed splitting of $\sim$ 44 G by [95]Mo indicate that it is due to a Mo-containing impurity of unknown nature. The splitting of the small amount of Mo(V) impurity near g = 2 by [95]Mo is also apparent.

natant by EPR spectroscopy reveals the presence of a paramagnetic species (Figure 2). The observed rhombic spectrum (g = 2.29, 2.15, 1.97; g_{av} = 2.14) is not characteristic of the Mo(III) species described above and lies far outside the range observed for Mo(V) complexes.[21] Substitution of [95]Mo gives no splitting or detectable broadening of the spectrum (Figure 2), even though the lines are relatively narrow ($\sim$ 45 G). This result suggests that the observed spectrum may not be due to a molybdenum species at all, but instead

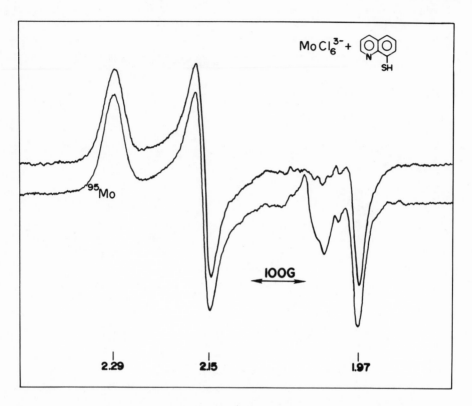

Figure 2. EPR spectra of the toluene supernatant from the reaction
of 8-mercaptoquinoline with $[NH_4]_3[MoCl_6]$ and $[NH_4]_3[^{95}MoCl_6]$. Conditions of EPR spectroscopy were as in Figure 1 except for the following: microwave power, 100 μW; magnetic field sweep rate, 250 G
min^{-1}; time constant, 1.0 sec; instrument gain, 5000 (^{95}Mo), 6300
(natural abundance Mo).

to an organic free radical. Indeed, highly anisotropic signals
with g_{av} = 2.08-2.09 have been observed in crystals of cysteine
hydrochloride irradiated at low temperature.[22] The stability of the
observed species in contact with aqueous media at room temperature,
the anomalously high microwave power required to saturate the signal (> 1 mW at 13 K), and the fact that no such signals have been
observed in previous work on Mo(V) complexes of 8-mercaptoquinoline[23]
all suggest that this signal may be due to a radical stabilized by
a diamagnetic molybdenum complex. An alternative interpretation is
that the signal originates from a mono- or polynuclear molybdenum
complex with one unpaired electron. At the moment, we have no basis
to distinguish between these alternatives.

Conclusions

At temperatures below 30 K, monomeric Mo(III) complexes based on an octahedral geometry exhibit broad, axial or slightly rhombic EPR signals, centered at g = 4 and 2. Even with the large line-widths, substitution of ^{95}Mo results in significant broadening of the signal with $A_\perp$ (^{95}Mo) on the order of 60 G. These data suggest that in enzymes such as nitrogenase and nitrate reductase, which display similar signals but with even narrower lines, substitution of ^{95}Mo should result in significant broadening or splitting of the signal. The failure to observe this effect in the ^{95}Mo-substituted Mo-Fe protein of nitrogenase[5],[7] is, therefore, further evidence that the characteristic EPR signal is <u>not</u> due to a simple Mo(III) site. Assignment of narrow signals at g $\sim$ 2, which are observable at > 77 K, to Mo(III) in enzymes containing only isolated molybdenum sites is questionable. Similar signals reported previously for Mo(III) complexes are probably due to polynuclear mixed valence complexes or monomeric Mo(V) impurities.

NEW Mo-Fe-S CLUSTERS

Preparation and Characterization of $Fe_4Mo_4S_{20}{}^{6-}$

Reaction of 4 equivalents of 2,4,6-trimethylpyridine hydro-chloride and $[Me_4N]_2MoS_4$ with 1 equivalent of $[Me_4N]_2[Fe_4S_4(S-t-Bu)_4]$ in acetonitrile/N-methylformamide solution results in the formation of $[Me_4N]_6[Fe_4Mo_4S_{20}]$, obtained as black microcrystals, mp. > 300° C. <u>Anal.</u> Calc. for $C_{24}H_{72}Fe_4Mo_4N_6S_{20}$: C, 17.02; H, 4.30; Fe, 13.19; Mo, 22.66; N, 4.96; S, 37.87. Found: C, 16.94; H, 4.46; Fe, 12.90; Mo, 22.32; N, 4.81; S, 37.65 (Spang Microanalytical Laboratories, Eagle Harbor, Michigan, U.S.A.). The reaction is most readily formulated as an exchange of $MoS_4{}^{2-}$ for <u>t</u>-BuS$^-$ on the intact $Fe_4S_4{}^{2+}$ core. We propose the structure shown schematically, (I), in which each iron of an Fe_4S_4 cluster is coordinated by a bidentate thio-molybdate ligand. Our evidence is as follows. Far-infrared spectra (Nujol mulls) show a splitting of the 472 cm^{-1} band of free $MoS_4{}^{2-}$ into three bands (492, 471, 435 cm^{-1}). The optical spectrum (Figure 3A) shows a shift of the lowest energy band of free $MoS_4{}^{2-}$ (467 nm) to longer wavelength (580 nm, ε = 1.40 x 10^4). Both of these effects are typical of compounds containing bidentate tetrathiometallate ions coordinated to first-row transition metals[24] and suggest strongly that FeS_2Mo units are present. Preliminary Mössbauer spectra at 4.2° K and in zero field give the following parameters (mm/sec): isomer shift (vs. metallic Fe), 0.44; quadrupole splitting, 1.08; linewidth, 0.45. The first two parameters are similar in magnitude to those observed for tetranuclear Fe-S clusters with simple thio-late ligands.[25] This result is somewhat surprising, since the quad-rupole splitting is normally quite sensitive to the local symmetry and should be affected by a change from four- to five-coordinate

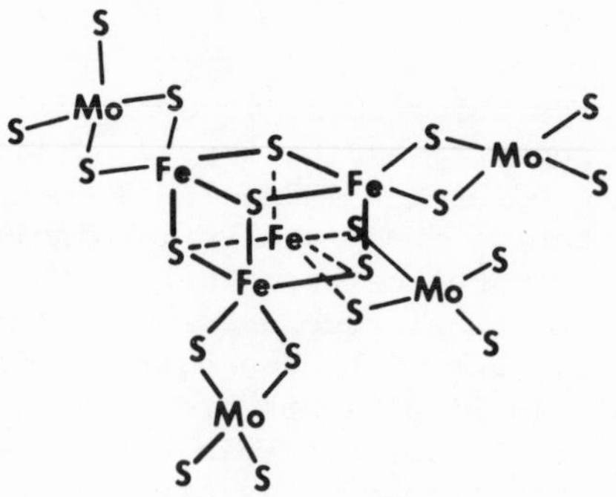

I

iron. The observed linewidths are a factor of two larger than
normal. Examination of the sample in high applied fields (ca. 50
kG) shows the system to be much more complex than the simple clusters, with at least two distinct iron environments. Further studies
are necessary and will be reported elsewhere.

Reactivity

The parent iron-sulfur tetramers are characterized by two kinds
of reactivity,[26] electron transfer and ligand exchange reactions.
Polarographic measurements on I in N,N-dimethylacetamide (0.05 M
Et_4NClO_4 supporting electrolyte, dropping mercury electrode) show
only a single reduction at -0.97 V (vs. SCE), before an irreversible
multi-electron reduction at $\sim$ 1.6 V. The irreversible nature of the
first, presumably one-electron, reduction may well be due to ad-
sorption of the sulfur-rich cluster on the mercury electrode; the
second is apparently due to reductive decomposition. Examination
of the reactivity of I with thiols indicates that the FeS_2MoS_2 unit
is kinetically rather inert. Thus, treatment of I with 8 equivalents
of $PhSH/Et_3N$ in MeCN gives no apparent reaction (cf. Fig. 3A,B). In
contrast, simple alkylthiolates are rapidly and quantitatively displaced from the Fe_4S_4 center by thiophenol under these conditions.[27]
At very high concentrations of thiophenol ($\geq$ 400 equivalents), optical spectra show that a slow reaction occurs to give a new species
in which the FeS_2Mo unit is apparently retained, but in which the
Fe_4S_4 core has been disrupted. The reaction product (II) has been
obtained in crystalline form as its tetraethylammonium salt. Preliminary crystallographic results identify the anion as the novel
binuclear complex, $[S_2MoS_2Fe(SPh)_2]^{2-}$; its chemical and physical
properties are being investigated.

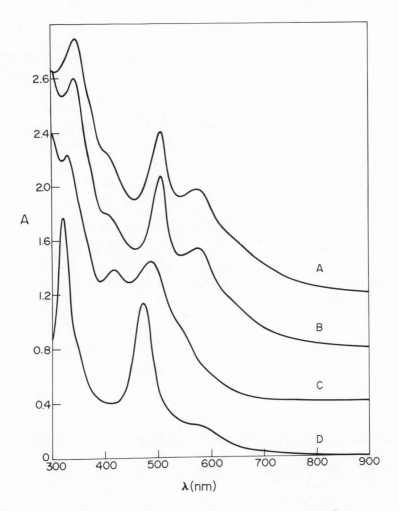

Figure 3. Optical spectra of: (A) 2.6 mM $(Et_4N)_6[Fe_4Mo_4S_{20}]$; (B) sample (A) after adding 8 equivalents of PhSH and Et_3N; (C) crystalline reaction product (II); (D) a mixture of 6.5 mM $(Et_4N)_2MoS_4$ and 1.6 mM $(Et_4N)_2[Fe_4S_4(SPh)_4]$. All spectra obtained in acetonitrile. Optical pathlengths: A,B: 0.20 mm; C: 0.50 mm; D: 0.10 mm.

Discussion

The results presented above indicate that compounds containing the FeS_2Mo linkage, in which the iron is part of an Fe_4S_4 unit, can be prepared and are quite stable. Structure characterization (by crystallography and EXAFS spectroscopy) should furnish data for compounds containing one iron as nearest neighbor to molybdenum. Together with the previously characterized $MoFe_3$ compounds, this information will place the interpretation of spectroscopic data on the FeMo-cofactor on even firmer ground. Synthetic work to date has been hampered by our inability to perform selective substitution at a single iron of an Fe_4S_4 cluster. The preparation of ligands which will permit this as well as the synthesis of complexes in which molybdenum bridges two or more Fe_4S_4 clusters is currently underway.

ACKNOWLEDGEMENTS

We thank Dr. B. H. Huynh and Dr. E. Münck of the Freshwater Biological Institute, University of Minnesota, Navarre, Minnesota, U.S.A. for the Mössbauer data, and Dr. W. E. Blumberg for helpful comments on the EPR spectra. This work was supported by grants from the Research Corporation, the National Science Foundation (CHE-7715990), and the USDA/SEA Competitive Grants Office (5901-0410-8-0175-0) (B.A.A.) and by grant GM 17170 from the National Institutes of Health and the College of Agriculture and Life Sciences, the University of Wisconsin, Madison (W.H.O.-J.).

REFERENCES

1. R. C. Bray and J. C. Swann, Molybdenum-Containing Enzymes, Structure and Bonding, 11:107 (1972).
2. E. I. Stiefel, Proposed Molecular Mechanism for the Action of Molybdenum in Enzymes: Coupled Proton and Electron Transfer, Proc. Natl. Acad. Sci. USA, 70:988 (1973).
3. (a) P. Forget and D. V. DerVartanian, The Bacterial Nitrate Reductases: EPR Studies on Nitrate Reductase A from Micrococcus denitrificans, Biochim. Biophys. Acta, 256:600 (1972); (b) D. V. DerVartanian and P. Forget, The Bacterial Nitrate Reductase. EPR Studies on the Enzyme A of Escherichia Coli K12, Biochim. Biophys. Acta, 379:74 (1975).
4. P. C. H. Mitchell and R. D. Scarle, Complexes of Molybdenum(III) with Sulphur-donor Ligands, J. Chem. Soc. Dalton Trans., 110 (1975).
5. W. H. Orme-Johnson, G. S. Jacob, M. T. Henzl, and B. A. Averill, Molybdenum in Enzymes, in:"Bioinorganic Chemistry-II", K. Raymond, ed., ACS Adv. Chem. Ser., 162:389 (1977).
6. (a) H. S. Jarrett, Paramagnetic Resonance in Trivalent Transition Metal Complexes, J. Chem. Phys., 27:1298 (1957); (b) A. J. Schoffman, Electron Spin Resonance Study of Molybdenum(III) Acetylacetonate and Its Oxidation, Ph.D. Thesis, Polytechnic Institute of Brooklyn, Dissertation Abstracts, 31:3311B (1970).
7. G. Palmer, J. S. Multani, W. C. Cretney, W. G. Zumft, and L. E. Mortenson, Electron Paramagnetic Resonance Studies on Nitrogenase, Arch. Biochem. Biophys., 153:325 (1972).
8. P. T. Pienkos, V. K. Shah, and W. J. Brill, Molybdenum Cofactors

from Molybdoenzymes and in vitro Reconstitution of Nitrogenase and Nitrate Reductase, Proc. Natl. Acad. Sci. USA 74:5468 (1977).

9. V. K. Shah and W. J. Brill, Isolation of an Iron-Molybdenum Cofactor from Nitrogenase, Proc. Natl. Acad. Sci. USA, 74:3249 (1977).

10. J. Rawlings, V. K. Shah, J. R. Chisnell, W. J. Brill, R. Zimmerman, E. Münck, and W. H. Orme-Johnson, Novel Metal Cluster in the Iron-Molybdenum Cofactor of Nitrogenase, J. Biol. Chem., 253:1001 (1978).

11. S. P. Cramer, W. O. Gillum, K. O. Hodgson, L. E. Mortenson, E. I. Stiefel, J. R. Chisnell, W. J. Brill, and V. K. Shah, The Molybdenum Site of Nitrogenase. 2. A Comparative Study of Mo-Fe Proteins and the Iron-Molybdenum Cofactor by X-Ray Absorption Spectroscopy, J. Am. Chem. Soc., 100:3814 (1978).

12. B. H. Huynh, E. Münck, and W. H. Orme-Johnson, Nitrogenase XI: Mössbauer Studies on the Cofactor Centers of the MoFe Protein from Azotobacter Vinelandii OP, Biochim. Biophys. Acta, 527: 192 (1979).

13. T. E. Wolff, J. M. Berg, C. Warrick, K. O. Hodgson, R. H. Holm, and R. B. Frankel, The Molybdenum-Iron-Sulfur Cluster Complex $[Mo_2Fe_6S_9(SC_2H_5)_8]^{3-}$. A Synthetic Approach to the Molybdenum Site in Nitrogenase, J. Am. Chem. Soc., 100:4630 (1978).

14. G. Christou, C. D. Garner, F. E. Mabbs, and T. J. King, Crystal Structure of Tris(tetra-n-butylammonium) Tri-μ-benzenethiolato-bis{tri-μ-sulphido-[μ_3-sulphido-tris(benzenethiolatoiron)]molybdenum}[Bu^n_4N]$_3$[{(PhSFe)$_3$MoS$_4$}$_2$(SPh)$_3$]; an Fe$_3$MoS$_4$ Cubic Cluster Dimer, J. Chem. Soc. Chem. Comm., 740 (1978).

15. W. G. Zumft, Isolation of Thiomolybdate Compounds from the Molybdenum-Iron Protein of Clostridial Nitrogenase, Eur. J. Biochem., 91:345 (1978).

16. R. G. Cavell and A. R. Sanger, Metal Complexes of Substituted Dithiophosphinic Acids. VI. Reactions of Difluorodithiophosphinic Acid with Chlorides and Oxychlorides of Chromium, Molybdenum, and Tungsten, Inorg. Chem., 11:2011 (1972).

17. E. Münck, H. Rhodes, W. H. Orme-Johnson, L. C. Davis, W. J. Brill, and V. K. Shah, Nitrogenase. VIII. Mössbauer and EPR Spectroscopy. The MoFe Protein Component from Azotobacter Vinelandii OP, Biochim. Biophys. Acta, 400:32 (1975).

18. (a) A. Nieuwpoort, New Compounds Containing Molybdenum and Diethyldithiocarbamate, J. Less-Common Metals, 36:271 (1974); (b) A. Nieuwpoort, H. M. Claessen, and J. G. M. van der Linden, Diphenyl Dithiocarbamato Complexes of Molybdenum and Tungsten, Inorg. Nucl. Chem. Lett., 11:869 (1975).

19. J. Gardner, N. Pariyadath, J. L. Corbin, and E. I. Stiefel, Molybdenum and Rhenium Complexes of Aromatic Amine Thiolate Ligands, Inorg. Chem., 17:897 (1978).

20. L. F. Lindoy, S. E. Livingstone, and T. N. Lockyer, Chelate Complexes of Molybdenum, Austr. J. Chem., 18:1549 (1965).

21. E. I. Stiefel, The Coordination and Bioinorganic Chemistry of Molybdenum, Progr. Inorg. Chem., 22:1 (1977).

22. (a) K. Akasaka, Paramagnetic Resonance in L-Cysteine Hydro-
     chloride Irradiated at 77°K, J. Chem. Phys., 43:1182 (1965);
     (b) H. C. Box, H. G. Freund, and E. E. Budzinski, Free-Radical
     Formation by Ultraviolet Irradiation in Single Crystals of
     Cysteine HCl, J. Chem. Phys., 45:809 (1966).
23. (a) G. R. Lee and J. T. Spence, Electron Spin Resonance Active
     Molybdenum(V) Complexes in Dimethylformamide, Inorg. Chem.,
     11:2354 (1972); (b) I. N. Marov, V. K. Belyaeva, Yu, N. Dubrov,
     and A. N. Ermakov, EPR Study of Thio-oxinato-complexes of Molyb-
     denum(V), Tungsten(V), and Copper(II), Russ. J. Inorg. Chem.,
     17:515 (1972).
24. E. Diemann and A. Müller, Thio and Seleno Compounds of the
     Transition Metals with the d^o Configuration, Coord. Chem. Rev.,
     10:79 (1973).
25. R. B. Frankel, B. A. Averill, and R. H. Holm, Mössbauer Proper-
     ties of Synthetic Analogs of Active Sites of the Iron-Sulfur
     Proteins, J. de Physique, Colloque C6, 35:C6-107 (1974).
26. R. H. Holm, Synthetic Approaches to the Active Sites of Iron-
     Sulfur Proteins, Acc. Chem. Res., 10:427 (1977).
27. L. Que, Jr., M. A. Bobrik, J. A. Ibers, and R. H. Holm, Synthe-
     tic Analogs of the Active Sites of Iron-Sulfur Proteins. VII.
     Ligand Substitution Reactions of the Tetranuclear Clusters
     $[Fe_4S_4(SR)_4]^{2-}$ and the structure of $[(CH_3)_4N]_2[Fe_4S_4(SC_6H_5)_4]$,
     J. Am. Chem. Soc., 96:4168 (1974).

MOLYBDO-IRON MIXED CLUSTER AS A Fe-Mo COFACTOR MODEL

Sei Otsuka and Masato Kamata

Department of Chemistry
Faculty of Engineering Science
Osaka University, Toyonaka, Osaka 560, Japan

INTRODUCTION

Recently, an iron-molybdenum cofactor (Fe-Moco) of low molecular weight (less than 5000 Dalton) was isolated from the iron-molybdenum protein of A. vinelandi and C. pasteurianum nitrogenases.[1,2] The molybdenum environment in the Fe-Moco was studied by EXAFS[3] (extended X-ray absorption fine structure) spectroscopy, which implicated molybdenum in a Mo-Fe-S cluster unit, possibly 1 or 2.[4] Apparently prompted by these findings, bi-cluster compounds, 3[5]

and 4[6], have been prepared, whose structures were established by X-ray analysis. We were able to synthesize a mono-cluster compound whose framework is believed to be similar to 1. This paper describes

229

it's characterization and catalytic activity for acetylene half-reduction.

SYNTHESIS

A DMF (20 ml) solution of $(t\text{-}BuS)_4Mo$ (0.10 mmol) and $[(n\text{-}Bu)_4N]_2[Fe_4S_4(SPh)_4]$ (0.11 mmol) was stirred at 90-100° overnight under a pure nitrogen atmosphere. The reaction mixture was filtered, and concentrated <u>in vacuo</u> to dryness. The residue was extracted with dried THF. The crude product (20 % yield based on the iron cluster) obtained from the THF extract was recrystallized from a THF/pyridine mixture to give an analytically pure sample of very dark brown needles, mp 145-6°. The elemental analysis (C, H, N, Mo) conforms well to the theoretical values for a mixed cluster of the formula $[(n\text{-}Bu)_4N]_2[(t\text{-}BuS)MoFe_3S_4(SPh)_3]$ (5). It is soluble in polar organic solvents like DMF, HMPA (hexamethylphosphoramide), N-methyl-pyrrolidone etc., but practically insoluble in dialkylethers and aromatic hydrocarbons.

The stoichiometry of formation of the mixed cluster may be expressed as follows:

$$(t\text{-}BuS)_4Mo + [Fe_4S_4(SPh)_4]^{2-} = [(t\text{-}BuS)MoFe_3S_4(SPh)_3]^{2-}$$
$$+ \frac{1}{n}[Fe(SPh)(S\text{-}t\text{-}Bu)_2]_n + \frac{1}{2}(t\text{-}Bu)\text{-}SS\text{-}(t\text{-}Bu)$$

The reduced form $[(n\text{-}Bu)_4N]_3[(t\text{-}BuS)MoFe_3S_4(SPh)_3]$ (6) was prepared by reducing a HMPA solution of 5 (0.1 mmol), containing (n-Bu)$_4$NBr (0.1 mmol), with sodium acenaphthylenide (0.12 mmol). The crude product obtained by precipitating with diethylether was washed with the ether and recrystallized from HMPA-THF to give an 80 % yield of extremely air-sensitive black needles.

SPECTROSCOPIC CHARACTERISTICS

The structural assignment is based on the spectroscopic properties (electronic, ir, ^{1}H nmr, epr, and Mössbauer) and electrochemical measurements (cyclic voltammogram) as follows.

The electronic spectra of 5 and its reduced form 6 are compared with those of the bi-cluster compound $[Mo_2Fe_6S_8(SPh)_9]^{3-}$, $[Fe_4S_4(SPh)_4]^{2-}$, and $[Fe_4S_4(SPh)_4]^{3-}$ in Table 1. The spectrum of 5 is characterized by an absorption at 440 nm and rather featureless sulfide to metal charge transfer bands in the uv region. The reduced form 6 shows a distinct maximum at 313 nm and an intense absorption around 370 nm (Table 1). The absorption in the 370-470 nm region appears to be characteristic of the cluster core $M_4S_4{}^{n+}$. A number of synthetic tetranuclear iron-sulfur clusters of cubane-like structure $[Fe_4S_4(SR)_4]^{2-}$ as well as the dimeric mixed clusters 4[6] exhibit a strong absorption in this region. The intensity of the visible absorption of 6 is stronger than that of 5. This trend contrasts with that observed for the ferredoxin proteins and the synthetic analogs[7,8] (Table 1).

Table 1.   Comparison of Electronic Spectra

Cluster	Solvent	λ max, nm (ε)	Ref.
$[(t\text{-}BuS)MoFe_3S_4(SPh)_3]^{2-}$	HMPA	293(sh, 19300) 356(sh, 11900) 440(9200)	This work
$[(t\text{-}BuS)MoFe_3S_4(SPh)_3]^{3-}$	HMPA	313(34000) 370(sh, 18000)	This work
$[Mo_2Fe_6S_8(SPh)_9]^{3-}$	DMSO	284(62300) 389(41200)	6
$[Fe_4S_4(SPh)_4]^{2-}$	DMSO	259(41400) 325(sh, 23000) 458(17600)	9
$[Fe_4S_4(SPh)_4]^{3-}$	DMSO	267(46100) 340(sh, 20200) 400(sh, 14800)	7

The ^{1}H nmr spectrum of $\underset{\sim}{5}$ measured at 25° in DMSO contains three broad signals assignable to the phenyl protons and several complex signals ascribable to the aliphatic protons. Although detailed assignments have not been made, it can be said that their normal chemical shifts reflect diamagnetism of the $MoFeS_4^{2+}$ core, derived presumably from antiferromagnetic coupling. Consistent with this, $\underset{\sim}{5}$ is essentially epr silent. The reduced mixed cluster $\underset{\sim}{6}$ in DMF shows a resonance centered near g=1.995. The epr spectrum of $\underset{\sim}{6}$ (77°K) lacks the S=3/2 resonances (g=4.3 and $\sim$3.6) characteristic to the iron-molybdenum protein or its cofactor[12].

A preliminary ^{57}Fe Mössbauer study was made for $\underset{\sim}{5}$. The spectra were recorded at 77, 195K and room temperature, employing the crys-

Table 2. Mössbauer Data[a] of $[(t-BuS)MoFe_3S_4(SPh)_3]^{2-}$ (5) and $[Fe_4S_4(SPh)_4]^{n-}$ (n=2,3)

Cluster	T, °K	δ(mm/s)[b]	ΔE_Q	$\Gamma 1/2$[c]	Av. Oxidn. State of Fe	Ref.
5	77	0.452	0.903	0.227		This work
	195	0.374	0.662	0.201		
	r.t.	0.337	0.565	0.193	~ 2.5	
$[Fe_4S_4(SPh)_4]^{2-}$	4.2	0.35	1.10	0.52	2.5	10
	296	0.34	0.54	0.65		
$[Fe_4S_4(SPh)_4]^{3-}$	77	0.61	1.63	0.53	2.25	11
	200	0.56	0.90	0.60		
		0.53	0.89	0.70		

a. Polycrystalline sample.   b. Referenced to metallic iron at room temperature.

c. Line widths of half height, mm/s.

talline sample. The data are compared with those reported for $[Fe_4$-
$S_4(SPh)_4]^{2-}$ and $[Fe_4S_4(SPh)_4]^{3-}$ in Table 2. In the temperature
range studied, one quadrupole doublet was observed. It remains quite
sharp ($\Gamma\frac{1}{2}$ < 0.23 mm/s), implying effective equivalence of the three
iron atoms. The isomer shift (δ mm/s) is markedly temperature-
dependent. Comparison with the four iron clusters indicates that
the average oxidation state of the iron atoms moves from 2.25 towards 2.5 as the temperature rises. This implies an increase in
electron density on the molybdenum. Such polarization could be a
factor assisting the electron transfer to substrates such as dinitrogen, acetylene and similar π-acids. This is of course a hypothetical proposal which will clearly require further investigation.

Table 3. Redox Potential Data in DMF at 25°

Compounds	2-/3-	3-/4-	4-/5-	Ref.
$[(t\text{-}BuS)MoFe_3S_4(SPh)_3]^{n-,a}$	-1.09	-1.74	-	This work
$[Mo_2Fe_6S_8(SPh)_9]^{n-}$	-	-0.87	-1.08	13
$[Fe_4S_4(SPh)_4]^{n-,b}$	-1.04	-1.75		9

a. Potentials vs. SCE. $[NEt_4]ClO_4$ as supporting electrolyte. The
potential value obtained by using Ag/0.01 M $AgNO_3$ electrode was corrected by 0.3 V.   b. $E_{1/2}$ values obtained by cathodic polarography.

ELECTROCHEMICAL PROPERTIES

The current-voltage curves for the redox processes in the mixed cluster anion 5 are characterized by two redox events which are quasi-reversible. A marked difference between the monomeric cluster and Garner's dimeric one 4 is that the second reduction of the latter occurs at a much lower potential (-1.08 V) than that of the former (Table 3). These redox levels are compared with those of $[Fe_4S_4(SPh)_4]^{n-}$ in Table 3.

CATALYSIS

The iron-molybdenum mixed cluster 5 alone and in combination with the ferredoxin model cluster $[Fe_4S_4(SPh)_4]^{2-}$ was tested for the electron-transfer hydrogenation of acetylene. Due to solubility problems and to the instability toward water of 5 and 6, the choice

Table 4. Half-Reduction of Acetylene with $NaBH_4$[a]

Catalyst	mmol	Turn over[b]
$(t-BuS)_4Mo$	(0.04)	4.4
$(t-BuS)_4Mo/Fd$[c]	(0.04/0.04)	8.1
Fd[c]	(0.08)	14.8
$MoFe$[d]$/Fd$[c]	(0.04/0.04)	15.9

a. C_2H_2 (14 mmol), $NaBH_4$ (7 mmol); Solvent = DMF (5 ml) + EtOH (10 ml); 30°, 5h.   b. C_2H_4 mol/cluster mol/5h.   c. Fd = $[Fe_4S_4(SPh)_4]^{2-}$.   d. MoFe = $[(t-BuS)MoFe_3S_4(SPh)_3]^{2-}$.

of the reaction medium was limited. The catalytic reaction was
carried out using $NaBH_4$ in alcohol-DMF or $Na_2S_2O_4$ in aq. DMF. Typical results are shown in Table 4. Most noteworthy is the observation
of selective half-reduction of acetylene effected by dithionite as
the electron source, however the rate is slow compared to redution
with BH_4^-. The turn-over is also low due to the apparent deterioration of the mixed cluster by hydrolysis. The use of micellar systems
(DMF/H_2O/detergent) slightyly improves the turn over (Table 5).
Further improvement of the reaction medium is under investigation.

Table 5. Half-Reduction of Acetylene with $Na_2S_2O_4$[a]

Cluster	(mmol)	Medium	(ml)	Turn over[b]
Fd[c]	(0.08)	DMF/H_2O	(10/2)	2.5
MoFe[d]/Fd[c]	(0.01/0.08)	"	"	2.8
"	"	"	" [e]	3.6
"	"	$CHCl_2CHCl_2/H_2O$	(5/5)[f]	2.9
"	"	"	" [e]	4.3
MoFe[d]	(0.08)	DMF/H_2O	(10/2)[f]	1.4
MoFe[d]/Fd[c]	(0.04/0.04)	"	" [f]	3.4

a. C_2H_2 (14 mmol), $Na_2S_2O_4$ (10 mmol); 30°, 5h.   b. C_2H_4 mol/cluster
mol/5h.   c. Fd=$[Fe_4S_4(SPh)_4]^{2-}$.   d. MoFe=$[(t\text{-}BuS)MoFe_3S_4(SPh)_3]^{2-}$.
e. $[C_{12}H_{25}NMe_3]Cl$ (1 mmol).   f. $C_{12}H_{25}C_6H_4SO_3Na$ (1 mmol).

REFERENCES

1. V.K. Shah and W.J. Brill, Isolation of an iron-molybdenum co-factor from nitrogenase, Proc. Natl. Acad. Sci. USA, 74:3249 (1977).

2. P.T. Pienkos, V.K. Shah, and W.J. Bill, Molybdenum cofactors from molybdo-enzymes and in vitro reconstitution of nitrogenase and nitrate reductase, Proc. Natl. Acad. Sci. USA, 74:5468 (1977).

3. S.P. Cramer, K.O. Hodgson, W.O. Gillum, and L.E. Mortenson, The molybdenum site of nitrogenase. Preliminary structural evidence from X-ray absorption spectroscopy, J. Am. Chem. Soc., 100:3398 (1978).

4. S.P. Cramer, W.O. Gillum, K.O. Hodgson, L.E. Mortenson, E.I. Stiefel, J.R. Chisnell, W.J. Brill, and V.K. Shah, The molyb-denum site of nitrogenase. 2. A comparative study of Mo-Fe proteins and the iron-molybdenum cofactor by X-ray absorption spectroscopy, J. Am. Chem. Soc., 100:3814 (1978).

5. T.E. Wolff, J.M. Berg, C. Warrick, K.O. Hodgson, R.H. Holm, and R.B. Frankel, The molybdenum-iron-sulfur cluster complex $[Mo_2Fe_6S_9(SC_2H_5)_8]^{3-}$. A synthetic approach to the molybdenum site in nitrogenase, J. Am. Chem. Soc., 100:4630 (1978).

6. C.D. Garner, G. Christou, S.R. Acott, F.E. Mabbs, R.E. Miller, T.J. King, M.G.B. Drew, C.E. Johnson, J.D. Rush, and G.W. Kenner, Synthesis and characterization of iron-molybdenum-sulfur cluster. This volume and references therein.

7. J. Cambray, R.W. Lane, A.G. Wedd, R.W. Johnson, and R.H. Holm, Chemical and electrochemical interrelationship of the 1-Fe, 2-Fe, and 4-Fe analogues of the active sites of iron-sulfur proteins, Inorg. Chem., 16:2565 (1977).

8. R. H. Holm and J.A. Ibers, Synthetic analogues of the active sites of iron-sulfur proteins, in: "Iron-sulfur Proteins," Vol. III, ed. W. Lovenberg, Academic Press, New York, 1977, p. 205.

9. B.V. Depamphilis, B.A. Averill, T. Herskovitz, S. Que, Jr., and R.H. Holm, Synthetic analogues of the active sites of iron-sulfur proteins. VI. Spectral and redox characteristics of the tetranuclear clusters $[Fe_4S_4(SR)_4]^{2-}$, J. Am. Chem. Soc. 96:4159 (1974).

10. R.B. Frankel, B.A. Averill and R.H. Holm, Mössbauer properties of synthetic analogs of active sites of the iron-sulfur proteins, J. de Phys., 35:C6-107 (1974).

11. E.J. Laskowski, R.B. Frankel, W.O. Gillum, G.C. Papaefthymiou, J. Renaud, J.A. Ibers and R.H. Holm, Synthetic analogues of the 4-Fe active sites of reduced ferredoxins. Electronic properties of the tetranuclear trianions $[Fe_4S_4-(SR)_4]^{3=}$ and the structure of $[(C_2H_5)_3(CH_3)N]_3[Fe_4S_4(SC_6H_5)_4]$, J. Am. Chem. Soc., 100:5322 (1978).

12. J. Rawlings, V.K. Shah, J.R. Chisnell, W.J. Brill, R. Zimmerman, E. Münck and W.H. Orme-Johnson, Novel metal cluster in the iron-molybdenum cofactor of nitrogenase, J. Biol. Chem., 253:1001 (1978).

13. G. Christou, C.D. Garner, F.E. Mabbs, A molybdenum derivative of a four-iron ferredoxin type center, Inorg. Chim. Acta., 28:L189 (1978).

REACTIONS AND PROPERTIES

OF MOLYBDENUM COMPLEXES

FORMATION OF AMMONIA FROM LIGATING DINITROGEN AT A SINGLE METAL

SITE AND ITS POSSIBLE RELEVANCE TO THE NITROGENASE REACTION

J. Chatt

A.R.C. Unit of Nitrogen Fixation, University of Sussex,
Falmer, Brighton, Sussex, BN1 9RQ, England

INTRODUCTION

The nitrogenase reaction has intrigued coordination chemists since 1930 when Bortels showed that Azotobacter in a nitrogen-free medium required traces of molybdenum for growth, although it would grow without molybdenum when it was supplied with ammoniacal nitrogen.[1] That discovery sparked a number of sporadic attempts to bring gaseous nitrogen into reaction with molybdenum compounds, or with dihydrogen in the presence of molybdenum compounds, during the intervening decades. Many claims of fixation, some quite substantial, resulted but rarely have they been substantiated.[2] Here, I shall report on attempts at a possible simulation of the nitrogenase reaction in a stoichiometric manner by the conversion of ligating dinitrogen to ammonia on a molybdenum(0) or tungsten(0) center with protic acid.

Nitrogenase, which causes the reduction of dinitrogen to ammonia in certain very primitive microorganisms, is a complex molybdo-iron enzyme. Although the biochemistry of nitrogen fixation is reasonably well understood at the descriptive level,[3,4] there is still very little definite information about the chemistry of the fixation reaction at the molecular level. This situation leaves plenty of room for speculation and such has been fairly rife. This paper, however, concentrates only on speculations from within my own research group. Circumstantial evidence suggests very strongly that the molybdenum atoms in nitrogenase provide the site or sites where molecular nitrogen is reduced to ammonia.[5] Two significant steps towards the elucidation of the nitrogenase reaction were made recently. The first concerns the environment of molybdenum in the enzyme, which has been inferred from X-ray

absorption spectroscopic studies.[6,7]   The second concerns the
detection by biochemists in my laboratory of an intermediate
dinitrogen hydride which is formed on the enzyme when it is actively
fixing nitrogen.[8]

The nitrogenase reaction

     The chemical reaction which occurs on nitrogenase is the
reduction of dinitrogen to ammonia.   Nitrogenase is not specific
for the reduction of dinitrogen;   dinitrogen oxide, azide ion,
cyanide ion, acetylene, methyl isocyanide, cyclopropene and, less
efficiently, methyl cyanide are all reduced, but not ethylene.
Dihydrogen is a competitive inhibitor and carbon monoxide a strong
inhibitor of nitrogenase action.   All of these are substances
which bind strongly to transition metal ions in low oxidation
states.   In the biological reaction, no free intermediate, such
as hydrazine, has been detected.   The reduction of dinitrogen is
accompanied by the production of dihydrogen in vitro and in vivo,
but certain aerobes, e.g. Azotobacter, reabsorb their dihydrogen
before it is evolved by means of a hydrogenase specifically for
that purpose.[9,10]

     Nitrogenase[3,4] is a brown molybdo-iron enzyme which can be
separated on a Sephadex column into a brown protein of molecular
weight around 220,000 and a yellow protein of molecular weight
around 68,000.   The molecular weights vary slightly depending on
the microorganism from which the nitrogenase has been extracted.
The large protein contains ideally two atoms of molybdenum and a
quantity variously stated as 24 to 34 atoms of iron.   The iron
content may depend on the source of the protein but now it is
generally thought that the number of iron atoms is nearer 34 than
24.   The small protein contains 4 atoms of iron and no molybdenum.
The iron atoms are accompanied by an approximately equal number of
sulphide ions.   In the larger protein, at least half of the iron
appears to be combined in cubane Fe_4S_4 ferredoxin-type clusters.
The smaller protein contains one such cluster.   These clusters
are bonded through four cysteine sulphur atoms into the protein,
one cysteine sulphur to each iron atom.[11]   The nitrogenase
proteins are sensitive to air, the larger one being less sensitive
than the smaller one which is destroyed immediately on the briefest
exposure to air.

     It appears that the catalytic reduction of dinitrogen to
ammonia occurs on the large protein and that the small protein is
a specific electron carrier to the large protein.   No reducing
agent or other electron carrier has been found to reduce the large
protein directly.   In vitro, in the presence of the monomagnesium
salt of ATP as energy source and around neutral pH with sodium
dithionite as an artificial reducing agent, the two proteins
convert dinitrogen into ammonia.   Dihydrogen is formed as a

by-product.   The transfer of an electron from the small protein to
the large protein is accompanied by the hydrolysis of the ATP to
ADP and inorganic phosphate, and a direct correspondence of one
electron transferred for each molecule of ATP hydrolysed has been
demonstrated.[12]   The monomagnesium salt of ADP is an inhibitor of
the nitrogenase reaction.   The Fe_4S_4 clusters probably provide the
electron storage system from which electrons pass to the active
site and hence, into a ligating dinitrogen molecule which
simultaneously picks up protons from its environment until it has
been completely reduced to ammonia.   The ammonia, presumably, is
then displaced by dinitrogen, and the cycle repeats.

The active site

    The molybdenum atoms are believed to provide the active site
or sites for the reduction of dinitrogen and it is the reaction on
the molybdenum atoms which particularly concerns us here.   Until
last year, all attempts to determine their environment in the
enzyme by spectroscopic means were unsuccessful.   Recent studies
of the X-ray absorption fine structure (EXAFS) spectrum of the
large protein now confirm the long entrenched belief that the
molybdenum is associated with sulphide ions in the same way as the
iron, and indicates that the molybdenum is surrounded by three or
four sulphur atoms at 2.35 ± 0.03 Å, one or two sulphur atoms at
2.49 ± 0.35 Å, and two or three iron atoms at 2.72 ± 0.05 Å.[7,8]
Two types of sulphur bridged structures containing iron and
molybdenum were proposed as satisfying the spectroscopic criteria,
but at that time, no such structures were known.   One suggestion
was an $MoFe_3S_4$ cluster analogous to the ferredoxin Fe_4S_4 type
cluster but having one iron atom replaced by molybdenum.   Now, two
compounds each containing two such clusters have been described,
$[Et_4N]_3[\{(FeSEt)_3MoS_4\}_2S(SEt)_2]$[13] and $[Bu^n_4N]_3[\{(FeSPh)_3MoS_4\}_2-$
$(SPh)_3]$.[14]   There is no evidence so far that either compound can
be induced to take up molecular nitrogen or to catalyse its
reduction to ammonia.

    Further evidence concerning the possible nature of the
molybdenum site has appeared from the isolation of a very unstable
cofactor (FeMoco) obtained by dimethylformamide extraction of the
large protein after treatment with acid.[15]   This cofactor
activates the inactive large protein obtained from mutants of
nitrogen-fixing bacteria which produce nitrogenase containing no
molybdenum.   It is the only molybdenum-containing substance which
is known to do so.   It is highly unstable in aqueous solution,
contains no protein or amino acids, apparently has a low molecular
weight, and contains molybdenum, iron and sulphur probably in the
ratio of 1:8:6.

    The enzyme has no esr absorption which can be attributed to
molybdenum, although there are a number attributed to the iron-

sulphur clusters.   The changes in these absorptions induced by oxidation or reduction of the proteins have been extensively investigated.[3,4]

Whether the molybdenum atoms in the enzyme occur as a pair or as independent atoms is an important matter in considering the chemical mechanism of the reduction process.   However, there is no direct evidence on the matter.   The large protein can be separated into two pairs of subunits.   The components of one pair have a molecular weight of about 60,000 each and the other of about 51,000 each.   The molybdenum presumably occurs as one atom in each component of one of these pairs, although they could lie together as a pair across the boundary between the matching subunits.   The enzyme as usually isolated is deficient in molybdenum and its activity is more nearly linearly related to the concentration of molybdenum than to the square or other such function.   This result suggests that the molybdenum atoms behave individually or that they are inserted into the enzyme in pairs.

CHEMISTRY OF NITROGEN FIXATION

In our study of the possible chemistry of the nitrogenase reaction, we assumed that it takes place on an individual molybdenum atom in the enzyme.   We, thus, conclude that reduction can take place through the formation of a ligating hydrazide(2-), $N-NH_2^{2-}$ or isodiazene, $N=NH_2$ intermediate.   These two would be indistinguishable when attached to a transition metal atom and which name is correct is purely semantic.   Other research groups associated with Professor A.E. Shilov[16] and Professor G.N. Schrauzer[17] have proposed other mechanisms for the reduction of dinitrogen to ammonia in protic media.   The first proposes that dinitrogen bridges two molybdenum atoms before it is reduced and the second that dinitrogen, side-bonded to one molybdenum atom, is reduced to diazene in the initial stage of reduction.   Neither has identified any intermediate stage of reduction from dinitrogen to ammonia in his model system.

Our work is concerned with the reduction of dinitrogen to ammonia in certain mononuclear complexes of molybdenum and tungsten.   In these, the dinitrogen is certainly bonded end-on to the metal.   The reduction takes place with a very high efficiency in relation to the reducing capacity of the system and we have also been able to isolate and characterize complexes containing the dinitrogen in the initial stages of its reduction.   These studies have led to a very satisfying chemical mechanism for the reduction process on a single metal atom, which might well model the reduction on the molecular level in the natural system.   However, there is no experimental evidence to exclude other possibilities such as the two referred to above.

Ammonia from mononuclear metal complexes

The only mononuclear dinitrogen complexes to undergo
substantial protonation on nitrogen in a protic medium are certain
bis-dinitrogen complexes of molybdenum or tungsten of the types
cis- or trans-$[M(N_2)_2L_4]$ (M = Mo or W; L = $PMePh_2$, PMe_2Ph, or
$\frac{1}{2}$ $Ph_2PCH_2CH_2PPh_2$).[18] They are prepared from molecular nitrogen
by the reduction of halo-complexes such as $MoCl_5$, $[MoCl_3(thf)_3]$,
$[WCl_4L_2]$ or $[WCl_4L_3]$ with Grignard magnesium in tetrahydrofuran
(thf) solution under a dinitrogen atmosphere in the presence of the
phosphine at room temperature, e.g., reaction (1). When the above

$$[MoCl_3(thf)_3] + 4PMe_2Ph \xrightarrow[thf]{Mg,N_2} cis-[Mo(N_2)_2(PMe_2Ph)_4] \qquad (1)$$

complexes in solution in organic solvents are treated with an
excess of mineral acid, one dinitrogen molecule is protonated at
the terminal nitrogen (designated as β) and the second molecule is
liberated. The reaction proceeds particularly efficiently in
methanol with sulphuric acid. When the metal is tungsten and the
complex contains monophosphines, protonation proceeds rapidly at
room temperature to give ammonia in yields of over 90% with about
5% of hydrazine and a little hydrogen. Ideally, the reaction
proceeds as reaction (2). The reaction cannot be rendered

$$[W(N_2)_2(PMe_2Ph)_4] \xrightarrow[MeOH]{acid} 2NH_3 + N_2 + W^{VI} \text{ products} \qquad (2)$$

catalytic or cyclic because the product, probably tungsten(VI)
oxide, cannot be recycled. When the metal is molybdenum, the
yield of ammonia is only 35%, presumably because molybdenum in its
lower oxidation states is less strongly reducing than tungsten.
But as its complexes are also less robust, they are more prone to
side reactions, such as the production of hydrido-complexes and,
thence, dihydrogen.

The initial step is undoubtedly the direct protonation of the
complex, probably at the β-nitrogen of one of the ligating
dinitrogen molecules (reaction 3). This protonation causes

$$[M(N_2)_2L_4] + H^+ \longrightarrow [M(N_2)(N_2H)L_4]^+ \xrightarrow{X^-} [MX(N_2H)L_4]^- + N_2 \qquad (3)$$

electron withdrawal from the metal, essentially an oxidation,
which will assist the liberation of the second dinitrogen ligand
and its replacement by the acid anion (reaction 3). This
replacement of a π-acceptor N_2 ligand by an anion, e.g., Cl^- or
HSO_4^-, which is a π-donor and carries negative charge, will repel
more electronic charge into the $\equiv N\equiv NH$ ligand resulting from the
first protonation, so assisting a second protonation.

The direct protonation of a dinitrogen ligand without displacement of any other ligand has been observed,[19] as shown in equation (4). It is interesting that the protonation occurs at

$$trans\text{-}[Mo(NCPr^n)(N_2)(dppe)_2] + 2H^+ \longrightarrow [Mo(NCPr^n)(N\text{-}NH_2)(dppe)_2]^{2+} \quad (4)$$

the terminal nitrogen atom rather than the carbon of the n-propyl cyanide. The above protonation is completely reversed by weak bases, such as triethylamine, to regenerate the original dinitrogen complex. The protonation at room temperature stops at the $N\text{-}NH_2$ stage as does that of the bis-dinitrogen complexes which contain dppe as the only phosphine ligands (reaction 5). In this case,

$$trans\text{-}[M(N_2)_2(dppe)_2] \xrightarrow[\substack{acid \\ HX}]{excess} trans\text{-}[MX(\text{=}N\text{-}NH_2)(dppe)_2]X \quad (5)$$

weak base will remove only one proton from the N_2H_2 ligand to give complexes of the iminonitrosyl or diazenido-ligand, N=NH (equation 6),[21] and this reaction provides the only means of

$$trans\text{-}[MX(N\text{-}NH_2)(dppe)_2]X \xrightarrow{NEt_3} [MX(N\text{=}NH)(dppe)_2] + NHEt_3X \quad (6)$$

preparing these N=NH complexes in a pure state. Both protons can be removed only in the presence of an uncharged ligand Q, e.g. N_2 or PhCN, which is capable of replacing the ligand X^-, so producing dinitrogen complexes of the type $[M(Q)(N_2)(dppe)_2]$.[22]

For protonation to proceed beyond the $N\text{-}NH_2$ stage in the bis-(dppe) complexes, it is necessary to carry out the reaction at higher temperatures to cause the displacement of some of the phosphine. However, the monophosphine complexes shed their phosphine ligands more easily and, in methanol/sulphuric acid, the reactions in the tungsten series proceed rapidly to ammonia at room temperature. The intermediate stages are then difficult to catch, but by careful manipulation and adding H_2SO_4 (two equivalents) to cis-$[W(N_2)_2(PMe_2Ph)_4]$ in tetrahydrofuran followed by the precipitation of the product after 30 mins with pentane, it is possible to obtain a substance of stoichiometry $W(N\text{-}NH_2)(PMe_2Ph)_3$-$(HSO_4)_2$ which shows the $N\text{-}NH_2$ group in its infra-red spectrum.[23] This material is very unstable and decomposes even in aprotic solvents such as methyl cyanide. In methanol, it degrades immediately to give nitrogen hydrides quantitatively (reaction 7).

$$W(N\text{-}NH_2)(PMe_2Ph)_3(HSO_4)_2 \xrightarrow{MeOH} NH_3(95\%) + N_2H_4(5\%) \quad (7)$$

In the reaction of methanolic sulphuric acid with
cis-$[W(N_2)_2(PMe_2Ph)_4]$, the ratio of ammonia to hydrazine depends
markedly on the solvent with the less protic solvents giving more
hydrazine.[18,24]   The extent of the reaction depends upon both the
tertiary phosphine and the acid used, with the halogen acids tending
to give complexes representing intermediate stages of reduction more
readily than sulphuric acid.   By the use of various acids,
different phosphines and solvents, it has been possible to isolate
complexes containing dinitrogen in monoprotonated and diprotonated
forms.   We have not been able to isolate any molybdenum complexes
containing a third protonation stage, but in the tungsten series,
a complex representing the third protonation stage was obtained as
a by-product.   This species was originally formulated as
$[WCl_3(NHNH_2)(PMePh_2)_2]$.[25]   However, a recent [15]N n.m.r. study
shows that the third proton is not attached to the α-nitrogen atom,
but is probably on the metal.[26]   Some typical examples of complexes
containing dinitrogen in the first, second and third protonation
stages of the complexes are shown in Table 1.   Those complexes
which contain monophosphines and dinitrogen degrade rapidly with
sulphuric acid in methanol to give ammonia.   In the case of the
tungsten complexes, they also give a little hydrazine.   Those
which contain monophosphines and N_2H_2 ligands give ammonia and
hydrazine in the same yields and ratios as are obtained from the

Table 1.   Some complexes of molecular nitrogen in the initial
           stages of reduction towards ammonia and hydrazine

Atomic grouping	Typical examples
(1) M≋N≡N	$[M(N_2)_2(dppe)_2]$, $[M(N_2)_2(PR_3)_4]$
(2) M⋰N⋮NH	$[MX(N_2H)(dppe)_2]$ cf. $[MX(NO)(dppe)_2]$
(3) M≡N⋯NH$_2$	$[MX(N_2H_2)(dppe)_2]^+$, $[MX_2(N_2H_2)(PR_3)_3]$
	$[MX(N_2H_2)(py)(PR_3)_3]^+$
(4) M≡N⋯NH$_2$ with H	$[WX_3H(N_2H_2)(PR_3)_n]$

dppe = $Ph_2PCH_2CH_2PPh_2$, PR_3 = PMe_2Ph or $PMePh_2$, M = Mo or W,

X = Cl or Br, py = pyridine, n = 2 or 3.

corresponding bis-dinitrogen complexes, suggesting that the N-NH$_2$
complexes lie on the direct route to ammonia.   However, the
complexes WCl$_3$H(NNH$_2$)(PR$_3$)$_2$ give a much higher ratio of hydrazine
(30%) on treatment with methanol/sulphuric acid,[23] which suggests
that these monohydride complexes are not intermediates on the way
to ammonia but along the side reaction to hydrazine.   It is
surprising that this hydride complex gives any ammonia because it
is a hydrazide(2-) complex of tungsten(VI) and has no non-bonding
d-electrons to effect reduction.   We shall return to this point
later.

A study of the protonation of cis-[W(N$_2$)$_2$(PMe$_2$Ph)$_4$] by [15]N
n.m.r. spectroscopy in thf with sulphuric acid (20 mols.) initially
shows the characteristic two line spectrum of the starting complex
([15]N$_\alpha$ at -61.2 ppm, [15]N$_\beta$ at -35.9 ppm relative to nitro-methane,[27]
J(NN) = 5.4(1)Hz, ^{2}J(NP) = 1.9(1)Hz, ^{3}J(NP) = 0.9(1), and ^{1}J(WN)
not resolved at 18.24 MHz and 30°C).   This spectrum is replaced
as the reaction proceeds by an equally characteristic spectrum of
the [15]N-[15]NH$_2$ ligand ([15]N$_\alpha$ at -78.2 ppm, [15]N$_\beta$ at -243.0 ppm with
no hydrogen coupling owing to exchange in acid medium).[26] The
product of the first protonation, which would contain the W-N=NH
grouping, never reaches sufficient concentration to be observed.
The N-NH$_2$ spectrum then shifts showing four further distinct stages
without any chemical change in the N-NH$_2$ group.   These stages are
not necessarily all on the way to ammonia since at least one may
be the hydride stage corresponding to atomic grouping (4) of
Table 1.   However, most of these spectral shifts probably record
the successive replacements of the monophosphines by the hydrogen
sulphate anion or solvent.   No further nitrogen intermediate in
the reaction reaches sufficient concentration to be detected until
the final product [15]NH$_4^+$ whose absorption appears at -360 ppm.
The N-NH$_2$ intermediate obviously represents the plateau of highest
stability on the route from ligating dinitrogen to ammonia in the
monophosphine tungsten complexes.   The course of the reaction in
the molybdenum series is still under study and not yet complete,
but it also shows the N-NH$_2$ stage as a persistent one.

The N-NH$_2$ ligand is strongly conjugated to the metal.   A
number of stable complexes containing that ligand have been isolated
during our protonation studies and six have had their structures
determined by X-rays.   In all cases, the M-N-N chain is linear,
the M-N bond being essentially triple and the N-N bond has an order
rather greater than one.   In the recently determined structure of
[WCl$_3$(NNH$_2$)(PMe$_2$Ph)$_2$], even the hydrogen atoms are resolved.[28]
One hydrogen atom is hydrogen bonded to the chlorine atom of an
adjacent molecule.   The W-N-NH$_2$ moiety is planar with W-N = 1.75(1)
Å, N-N = 1.30(2) Å, N-H^1 = 0.784(10) Å, N-H^2 = 1.083(12) Å,
H^2...Cl = 2.27 Å, < WNN = 179°, and < HNH = 143°.   The shape and
bond distances clearly show that the N-NH$_2$ system is strongly

conjugated to the metal, which is undoubtedly why the third
protonation does not occur at the N_α-atom, but rather, we suppose,
at the N_β-atom.   This supposition leads to a chemically satisfying
mechanism for the degradation of dinitrogen to ammonia.[29]

Chemical mechanism of degradation of dinitrogen to ammonia

    The protonation of the β-nitrogen atom of the $N-NH_2$ ligand
will finally reduce the N-N bond order to unity and at the same
time by putting a positive charge on the β-nitrogen atom induce
heterolytic cleavage of the N-N bond as in reaction sequence (8).

$$M \equiv N \equiv NH_2 + H^+ \longrightarrow M \equiv N - NH_3^+ \longrightarrow M^+ \equiv N + NH_3 \qquad (8)$$

In this way, the hydrazide(2-) ligand is protonated and reduced to
a nitrido-ligand and one molecule of ammonia.   The nitrido-ligand
will readily hydrolyze from the metal to give the second molecule
of ammonia.   The conditions necessary to protonate the $N-NH_2$ ligand
to $N-NH_3^+$, which doubtless degrades rapidly, are just those
necessary to hydrolyze the resulting nitride and so it is not
surprising that no intermediate can be seen between the $N-NH_2$ stage
and the appearance of the ammonium salt by the somewhat insensitive
method of ^{15}N n.m.r.   Although no nitride complex has been isolated
from the protonation studies, a nitride complex of the type expected,
$[MoCl_2N(PMe_2Ph)_2]$, has been shown to give ammonia quantitatively on
hydrolysis in methanolic sulphuric acid.[30]

    The hydrido-hydrazido(2-)-complexes (item 4 in Table 1), which
appear to be a by-product at the third stage of protonation of the
dinitrogen complexes, may be responsible for the production of
hydrazine from these reactions.   Complexes of this type can be
isolated in 70% yields if the appropriate dinitrogen complex
$[W(N_2)_2L_4]$ is protonated with a halogen acid (HX) in dichloromethane.
Four such complexes of general formula $[WHX_3(\equiv N\equiv NH_2)L_3]$ (X = Br or
Cl; $L = PMe_2Ph$ or $PMePh_2$) have been isolated.   Since they are
complexes of tungsten(VI) and have no d-electrons to effect
reduction of the nitrogen-containing ligand, they would be expected
to hydrolyze directly to hydrazine and dihydrogen (reactions 9 and
10).   The fact that only 30% of the nitrogen appears as hydrazine

$$[WHX_3(\equiv N\equiv NH_2)L_3] + 2HX \longrightarrow WHX_5 + N_2H_4 + 3L \qquad (9)$$

$$WHX_5 + HX \longrightarrow WX_6 + H_2 \qquad (10)$$

and the remainder as ammonia suggests that, as in the hydrazido(2-)
complexes mentioned earlier, the hydrazido(2-) ligand here is
strongly conjugated to the metal and that protonation occurs at the

β-nitrogen atom, which withdraws the electrons of the W-H bond to effect reduction as in reaction sequence (11).

$$\underset{\underset{W \;\equiv\; N \;\cdots\; NH_2}{|}}{H} + H^+ \longrightarrow \underset{\underset{W \;\equiv\; N - NH_3^{+}}{|}}{H} \longrightarrow \underset{\underset{W \;\equiv\; N + NH_3}{}}{H^+} \tag{11}$$

Hydrazine is not obtained during the nitrogenase reaction nor is it obtained, except in minute traces, from the protonation of the molybdenum complexes $[Mo(N_2)_2(PR_3)_4]$. Nevertheless, because of the much greater yield of ammonia from the tungsten complexes, it seems that these represent the better model for the nitrogenase reaction.

Proposed catalytic cycle and mechanism for the nitrogenase reaction

The above considerations lead to a very plausible mechanism for the reduction of dinitrogen end-bonded to a molybdenum atom in the enzyme (Scheme 1). It is not proposed that the metal atom

Scheme 1.   Proposed mechanism for the nitrogenase reaction

Probably in
3 steps

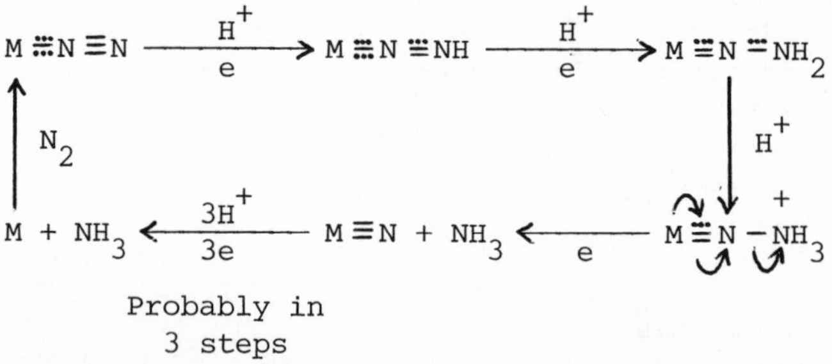

switches oxidation state by six units as in our tungsten complexes, but that in the enzyme, the reducing electrons are fed from the Fe_4S_4 cluster systems through the molybdenum atom into dinitrogen. Of course, in our mononuclear complexes, all six electrons had to be present in the center metal atom so that the protonation reaction could provide two molecules of ammonia from one dinitrogen molecule. The fact that our molybdenum complexes gave only a 34% yield of ammonia as compared with over 90% in the case of the tungsten complexes is also no argument that the Scheme is a poor model for the nitrogenase reaction. Molybdenum complexes are much more labile than those of tungsten and it could well be that molybdenum coordinated to the more rigid framework of the protein will behave somewhat like tungsten surrounded by monodentate ligands. Our tungsten complexes produced some hydrazine, whereas nitrogenase does not. We have seen that this hydrazine probably arises because the metal has been protonated to produce a hydrido-hydrazido(2-)-complex. It may be that on the molybdenum of the enzyme such an error of protonation cannot occur, or if it does occur, the resultant ligating hydride ion, which will be more labile, reacts with hydrogen ion to produce dihydrogen so correcting the protonation error.   This latter suggestion provides a possible explanation of the production of dihydrogen together with ammonia during the nitrogenase reaction.   It may provide a mechanism to prevent nitrogenase producing free hydrazine.

We have shown that the $N-NH_2$ stage of the reduction is the most stable, partially-reduced state during the protonation of our dinitrogen complexes.   The biochemists in my group have looked for it in the functioning enzyme.[8]   The enzymic reaction was rapidly quenched with acid or alkali and the resultant solution tested for hydrazine, which was found to be present in the same quantity whether acid or alkali was used.   When the enzyme functioned under argon or was reducing cyanide ion, no hydrazine was found.   It was found only when the enzyme was actually fixing dinitrogen and so must have arisen from some dinitrogen hydride intermediate which was bound to the enzyme when the reaction was quenched.   It was also shown that the rate of ammonia production was proportional to the concentration of the hydrazine-producing intermediate.   It is interesting that when our $N-NH_2$ complexes of tungsten were treated with strong alkali or strong acid to give ammonia and hydrazine the quantities of ammonia and hydrazine obtained were the same whether acid or alkali was used.

The fact that the mechanism of Scheme 1 makes good chemical sense and that both the functioning enzyme and our tungsten complexes give the same quantity of hydrazine with acid as with alkali does not prove that Scheme 1 provides the mechanism of the reduction of dinitrogen on nitrogenase.   Its main virtue is that the first two stages of the protonation in our dinitrogen complexes are well established by characterized examples.   The structure of the $M=N-NH_2$ grouping indicates a strong triple bond between the metal and its attached nitrogen atom, which should encourage the next stage of protonation at the terminal nitrogen.   Logically, such protonation by putting a positive charge on the terminal nitrogen and reducing still further the N-N bond order should assist heterolytic splitting of the N-N bond to give ammonia and a ligating nitride ion.   This species, in its turn, should easily be removed by protonation and reduction to provide the second molecule of ammonia.   The proposed mechanism is plausible and most of its steps have been realized as individual chemical reactions.   To what extent these reactions are those which occur during the reduction of dinitrogen on the active site in nitrogenase is still an open question.

REFERENCES

1.   H. Bortels, Molybdän als Katalysator bei der biologischen
     Stickstoff-bindung, Arch. Microbiol., 1:333 (1930).

2.   J. Chatt and G.J. Leigh, Nitrogen Fixation, Chem. Soc. Rev.,
     1:12 (1972).

3.   W.G. Zumft and L.E. Mortenson, Biochemistry of nitrogen
     fixation, Biochim. Biophys. Acta, 416:1 (1975).

4.   W.H. Orme-Johnson and L.C. Davies, Current topics and
     problems in enzymology of nitrogenase, in: "Iron-Sulfur
     Proteins", W. Lovenberg, ed., Academic Press, New York
     and London (1977) p.15.

5.   B.E. Smith, The structure and function of nitrogenase: a
     review of the evidence for the role of molybdenum,
     J. Less-Common Metals, 54:465 (1977).

6.   S.P. Cramer, K.O. Hodgson, W.O. Gillum and L.E. Mortenson,
     The molybdenum site of nitrogenase. Preliminary structural
     evidence from X-ray absorption spectroscopy, J. Am. Chem.
     Soc., 100:3398 (1978).

7.   S.P. Cramer, W.O. Gillum, K.O. Hodgson, L.E. Mortenson,
     E.I. Stiefel, S.R. Chisnell, W.J. Brill and V.K. Shah,
     The molybdenum site of nitrogenase. 2. A comparative
     study of Mo-Fe proteins and the iron-molybdenum cofactor
     by X-ray absorption spectroscopy, J. Am. Chem. Soc.,
     100:3814 (1978).

8.   R.N.F. Thorneley, R.R. Eady and D.J. Lowe, Biological nitrogen
     fixation by way of an enzyme-bound dinitrogen-hydride
     intermediate, Nature, 272:557 (1978).

9.   H.J. Evans, T. Ruiz-Argüeso, N.T. Jennings and H. Hanus,
     Energy coupling efficiency of symbiotic nitrogen fixation,
     in "Genetic Engineering for Nitrogen Fixation",
     A. Hollaender, ed., Plenum Press, New York and London
     (1977) p.333.

10. C.C. Walker and M.G. Yates, The hydrogen cycle in nitrogen-
     fixing Azotobacter chroococcum, Biochemie, 60:225 (1978).

11. W.H. Orme-Johnson, L.C. Davis, M.T. Henzl, B.A. Averill,
     N.R. Orme-Johnson, E. Münck and R. Zimmerman, Components
     and pathways in biological nitrogen fixation, in: "Recent
     Developments in Nitrogen Fixation", W. Newton, J.R. Postgate
     and C. Rodriguez-Barrueco, eds., Academic Press, London and
     New York (1977) p.131.

12. R.R. Eady, D.J. Lowe, R.N.F. Thorneley, Nitrogenase of
    Klebsiella pneumoniae: a pre-steady state burst of ATP
    hydrolysis is coupled to electron transfer between the
    component proteins, FEBS Lett., 95:211 (1978).

13. T.E. Wolf, J.M. Berg, C. Warrick, K.O. Hodgson, R.H. Holm,
    and R.B. Frankel, The molybdenum-iron-sulfur cluster
    complex $[Mo_2Fe_6S_9(SC_2H_5)_8]^{3-}$. A synthetic approach to
    the molybdenum site in nitrogenase, J. Am. Chem. Soc.,
    100:4630 (1978).

14. G. Christou, C.D. Garner, F.E. Mabbs and T.J. King, Crystal
    structure of tris(tetra-n-butylammonium)tri-μ-benzene-
    thiolatobis{tri-μ-sulphido-[μ_3-sulphido-tris(benzenethio-
    latoiron)]molybdenum}$[Bu^n_4N]_3[\{(PhSFe)_3MoS_4\}_2(SPh_3]$; an
    Fe_3MoS_4 cubic cluster dimer, J. Chem. Soc. (Chem. Comm.),
    740 (1978).

15. V.K. Shah and W.J. Brill, Isolation of an iron-molybdenum
    cofactor from nitrogenase, Proc. Natl. Acad. Sci., U.S.A.,
    74:3249 (1977).

16. A.E. Shilov, Dinitrogen fixation in protic media: a
    comparison of biological dinitrogen fixation with its
    chemical analogues, in: "Biological Aspects of Inorganic
    Chemistry", A.D. Addison, W.R. Cullen, D. Dolphin and
    B.R. James, eds., Wiley-Interscience, New York, Sydney,
    London, Toronto (1977) p.197.

17. G.N. Schrauzer, Nonenzymatic simulation of nitrogenase
    reactions and the mechanism of biological nitrogen
    fixation, Angew. Chem. Int. Ed., 14:514 (1975); for later
    references see G.N. Schrauzer, P.R. Robinson, E.L. Moorhead,
    B.J. Weathers, E.A. Ufkes and T.M. Vickery, The chemical
    evolution of a nitrogenase model. 14. Stoichiometric
    reactions of complexes of molybdenum(V), molybdenum (IV),
    and molybdenum(III) with acetylene and nitrogen, J. Am.
    Chem. Soc., 99:3657 (1977).

18. J. Chatt, A.J. Pearman and R.L. Richards, The reduction of
    mono-coordinated molecular nitrogen to ammonia in a protic
    environment, Nature, 253:39 (1975); J. Chatt, A.J. Pearman
    and R.L. Richards, Conversion of dinitrogen in its
    molybdenum and tungsten complexes into ammonia and
    possible relevance to the nitrogenase reaction, J. Chem.
    Soc., Dalton Trans., 1852 (1977).

19. J. Chatt, G.J. Leigh, H. Neukomm, C.J. Pickett and D.R. Stanley,
     Redox potential-structural relationships in metal complexes.
     Part II. The influence of trans-substituents upon the redox
     properties of certain dinitrogen complexes of molybdenum and
     tungsten and some carbonyl analogues: inner-sphere vs outer-
     sphere electron-transfer in the alkylation of coordinated
     dinitrogen, J. Chem. Soc., Dalton Trans. In the press.

20. J. Chatt, G.A. Heath and R.L. Richards, Diazene-N-(di-imide)
     and hydrazido(2-)-N (aminoimido) complexes: the addition of
     acids to dinitrogen complexes, J. Chem. Soc., Dalton Trans.,
     2074 (1974).

21. J. Chatt, A.J. Pearman and R.L. Richards, Diazenido-complexes
     of molybdenum and tungsten, J. Chem. Soc., Dalton Trans.,
     1520 (1976).

22. J. Chatt, G.J. Leigh and C.J. Pickett. To be published.

23. J. Chatt, M.E. Fakley and R.L. Richards. To be published.

24. M. Hidai, Y. Mizobe, T. Takahashi and Y. Uchida, Preparation
     and properties of molybdenum and tungsten dinitrogen
     complexes 9. Conversion of ligating dinitrogen into
     hydrazine, Chem. Lett., 1187 (1978).

25. J. Chatt, A.J. Pearman and R.L. Richards, The preparation and
     oxidation, substitution, and protonation reactions of
     trans-bis(dinitrogen)tetrakis(methyldiphenylphosphine)-
     tungsten, J. Chem. Soc., Dalton Trans., 2139 (1977).

26. J. Chatt, M.E. Fakley, R.L. Richards, J. Mason and
     I.A. Stenhouse. To be published.

27. J. Chatt, M.E. Fakley, R.L. Richards, J. Mason and
     I.A. Stenhouse, Nitrogen-15 nuclear magnetic resonance
     spectroscopic studies of dinitrogen complexes of molybdenum
     and tungsten, J. Chem. Res. (S), 44 (1979).

28. J. Chatt, M.E. Fakley, P.B. Hitchcock, R.L. Richards,
     N.T. Luong-Thi and D.L. Hughes, Hydrazido(2-) (or isodiazene)
     complexes of tungsten, J. Organomet. Chem. In the Press.

29. J. Chatt, J.R. Dilworth and R.L. Richards, Recent advances in
     the chemistry of nitrogen fixation, Chem. Rev. 78:589 (1978).

30. M.W. Bishop, J. Chatt, J.R. Dilworth, M.B. Hursthouse and
     M. Motevalle, The preparation, structures and reactivity
     of complexes with molybdenum-nitrogen multiple bonds,
     J. Less-Common Metals, 54:487 (1977).

REACTIONS OF HYDRAZIDO(2-) COMPLEXES DERIVED FROM MOLYBDENUM AND TUNGSTEN DINITROGEN COMPLEXES

Masanobu Hidai, Yasushi Mizobe, Tamotsu Takahashi,
Maki Sato, and Yasuzo Uchida

Department of Industrial Chemistry, University of Tokyo,
Hongo, Tokyo 113, Japan

INTRODUCTION

Since the discovery of the first dinitrogen complex of molybde-num, trans-[Mo(N$_2$)$_2$(dpe)$_2$] (dpe = Ph$_2$PCH$_2$CH$_2$PPh$_2$),[1] many dinitrogen complexes of molybdenum and tungsten have been prepared and their chemical properties have been extensively studied for possible rele-vance to nitrogen fixation as well as to open a new chemistry of dinitrogen.[2] Until now, many hydrazido(2-) complexes have been isolated as an intermediate stage in the reduction of dinitrogen. The hydrazido(2-) complexes, [MF(NNH$_2$)(dpe)$_2$][BF$_4$] (M = Mo or W), are readily obtained by the reaction of trans-[M(N$_2$)$_2$(dpe)$_2$] with fluoroboric acid.[3,4] On the other hand, treatment of [M(N$_2$)$_2$(PMe$_2$-Ph)$_4$] (M = Mo or W) with aqueous HX (X = Cl, Br, or I) in methanol affords a series of the hydrazido(2-) complexes [MX$_2$(NNH$_2$)(PMe$_2$Ph)$_3$] in good yields.[5] The X-ray structure of the former hydrazido(2-) complex (M = Mo) shows that the Mo-N-N linkage is essentially linear. The N-N bond distance indicates a bond order greater than unity while the Mo-N bond distance is consistent with considerable multiple bonding between the metal and nitrogen.[4] This result is interpreted in terms of a combination of two resonance structures (i) and (ii).

$$M \equiv N - NH_2 \quad \longleftrightarrow \quad M^- \rightleftharpoons N = N^+H_2$$

$$\text{(i)} \qquad\qquad\qquad\qquad \text{(ii)}$$

In this paper, we describe the chemical behavior of these hydrazido-(2-) complexes towards proton and carbonyl compounds.

REACTIONS OF HYDRAZIDO(2-) COMPLEXES WITH HCl IN 1,2-DIMETHOXYETHANE

Previously, Chatt and his coworkers[6] reported that the complexes

<u>cis</u>-$[M(N_2)_2(PMe_2Ph)_4]$ or <u>trans</u>-$[M(N_2)_2(PPh_2Me)_4]$ (M = Mo or W) give, on treatment with sulfuric acid in methanol at room temperature, followed by base distillation for M = Mo, high yields of ammonia, together with a little hydrazine for M = W and a trace for M = Mo. The yield of ammonia is essentially 2 mol/metal atom for M = W, but only ca. 0.66 mol/metal atom for M = Mo.

We have recently found[7] that treatment of <u>cis</u>-$[M(N_2)_2(PMe_2Ph)_4]$ or $[MX_2(NNH_2)(PMe_2Ph)_3]$ with HCl gas in 1,2-dimethoxyethane produces hydrazine in preference to ammonia as shown in Table 1. When methanol replaces 1,2-dimethoxyethane as solvent, the yield of hydrazine decreases remarkably and the yield of ammonia greatly increases. It is of great interest to note that the ligating dinitrogen on molybdenum is converted into hydrazine in moderate yields in these reactions and more than one nitrogen atom per molybdenum atom seems to be protonated to give hydrazine and ammonia. This result is in sharp contrast to the protonation reactions reported previously,[6] which were interpreted in terms of disproportionation of the N_2H_2 ligand of the hydrazido(2-) stage of reduction (equation 1) for the molybdenum dinitrogen complexes, which accounted for formation of only ca. 0.66 mol of ammonia per molybdenum atom. However, the results obtained here indicate that there is no substantial difference between

$$3N_2H_2 \longrightarrow 2NH_3 + 2N_2 \tag{1}$$

tained here indicate that there is no substantial difference between the protonation reactions of the molybdenum and tungsten dinitrogen complexes and that the N_2H_2 ligand in both complexes is further protonated to produce predominantly hydrazine on treatment with HCl in 1,2-dimethoxyethane.

In the course of our studies on these protonation reactions, several new complexes containing the MN_2H_3 moiety have been isolated. Treatment of a suspension of the hydrazido(2-) complex $[WBr_2(NNH_2)-(PMe_2Ph)_3]$ in 1,2-dimethoxyethane with one molar equivalent of HCl gas at room temperature affords an orange-yellow crystalline compound in a good yield. Analytical data are consistent with the formula $WClBr_2(N_2H_3)(PMe_2Ph)_3$. The infrared spectrum of a chloroform solution of this complex exhibits a weak band at 3320 cm^{-1} and a strong band at 2950 cm^{-1} assigned to ν(N-H). The ^{1}H nmr spectrum has resonances in CD_2Cl_2 at 9.9 (broad singlet, intensity 2H) and 9.8 ppm (quartet, intensity 1H, J_{P-H} = ca. 82 Hz), which are assigned to W-NH-N<u>H</u> (β) and W-N<u>H</u>-NH (α) protons, respectively. The β-N-H exchanges immediately on addition of D_2O, whereas the α-N-H, if it exchanges, does so[8] very slowly. Slow exchange is characteristic of such α-protons.[9] Preliminary X-ray analysis of the complex shows that one of the two bromide anions originally coordinated to tungsten is displaced from the inner coordination sphere in the reaction and the chloride anion is instead bound to the metal as

Table 1.   Yields of ammonia and hydrazine upon treatment with acids

Compound	Solvent	Acid	NH_3[a]	N_2H_4[a]
cis-[Mo(N$_2$)$_2$(PMe$_2$Ph)$_4$][b]	MeOH	H$_2$SO$_4$	0.64	trace
cis-[Mo(N$_2$)$_2$(PMe$_2$Ph)$_4$]	DME[c]	HCl	0.31	0.32
[MoCl$_2$(NNH$_2$)(PMe$_2$Ph)$_3$]	DME	HCl	0.24	0.52
[MoBr$_2$(NNH$_2$)(PMe$_2$Ph)$_3$][b]	DME	HCl	0.18	0.45
cis-[W(N$_2$)$_2$(PMe$_2$Ph)$_4$]	MeOH	H$_2$SO$_4$	1.86	0.02
cis-[W(N$_2$)$_2$(PMe$_2$Ph)$_4$]	DME	HCl	0.22	0.63
cis-[W(N$_2$)$_2$(PMe$_2$Ph)$_4$]	MeOH	HCl	0.64	0.33
[WBr$_2$(NNH$_2$)(PMe$_2$Ph)$_3$]	DME	HCl	0.05	0.64
[WClBr(NHNH$_2$)(PMe$_2$Ph)$_3$]Br	DME	H$_2$SO$_4$	0.45	0.30
[WClBr(NHNH$_2$)(PMe$_2$Ph)$_3$]Br	DME	HCl	0.50	0.55
[WBr$_2$(N$_2$CMe$_2$)(PMe$_2$Ph)$_3$][d]	CH$_2$Cl$_2$	HBr	Nil	0.60
[MoCl$_2$(N$_2$CMePh)(PMe$_2$Ph)$_3$]	CH$_2$Cl$_2$	HCl	0.21	0.12
[MoCl$_2$(N$_2$CMePh)(PMe$_2$Ph)$_3$]	DME	HCl	0.22	0.27

[a] mmol/mmol M.   [b] Reference 6.   [c] DME = 1,2-dimethoxyethane.   [d] Reference 17; acetone azine was produced in ca. 27% yield.

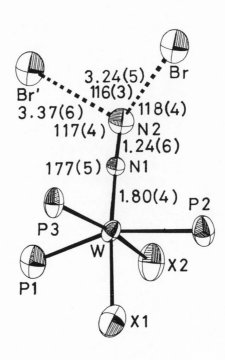

Fig. 1.   Perspective view of [WClBr(NHNH$_2$)(PMe$_2$Ph)$_3$]Br. The shapes
          of the atoms in this drawing represent 50% probability
          contours of thermal motions. The positions indicated by
          X are occupied by Cl or Br atom with the same probability.

shown in Fig. 1. Furthermore there is a strong hydrogen-bonding
interaction between the NNH$_2$ group and two bromide anions, one of
which is in another complex cation. The very low conductivity
observed for a solution of this complex in CH$_2$Cl$_2$ indicates that
strong hydrogen-bonding exists even in the solution, causing the
shift of the NH$_2$ resonance to low field and the ν(N-H) to very low
frequencies with strong intensity (vide supra). Thus, the most
reasonable formulation of this complex is a hydrazido(1-) complex
[WClBr(NHNH$_2$)(PMe$_2$Ph)$_3$]Br. Very recently, Chatt and his coworkers[5]
reported a similar strong hydrogen-bonding interaction between the
NNH$_2$ group and the halide anions in hydrazido(2-) complexes such as
[WBr(NNH$_2$)(NC$_5$H$_4$Me-4)(PMe$_2$Ph)$_3$]Br and [W(NNH$_2$)(quin)(PMe$_2$Ph)$_3$]I.
However, we cannot at present exclude the possibility of a hydra-
zinium complex [WClBr(NN$^+$H$_3$)(PMe$_2$Ph)$_3$]Br in the solid state since
the W-N-N linkage is rather linear, perhaps indicating three hydrogen atoms bonded to the terminal nitrogen atom.

In this complex, one of the two bromide anions is very labile and, by reaction with an excess of NaBPh$_4$ in tetrahydrofuran-CH$_2$Cl$_2$, another hydrazido(1-) complex [WClBr(NHNH$_2$)(PMe$_2$Ph)$_3$][BPh$_4$] is obtained. The complex is characterized by three bands in the infrared spectrum at 3320, 3200, and 3105 cm^{-1} assigned to ν(N-H). The hydrazido(1-) complex, [WCl$_3$(NHNH$_2$)(PPh$_2$Me)$_2$], which has recently been prepared by the reaction of trans-[W(N$_2$)$_2$(PPh$_2$Me)$_4$] with HCl in CH$_2$Cl$_2$ exhibits similar bands in the same region.[8] The ^{1}H nmr spectrum of the [BPh$_4$] complex gives rise to resonances at 3.6 (broad singlet, intensity 2H) and 9.3 ppm (quartet, intensity 1H, $^3J_{P-H}$ = ca. 82 Hz) which are assigned to β and α protons, respectively. Since the [BPh$_4$] anion is not capable of strong hydrogen-bonding, the NH$_2$ resonance appears at relatively high field.

The hydrazido(1-) complexes obtained here readily condense with aldehydes and ketones to afford diazoalkane or protonated diazo-alkane complexes which will be described in a later publication.[10] Treatment of the hydrazido(1-) complex [WClBr(NHNH$_2$)(PMe$_2$Ph)$_3$]Br with HCl or sulfuric acid in 1,2-dimethoxyethane produces hydrazine in preference to ammonia as shown in Table 1. This result is in marked contrast to the predominant formation of ammonia by treatment of the hydrazido(2-) complex [WBr$_2$(NNH$_2$)(PMe$_2$Ph)$_3$] with sulfuric acid in methanol.[11] The MNHNH$_2$ stage of reduction seems to be on the route to hydrazine. Further investigations are now in progress to understand subtle factors which determine the reduction course of ligating dinitrogen to ammonia or hydrazine.

CONDENSATION REACTIONS OF HYDRAZIDO(2-) COMPLEXES WITH ALDEHYDES AND KETONES

New types of catalytic reactions which produce organo-nitrogen compounds from dinitrogen and a relatively cheap organic feedstock are of potential industrial importance and will open new areas of dinitrogen chemistry. The catalytic cycles of these reactions will involve the formation of carbon-nitrogen bonds from ligating dinitrogen as an important elementary step. Aroyl, acyl, and alkyl halides have been shown to react with the dinitrogen complexes trans-[M(N$_2$)$_2$(dpe)$_2$] (M = Mo or W) to afford complexes containing organonitrogen ligands.[12,13] However, these reactions do not seem to be applicable to the dinitrogen complexes containing mono(tertiary) phosphines [M(N$_2$)$_2$(L)$_4$] (M = Mo or W; L = PMe$_2$Ph or PPh$_2$Me).[14] We have recently found another important route to the formation of carbon-nitrogen bonds from ligating dinitrogen which is applicable to complexes of both di(tertiary) or mono(tertiary) phosphines.[15] This route involves the condensation of hydrazido(2-) complexes with aldehydes and ketones (RR'C=O) to give hydrazone-type or diazoalkane complexes containing the M≡N-N=CRR' moiety. These condensation reactions are markedly accelerated by catalytic amounts of acids as is usually found in the condensation reactions of carbonyl compounds with hydrazine.

The hydrazido(2-) complexes $[MF(NNH_2)(dpe)_2][BF_4]$ condense with a variety of aldehydes and ketones to afford diazoalkane complexes (equation 2).[16] These complexes give rise to strong $\nu(C=N)$ bands

$$\underline{trans}\text{-}[M(N_2)_2(dpe)_2] \xrightarrow{HBF_4} [MF(NNH_2)(dpe)_2][BF_4]$$

$$\xrightarrow{RR'C=O} [MF(NN=CRR')(dpe)_2][BF_4] \quad (M = Mo \text{ or } W) \quad (2)$$

in the 1510-1595 cm^{-1} region of the infrared spectra. The ^1H nmr spectra show that the substituents R and R' on the nitrogen-bearing carbon are in inequivalent positions. Thus, diazomethane hydrogens in the complex $[MoF(N_2CH_2)(dpe)_2][BF_4]$ show a pair of doublets at 4.9 and 6.2 ppm. In these diazoalkane complexes, the diazoalkane ligand and fluoride anion are $\underline{trans}$ to each other. It is well known that many α,β-unsaturated aldehydes and ketones and 1,3-diketones react with hydrazine to give cyclic organo-nitrogen compounds. However, treatment of crotonaldehyde, cinnamaldehyde, and acetylacetone with the above hydrazido(2-) complexes give only the usual hydrazone-type complexes retaining a double bond or a carbonyl group.

The hydrazido(2-) complexes containing mono(tertiary) phosphines $[MX_2(NNH_2)(PMe_2Ph)_3]$ also condense with a variety of aldehydes and ketones (equation 3).[7,17] The ^1H nmr spectra show two triplets and

$$\underline{cis}\text{-}[M(N_2)_2(PMe_2Ph)_4] \xrightarrow{\text{aqueous HX in MeOH}} [MX_2(NNH_2)(PMe_2Ph)_3]$$

$$\xrightarrow{RR'C=O} \qquad (3)$$

(M = Mo or W; X = Cl, Br, or I)

one doublet, each with an intensity of 6H, which are characteristic of virtually coupled $\underline{meridional}$ phosphine methyls in a complex lacking a plane of symmetry containing the phosphines.

The reaction of $[WBr_2(N_2CMe_2)(PMe_2Ph)_3]$ with $Li[AlH_4]$ in ether produces Pr^iNH_2 (93-95%) and ammonia (55-60%).[17] The same diazoalkane complex also reacts with an excess of HBr in CH_2Cl_2 to give hydrazine and acetone azine, $Me_2C=N-N=CMe_2$, in moderate yields.[17] When the diazoalkane complexes of molybdenum $[MoX_2(N_2CRR')(PMe_2Ph)_3]$ are treated with an excess of HCl in CH_2Cl_2 or 1,2-dimethoxyethane, hydrazine and ammonia are produced as shown in Table 1, together with a little of the azines.

REFERENCES

1. (a) M. Hidai, K. Tominari, Y. Uchida, and A. Misono, A trans-
        dinitrogen complex of molybdenum, Chem. Commun. 1392 (1969).
    (b) M. Hidai, K. Tominari, and Y. Uchida, Preparation and proper-
        ties of dinitrogen-molybdenum complexes, J. Amer. Chem. Soc.
        94: 110 (1972).
2. J. Chatt, J. R. Dilworth, and R. L. Richards, Recent advances
    in the chemistry of nitrogen fixation, Chem. Rev. 78: 589
    (1978).
3. J. Chatt, A. J. Pearman, and R. L. Richards, Diazenido-complexes
    of molybdenum and tungsten, J. Chem. Soc. Dalton Trans. 1520
    (1976).
4. M. Hidai, T. Kodama, M. Sato, M. Harakawa, and Y. Uchida,
    Preparation and properties of dinitrogen-molybdenum complexes.
    3. Preparation and molecular structure of 1-(η-hydrazido(2-))-
    fluorobis[1,2-bis(diphenylphosphino)ethane]molybdenum tetra-
    fluoroborate-dichloromethane solvate, Inorg. Chem. 15: 2694
    (1976).
5   J. Chatt, A. J. Pearman, and R. L. Richards, Hydrazido(2-)
    complexes of molybdenum and tungsten formed from dinitrogen
    complexes by protonation and ligand exchange, J. Chem. Soc.
    Dalton Trans. 1766 (1978).
6. J. Chatt, A. J. Pearman, and R. L. Richards, Conversion of Di-
    nitrogen in its molybdenum and tungsten complexes into
    ammonia and possible relevance to the nitrogenase reaction,
    J. Chem. Soc. Dalton Trans. 1852 (1977).
7. M. Hidai, Y. Mizobe, T. Takahashi, and Y. Uchida, Preparation
    and properties of molybdenum and tungsten dinitrogen complex-
    es 9. Conversion of the ligating dinitrogen into hydrazine,
    Chem. Lett. 1187 (1978).
8. J. Chatt, A. J. Pearman, and R. L. Richards, The preparation
    and oxidation, substitution, and protonation reactions of
    trans-bis(dinitrogen)tetrakis(methyldiphenylphosphine)-
    tungsten, J. Chem. Soc. Dalton Trans. 2139 (1977).
9. The complex crystallized in the monoclinic space group $P2_1/c$
    with cell dimensions, a = 15.40(1) Å, b = 20.65(2) Å, c =
    9.72(1) Å, and β = 97.16(2)°. The final residuals were R =
    0.13 and Rw = 0.17 using full-matrix least-squares with
    anisotropic thermal parameters for W, Br, Cl, and P, and
    isotropic thermal parameters for N and C, respectively.
10. M. Hidai, T. Takahashi, and Y. Uchida, Unpublished results.
11. J. Chatt, A. J. Pearman, and R. L. Richards, Diazenido (imino-
    nitrosyl)(N_2H), hydrazido(2-)(N_2H_2), and hydrazido(1-)(N_2H_3)
    ligands as intermediate in the reduction of ligating dinitro-
    gen to ammonia, J. Organometal. Chem. 101: C45 (1975).
12. J. Chatt, A. A. Diamantis, G. A. Heath, N. E. Hooper, and G. J.
    Leigh, Reactions of ligating dinitrogen to form carbon-
    nitrogen bonds, J. Chem. Soc. Dalton Trans. 688 (1977).
13. T. Tatsumi, M. Hidai, and Y. Uchida, Preparation and properties

of dinitrogen-molybdenum complexes. 2. dinitrogen(organo-nitrile) complexes of molybdenum, Inorg. Chem. 14: 2530 (1975).

14. P. C. Bevan, J. Chatt, G. J. Leigh, Private communication.

15. M. Hidai, Y. Mizobe, and Y. Uchida, Preparation and properties of dinitrogen-molybdenum complexes. 5. Conversion of the coordinated dinitrogen into a hydrazone-type N_2CRR' ligand, J. Amer. Chem. Soc. 98:7824 (1976).

16. M. Hidai, Y. Mizobe, M. Sato, T. Kodama, and Y. Uchida, Preparation and properties of molybdenum and tungsten dinitrogen complexes. 8. Diazoalkane complexes of molybdenum and tungsten, J. Amer. Chem. Soc. 100: 5740 (1978).

17. P. C. Bevan, J. Chatt, M. Hidai, and G. J. Leigh, Diazoalkane complexes of tungsten from the condensation of hydrazido complexes with ketones, J. Organomet. Chem. 160: 165 (1978).

MODEL STUDIES FOR MOLYBDENUM ENZYMES: STRUCTURES, SPECTRA,

ELECTROCHEMISTRY AND REACTIONS OF OXO Mo(VI), (V), AND (IV) COMPLEXES

J. T. Spence, M. Minelli, and C. A. Rice
Department of Chemistry and Biochemistry
Utah State University, Logan, Utah 84322, U. S. A.

N. C. Chasteen and M. Scullane
Department of Chemistry
University of New Hampshire, Durham, New Hampshire 03824
U. S. A.

INTRODUCTION

Molybdenum is now well established as a necessary cofactor for a variety of enzymes which catalyze a number of important biological reactions,[1,2] including the oxidation of purines, aldehydes and sulfite, and the reduction of dinitrogen, nitrate and carbon dioxide. With the possible exception of nitrogenase, the presence of Mo(VI) is assumed in the oxidized form of the molybdoenzymes.[1] All the enzymes except nitrogenase give esr signals attributed to monomeric Mo(V) during catalysis.[1,2] While direct evidence is difficult to come by, it appears Mo(IV) occurs in the reduced state of these enzymes, again with the possible exception of nitrogenase.[1,2] Recent EXAFS results from xanthine oxidase indicate the presence of oxo groups as well as thiol ligands at the Mo(VI) site;[3] a terminal Mo(VI)=S group has also been suggested, on the basis of esr evidence, for this enzyme.[4] In view of these results and the well known fact that oxo species dominate the coordination chemistry of the +6, +5, and +4 oxidation states of molybdenum,[2] oxo complexes of these oxidation states are promising models for the molybdenum sites of the enzymes, particularly those with sulfur ligands.

We have prepared a number of oxo complexes of Mo(VI), (V), and (IV) with a variety of ligands and characterized them by analysis, electronic, infrared and esr spectroscopy, cyclic voltammetry and coulometry. In a few cases, unambiguous structures have been determined by X-ray crystallography. We have also prepared some of the W(VI) and W(V) analogues for comparative purposes. These results are summarized here.

MOLYBDENUM(VI) COMPLEXES

There are few Mo(VI) complexes which are stable at biological pH, due to hydrolysis to MoO_4^{2-}. In nonaqueous aprotic solvents (e.g., DMF, CH_3CN), however, many dioxomolybdenum(VI) complexes are quite stable and easily prepared. All reported complexes have the cis dioxo arrangement, a consequence of the well known trans effect of the oxo group.[2] We have prepared several new complexes of this type in order to investigate the relationship between structure and reduction potential (Table 1).[5] These Mo(VI) compounds are all monomeric except the sap (salicylaldehyde o-hydroxyanil) complex. IR data indicate it to be a dioxo bridged dimer with an unusual structure in which each Mo(VI) is bonded to only one terminal oxo group. The two new complexes (Table 1) with tetradentate amino-ethanethiol ligands are especially interesting, since they contain both thiol and amino coordinating groups in a saturated framework. Both groups are likely candidates for ligands at the molybdenum site of the enzymes.

The reduction potentials for these complexes are quite negative (Table 1) and may in fact be too low for oxidation of purines and SO_3^{2-}. Because of difficulties in comparing potentials between different solvents and the lack of reversibility of some of the systems (e.g., model Fe/S complexes have reduction potentials in DMF 0.10 to 1.00 volt more negative than Fe/S proteins in H_2O) these thermodynamic conclusion are tentative. It is also worth noting that the reduction potential of $MoO_2(tox)_2$ (tox= 8-mercaptoquinoline) is more positive, and its reduction much more reversible, than the corresponding 8-hydroxyquinoline complex ($MoO_2(ox)_2$), suggesting, again, that complexes with thiol ligands are promising models. While the reductions reported in Table 1 have been investigated in detail for only $MoO_2(ox)_2$,[6] the values of the potentials indicate the products of reduction are not oxo-Mo(V) monomers (see below), but Mo(V) oxobridged dimers or

$Mo_2O_4(sap)_2$ sap

TABLE 1

CYCLIC VOLTAMMETRIC DATA FOR Mo(VI) AND W(VI) DIOXO COMPLEXES[a]

Complex	E_{P_c} [b]	E_{P_a} [b]	ΔE_p
$MoO_2(ox)_2$	-1.220	-0.970	0.250
$MoO_2(tox)_2$	-1.000	-0.905	0.095
$MoO_2(mee)$	-1.315	-1.215	0.100
$MoO_2(mpe)$	-1.445	-0.890	0.555
$MoO_2(sap)$	-1.155	------[c]	-----
$MoO_2Cl_2(phen)$[d]	-1.035	------[c]	-----
$MoO_2(acac)_2$[d]	-1.035	------[c]	-----
$WO_2(ox)_2$	-1.480	-0.005	-1.475
$WO_2(tox)_2$	-1.315	------[c]	-----
$WO_2Cl_2(phen)$	-1.215	------[c]	-----
$WO_2Cl_2(bpy)$[d]	-1.215	------[c]	-----
$WO_2(sal_2en)$	-1.170	-0.555	-0.615

a) Scan rate 0.100 volts/sec, 0.10 M TEACl in DMF
b) Volts vs. SCE; E_{P_c} = cathodic peak, E_{P_a} = anodic peak
c) No well defined anodic wave
d) Unstable in DMF

LIGAND ABBREVIATIONS: ox = 8-hydroxyquinoline, tox = 8-mercapto-
quinoline, mee = N,N'-dimethyl-N,N'-bis(2-mercaptoethyl)ethylene-
diamine, mpe = N,N'-bis(2-mercapto-2-methylpropyl)ethylenediamine,
sap = salicylaldehyde o-hydroxyanil, acac = acetylacetone, bpy =
α,α'-bipyridyl, phen = o-phenanthrolene, sal_2en = disalicylaldehyde
ethylenediimine

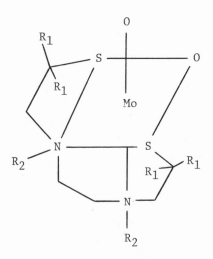

$$MoO_2(mpe), \quad R_1 = CH_3, \quad R_2 = H$$
$$MoO_2(mee), \quad R_1 = H, \quad R_2 = CH_3$$

mpe = N,N'-bis(2-mercapto-2-methylpropyl)ethylenediamine
mee = N,N'-dimethyl-N,N'-bis(2-mercaptoethyl)ethylenediamine

Mo(IV) complexes.[6] Further work is necessary to confirm this conclusion, but it appears that oxo-Mo(V) complexes are unobtainable by electrochemical reduction of the dioxo-Mo(VI) complexes under the conditions used.

Since tungsten will replace molybdenum in some enzymes (and may in fact, be present in formate dehydrogenase from C. thermoaceticum[7]), although usually producing inactive species,[1] a comparison of properties of W(VI) complexes with the corresponding Mo(VI) complexes is of interest. We have synthesized some dioxotungsten(VI) complexes, and measured their reduction potentials in DMF (Table 1). In all cases, the W(VI) complexes have more negative reduction potentials than the corresponding Mo(VI) complexes, which may possibly account for the general lack of activity of tungsten-substituted enzymes.

While not certain, it is highly probable that mononuclear molybdenum is present in all molybdoenzymes (except perhaps nitrogenase),[1] because esr signals, which may be attributed to monomeric Mo(V) (a d^1 system), appear during catalysis,[1] and all Mo(V) dimers (oxo or sulfido bridged) are esr inactive.[2] While much work has been reported concerning dimers, e.g., $[Mo_2O_4(cysteine)_2]^{2-}$,[2] the biological relevance of this bridged structure now seems minimal. A fruitful way to obtain structural information about the metal centers of enzymes is to compare their esr parameters with those of model complexes of known structure. There are, however, few well characterized monomeric Mo(V) complexes reported in the literature, particularly with ligands expected at the enzymic molybdenum sites.[2,8] Early esr work pointed to the thiol group of cysteine as a likely ligand; only recently, however, have sulfur-containing complexes with reasonably well characterized structures been reported. Of particular interest is the oxo-free complex, $Mo(Et_2NCS_2)[(HN)SC_4H_4)]_2$, which shows both ^{1}H and ^{14}N superhyperfine splitting in its esr spectrum, and has $\langle g \rangle$ and $\langle A \rangle$ values fairly close to those of the xanthine oxidase very rapid Mo(V) signal.[9] The X-ray crystal structure of this complex shows it to have a distorted trigonal prismatic geometry, with a short Mo---H(N) distance (possibly accounting for the large ^{1}H splitting).[10] The only other Mo(V) complex reported in the literature of known structure containing coordinate sulfur is $MoOCl_3(SPPh_3)$, which has a distorted square pyramidal geometry and esr parameters differing significantly from the xanthine oxidase signal.[11] The complex $MoO[S_2P(\underline{i}Pr)_2](OSC_6H_4)$ has esr parameters close to those of the enzyme.[9] Very recently, $MoO(SAr)_4^-$ (Ar = phenyl, alkyl-substituted phenyl) complexes having $\langle g \rangle$ values higher and $\langle A \rangle$ values lower than xanthine oxidase have been reported.[12]

Recently, we have developed relatively simple methods for the synthesis of monomeric Mo(V) oxo complexes of the formula $MoOCl_3L$, $MoOClL_2$ and MoOClL, where L is either a bi-, tri- or tetradentate ligand with O, N, or S coordinating groups, and we have prepared a number of these complexes with a variety of ligands (Table 2).[13] These complexes are unstable in H_2O, dimerizing rapidly to oxobridged, esr-inactive dimers; they are, however, quite stable in dry aprotic solvents (DMF, CH_3CN, $CHCl_3$, etc.).

Of particular interest are the complexes $MoOCl(tox)_2$ (tox = 8-mercaptoquinoline), MoOCl(mee) and MoOCl(mpe) (see Tables 1 and 2 for ligand abbreviations). The structure of $MoOCl(tox)_2$ has now been determined by X-ray crystallography.[13] Contrary to

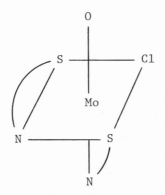

$\underline{cis}$-$MoOCl(tox)_2$
tox = 8-mercaptoquinoline

TABLE 2

ESR PARAMETERS FOR Mo(V)-OXO COMPLEXES

Complex	Probable Equatorial Ligands[a]	g_{xx}[b]	g_{yy}[b]	g_{zz}[b]	$A_{x'x'}$[c]	$A_{y'y'}$[c]	$A_{z'z'}$[c]	$\langle g \rangle$[b]	$\langle g_o \rangle$[b]	$\langle A \rangle$[b]	$\langle A_o \rangle$[b]
$(NH_4)_2MoOCl_5$	Cl_4	1.938	1.938	1.970	34.0	34.0	74.5	1.949	1.946	47.2	46.6
cis-MoOCl$_3$(phen)	N_2Cl_2	1.941	1.929	1.971	30	38	71	1.947	1.947	46.3	46.6
cis-MoOCl$_3$(bpy)	N_2Cl_2	1.944	1.931	1.971	29	35	73	1.949	1.948	45.6	46.0
trans-MoOCl$_3$(bpy)	NCl_3	1.968	1.953	1.939	40.0	11.4	71.5	1.953	1.952	41.0	42.9
trans-MoOCl(acac)$_2$	O_4	1.950	1.940	1.927	33.4	36.2	77.6	1.939	1.938	49.1	49.0
cis-MoOCl(ox)$_2$	NO_2Cl	1.939	1.953	1.970	46.5	9.1	74.2	1.954	1.954	43.3	43.2
cis-MoOCl(tox)$_2$	NS_2Cl	1.947	1.952	2.003	39.0	15.5	59.2	1.967	1.967	37.9	37.9
trans-MoOCl(sal$_2$phen)[d]	N_2O_2	1.955	1.926	1.947	25.0	33.4	71.5	1.943	1.940	43.3	44.1
cis-MoOCl(mee)	NS_2Cl	1.944	1.955	2.006	30.8	21.5	60.8	1.968	1.966	37.7	37.8
cis-MoOCl(mpe)	NS_2CL	1.943	1.958	2.011	22.5	36.0	57.5	1.971	1.969	38.7	38.0
trans-(Et$_4$N)MoOCl$_2$(sap)	O_2Cl_2	1.949	1.946	1.923	34.5	32.5	74.2	1.939	1.938	47.7	47.1
Xanthine Oxidase[e]	OS_3	1.951	1.956	2.025	35	22	38	1.977		31.7	

[a] Equatorial to the MoO^{3+} group.
[b] Errors in g nominally ± 0.001. $\langle g_o \rangle$ is room temperature solution value, $\langle g \rangle = (g_{xx} + g_{yy} + g_{zz})/3$.
[c] Units of 10^{-4} cm^{-1}. Errors nominally ± 0.5 x 10^{-4} cm^{-1} except for the A values less than 20 x 10^{-4} cm^{-1} which have errors nominally ± 1 x 10^{-4} cm^{-1}. $\langle A_o \rangle$ is room temperature solution value, $\langle A \rangle = (A_{x'x'} + A_{y'y'} + A_{z'z'})/3$.
[d] Sal$_2$phen = disalicylaldehyde o-phenylenediimine.
[e] Reference 1.

earlier reports based on esr work in solution,[14] the Cl is _cis_ to
the oxo group. This complex has been cited as a possible model for
the molybdenum site of xanthine oxidase.[1] Its esr parameters,
however, differ somewhat from those of the enzymic Mo(V) signal,
particularly in the value of $\langle A \rangle$ (Table 2). The MoOCl(mee) and
MoOCl(mpe) complexes are the first reported Mo(V) complexes with
thiol ligands in a saturated framework.[15] While the X-ray structures
have not yet been obtained, an analysis of both X and Q band esr
spectra for these complexes reveals that the principal magnetic axes
are non-coincident in the xy plane. The presence of non-coincident
g_{xx} and A_{xx}, and g_{yy} and A_{yy} tensor components requires that there
be no mirror plane perpendicular to the xy plane or proper rotation
axis contained within it. Thus, if we take the z direction of the
magnetic tensors as the Mo-O bond, as is usually the case for
molybdenyl complexes, then the _trans_ isomer (_trans_ refers to the
position of Cl⁻ with respect to oxo) is highly unlikely since it
contains a mirror plane. Furthermore, the esr parameters for
MoOCl(tox)$_2$ are remarkably similar to those of these complexes.[15]
Thus, a _cis_-Cl structure (as shown) for both complexes seems
probable.

The xanthine oxidase signal exhibits proton superhyperfine
splitting[1] and esr signals of oxo-free sulfur complexes, such as
Mo(S$_2$CNEt$_2$)[(HN)SC$_6$H$_4$]$_2$, exhibit both proton and nitrogen
superhyperfine splitting.[9] The MoOCl(mpe) complex, however, gave
no evidence for such splitting over a wide temperature range, from
ambient to frozen, in both DMF and CHCl$_3$.[15] This lack of
superhyperfine splitting may have its origin in differences in
geometry or in the absence in MoOCl(mpe) of extensive delocalization
of the Mo(V) electron into an aromatic system as is possible with

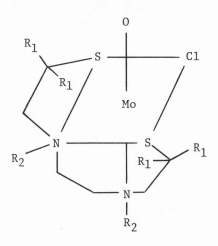

MoOCl(mpe)   R$_1$ = CH$_3$, R$_2$ = H
MoOCl(mee)   R$_1$ = H, R$_2$ = CH$_3$

$Mo(S_2CNEt_2)[(HN)SC_6H_4)]_2$. While the MoOCl(mpe) and MoOCl(mee) complexes have g values quite close to those of the xanthine oxidase very rapid signal[1] (Table 2), there are significant differences in the hyperfine splittings (A values). This difference, combined with the lack of proton superhyperfine splitting for MoOCl(mpe), suggests the Mo(V) site of xanthine oxidase differs from these complexes in some important way.

In order to obtain data concerning the relation of structure and reduction potentials, we have investigated the electrochemistry of monomeric oxomolybdenum(V) complexes in DMF.[5] Using cyclic voltammetry and coulometry, the reduction potentials and number of electrons involved in reduction were determined (Table 3). All complexes, with the exception of $MoOCl_2(sap)^-$, are facilely reduced to Mo(IV) species, some reversibly. $MoOCl_2(sap)^-$ is apparently reduced to a Mo(III) complex. In the voltage range used (+0.05 to -2.50 volts, vs. SCE), no oxidation peaks were observed, indicating oxomolybdenum(V) complexes are not easily oxidzed electrochemically in nonaqueous solvents, most likely due to the requirement of a cis dioxo structure in the Mo(VI) product. Perhaps of most importance, in those cases where reversible or quasireversible reductions are observed, the results indicate it is not possible to reduce dioxomolybdenum(VI) complexes to Mo(V) complexes; instead, either Mo(V) oxobridged dimers or Mo(IV) complexes are obtained.[6] This observation is in general agreement with a current mechanism for xanthine oxidase which postulates a similar relation between the potentials for Mo(VI)/(V)/(IV) in the absence of substrate.[16] Recent measurements of the reduction potentials for Mo(VI)/(V) and Mo(V)/IV) for xanthine oxidase and xanthine dehydrogenase indicates they are essentially equal.[17,18] Thus, given an equilibrium distribution of electrons, only part of the molybdenum should be observed as Mo(V), which possibly accounts for the levels of Mo(V) actually detected for most enzymes under these conditions.[1]

Again, for comparative purposes, we have prepared some corresponding oxotungsten(V) complexes and measured their reduction potentials. As with the W(VI) complexes, these potentials are lower for W(V) than Mo(V) (Table 3).

MOLYBDENUM(IV) COMPLEXES

While direct evidence is difficult to come by because Mo(IV) is an esr-inactive d^2 system, it appears Mo(IV) occurs in the reduced state of the molybdenum-containing oxidases and nitrate reductase.[1] The coordination chemistry of Mo(IV) has not been adequately explored.[2] A number of Mo(IV) complexes containing an $MoOS_4$ core are known, and may be useful as models for the reduced state of molybdoenzymes.[2] $MoO(Et_2NCS_2)_2$ forms stable adducts with

TABLE 3

ELCTROCHEMICAL DATA FOR Mo(V) AND W(V) OXO COMPLEXES[a,b]

Compound	E_{p_c}	E_{p_a}	ΔE_p	n(electrons/molecules)
cis-MoOCl$_3$(bpy)	-0.270	-0.200	0.070	0.77 ± .04
trans-MoOCl$_3$(bpy)*	-0.250	-0.180	0.070	0.86 ± .09
cis-MoOCl$_3$(phen)	-0.260	-0.189	0.071	1.00 ± .14
cis-MoOCl(ox)$_2$	-0.520	-0.445	0.075	0.93 ± .04
cis-MoOCl(tox)$_2$*	-0.415	-0.240	0.655	0.96 ± .01
trans-MoOCl(sal$_2$phen)	-0.535	-0.390	0.145	0.98 ± .02
cis-MoOCl(mee)	-0.910	-0.215	0.695	0.99 ± .01
cis-MoOCl(mpe)	-0.495	-0.340	0.155	0.84 ± .01
trans-(Et$_4$N)MoOCl$_2$(sap)*	-1.055	+0.050[c]	1.005	1.74 ± .09
trans-MoOCl(acac)$_2$	-0.700	-0.630	0.070	0.91 ± .05
(Et$_4$N)MoO(SPh)$_4$[d]	-0.845	-0.780	0.065	0.84 ± .05
(Et$_4$N)MoO(S$_2$C$_6$H$_3$CH$_3$)$_2$	-0.480	-0.370	0.100	0.87 ± .05
WOCl(ox)$_2$	-0.793	-0.733	0.060	1.00 ± .05
WOCl$_3$(bpy)	-0.550	-0.455	0.095	0.95 ± .05
WOCl$_3$(phen)	-0.635	-0.445	0.191	1.17 ± .10
WOCl(sal$_2$en)	-1.063	-0.833	0.230	0.94 ± .10
WOCl(tox)$_2$	-0.613	-0.443	0.170	1.02 ± .10
WOCl(sal$_2$phen)	-0.603	-0.443	0.160	1.04 ± .10
WOCl(acac)$_2$	-0.555	-0.473	0.082	0.90 ± .05

[a]Cyclic voltammograms, scan rate 0.100 V/sec, 0.10 M TEACL in DMF
[b]Volts vs. SCE
[c]Poorly defined anodic peak
[d]Unstable in DMF
*Structures established by X-ray crystallography

compounds containing -N=N- bonds, which may resemble the molybdenum-bound N_2 of nitrogenase.[19]

In order to obtain the three members of the series (Mo(VI), (V), and (IV)) we have been working for the past year on the synthesis of oxomolybdenum(IV) complexes. In general, Mo(IV) species are more difficult to prepare and more sensitive to oxygen and water than their oxomolybdenum(V) counterparts. Our results are found in Table 4. While three of the Mo(IV) complexes, as indicated by their electronic spectra and reduction potentials, are identical in DMF/TEACl solution with the Mo(IV) complexes obtained by coulometric reduction of the corresponding Mo(V) complexes, the 8-hydroxyquinoline complex appears to be a mixture of isomers, possibly cis and trans forms. The anodic cyclic voltammogram of $MoO(ox)_2(PPh_2CH_3)$ in DMF with TEACl as electrolyte shows two oxidation peaks and no reduction peaks. Upon scanning in an anodic direction, followed by scanning in a cathodic direction, and scanning again in an anodic direction to the starting potential, two cathodic peaks and two new, more negative, oxidation peaks are observed. One of the cathodic peaks and one of the new anodic peaks correspond to those found for $MoOCl(ox)_2$ (a Mo(V) complex, Table 3). After standing for about 30 minutes, the two new oxidation peaks appear in the anodic scan. Coulometric oxidation at +0.03 v indicates only 1 e$^-$ per molecule is involved, and the resulting solution has a cyclic voltammogram with two reduction peaks coupled to two oxidation peaks. Esr results show no ^{31}p superhyperfine splitting in the oxidized (Mo(V)) solution, indicating the PPh_2CH_3 has been displaced. The initial oxidation peaks for $MoO(ox)_2(PPh_2CH_3)$ are probably due to two isomers, each with a coordinated PPh_2CH_3 ligand. Upon 1 e$^-$ oxidation, two isomers are produced in which the PPh_2CH_3 has been displaced (probably by Cl$^-$). One of these oxidized species is cis-$MoOCl(ox)_2$ while the other may be the trans isomer (see Figure 1).

The results indicate that oxomolybdenum(V) and oxomolybdenum(IV) complexes form, in many cases, reversible systems while the dioxomolybdenum(VI) species is not electrochemically obtainable from either. Thus, a simple electron transfer series for the three oxidation states in nonaqueous solvents does not exist.

REDUCTION OF NITRATE

Earlier work on NO_3^- reduction by Mo(V) species in aqueous solution indicated Mo(V) dimers are inactive, while small amounts of Mo(V) monomers, in equilibrium with the dimers, reduce NO_3^- to NO. Recent work, both in this laboratory[13] and by Garner, et al.,[20] has shown that some Mo(V) monomers facilely reduce NO_3^- in a one-electron step to NO_2 in nonaqueous solvents (DMF, CH_2Cl_2). Upon addition of H_2O, NO_2 disproportionates into NO_2^- and NO_3^-, thus achieving the same product as the enzyme. Garner's group used

Table 4.   Electrochemical Properties of Oxo-Mo(IV) Complexes

	Ep_a [a]	Ep_c	ΔE_p	n [b]
$MoO(tox)_2$ (PPh_2CH_3)	+0.262	−0.420	0.694	0.98 ± .09
$MoO(ox)_2$ (PPh_2CH_3)	+0.252			0.87 ± .05
	+0.037			0.05 ± .02
	−0.218	−0.798	0.580	0.70 ± .05
	−0.453	−0.523	0.070	0.44 ± .05
$MoOCl_2(phen)(PPh_2CH_3)$	−0.205	−0.280	0.075	0.96 ± .02
$MoOCl_2(bpy)(PPh_2CH_3)$	−0.213	−0.298	0.085	0.95 ± .05

[a] Cylcic voltammograms, volts vs. SCE, 0.10 M TEACl, DMF.
[b] Electrons/molecule (oxidation).

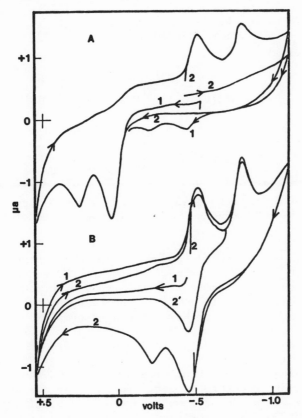

Figure 1.   Cyclic voltammongrams (vs.SCE) for $MoO(ox)_2(PPh_2CH_3)$.
A, before and B after oxidation at +0.30 V vs. SCE. 1. Anodic scans.
2. Cathodic scans.   5 x 10^{-4} M in DMF, 0.10 M TEACl; Pt electrodes.

$MoOCl_3L_2$ complexes (L = substituted phosphine oxides) and data from
this laboratory was obtained with a variety of complexes (Table 5).
The proposed mechanisms of reduction are similar. Dissociation of
a ligand (in our complexes, Cl^-) is followed by the formation of
a NO_3^- complex in a position cis to the oxo group, allowing orbital
overlap between the π orbitals on oxygen of the NO_3^- and the Mo(V)
d_{xy} orbital. Oxo group transfer then occurs, forming the products,
NO_2 and dioxomolybdenum(VI). This mechanism (oxo group transfer
coupled to the loss of an anionic ligand at a hydrophobic site) may
be operative in the nitrate reductases. A general mechanism
developed to explain our results takes the form of equations (1-3),
with

$$Mo(V)OCl \underset{k_{-1}}{\overset{k_1}{\rightleftharpoons}} Mo(V)O^+ + Cl^- \qquad (1)$$

$$Mo(V)O^+ + NO_3^- \underset{k_{-2}}{\overset{k_2}{\rightleftharpoons}} Mo(V)O(NO_3) \qquad (2)$$

$$Mo(V)O(NO_3) \overset{k_3}{\longrightarrow} Mo(VI)O_2 + NO_2 \qquad (3)$$

only the relevant parts of the complexes shown. The complete rate
law is:

$$-dMo(V)/dt = k_1k_2k_3[Mo(V)][NO_3^-]/(k_{-1}(k_{-2}+k_3)[Cl^-]+k_2k_3[NO_3^-])$$

The differences in experimental rate laws between the complexes are
accounted for by differences in the relative values of the rate
constants.[13]

     Recently, we have found some complexes which react rapidly with
both NO_2 and NO_2^- in DMF, giving a mixture of NO and N_2O, thus
proceeding beyond the nitrite stage of the enzyme reduction
(Table 5). Preliminary kinetic data indicate, however, that the
rate-determining step(s) for hexacoordinate complexes containing Cl^-
is the same as with the former group; the rates of reaction with
NO_2 and NO_2^- are fast and dissociation of Cl^- from the Mo(V) complex
is a slow step in most cases. It is not at all evident at this
time why certain complexes react rapidly with NO_2 (and NO_2^-), while
others of similar structure do not. In the case of pentacoordinate
$MoO(SPh)_4^-$, dissociation of a monodentate ligand, allowing
coordination of NO_3^- at a position cis to the oxo group, may occur.
Upon standing, even in the solid state, this complex undergoes
decomposition.

TABLE 5

REDUCTION OF NO_3^- BY Mo(V) AND Mo(IV) OXO COMPLEXES[a]

Complex	Product	k_1 $(sec^{-1} \times 10^4)$	$\dfrac{k_1 k_2 k_3}{k_{-1}(k_{-2} + k_3)}$ $(sec^{-1} \times 10^4)$	$k_2 k_3/k_{-2}$ $(sec^{-1} \times 10^4)$
$(NH_4)_2MoCl_5$	NO_2			$1.63 \pm .15^c$
$(NH_4)MoOCl_4(CH_3CN)$	NO_2			$1.47 \pm .06$
cis-$MoOCl(phen)$	NO_2	35.2 ± 4.3^b	$10.0 \pm .2$	
cis-$MoOCl(bpy)$	NO_2	24.4 ± 2.4	$12.0 \pm .3$	
trans-$MoOCl(bpy)$	NO_2	9.9 ± 4.0	$1.75 \pm .10$	
cis-$MoOCl(ox)_2$	NO_2	16.2 ± 4.4	$3.22 \pm .20$	
cis-$MoOCl(tox)_2$	NO_2	$1.86 \pm .20$		
cis-$MoOCl(CH_3OH)(sap)$	NO_2		$15.8 \pm .17$	
trans-$MoOCl(salphen)$		no reaction		
cis-$MoOCl(acac)_2^d$	NO_2	$2.48 \pm .38$		
$(Et_4N)MoO(SPh)_4$	NO, N_2O			
$MoO(ox)_2$	NO, N_2O			
$MoO(tox)_2$	NO, N_2O			

[a] DMF, 25°.
[b] $\Delta H^{\ddagger} = 12.2 \pm .2$ kcal mole^{-1}, $\Delta S^{\ddagger} = -27.6 \pm .6$ cal mole^{-1}.
[c] $\Delta H^{\ddagger} = 18.3 \pm .4$ kcal mole^{-1}, $\Delta S^{\ddagger} = -14 \pm 2$ cal mole^{-1}.
[d] 35°

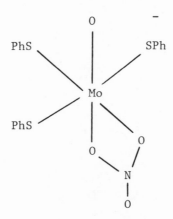

$$[MoO(SPh_3)(NO_3)]^-$$

Since a two-electron reduction of NO_3^- to NO_2^- by the enzyme seems more likely than the one-electron reduction observed with the Mo(V) complexes, reaction of NO_3^- with Mo(IV) complexes in DMF is under investigation. Preliminary results indicate the Mo(IV) complexes react more rapidly with NO_2^- than with NO_3^-, again producing NO and N_2O. In fact, all the complexes we have tried (both Mo(V) and Mo(IV)) react more rapidly with NO_2^- than with NO_3^-. Only in those Mo(V) cases involving one-electron reduction of NO_3^- to NO_2, in which the original complex does not react further with NO_2 to produce NO_2^-, is NO_2^- (produced by disproportionation of NO_2 when H_2O is added to the system) the final product. Since the enzyme does not react with NO_2^-, the relevancy of these investigations to the mechanism of action of nitrate reductase is somewhat questionable.[21]

While model studies have contributed much to our current understanding of the bioinorganic chemistry of molybdenum, much remains to be learned. The complete set of ligands at the molybdenum site of any enzyme is still unknown. No Mo(VI) models having the necessary thermodynamic and kinetic properties to oxidize purines or SO_3^{2-} have been developed. No generally satisfactory Mo(V) model which mimics all the esr parameters of the enzymes has appeared. Protein structures, which stabilize Mo(VI), monomeric Mo(V) and Mo(IV) against hydrolysis, dimerization and oxidation in an essentially aqueous environment, are not yet known. Mechanisms for electron transfer between cofactors at considerable distance in the enzyems have not been elucidated. The relationship between the Mo cofactor(s) and the apoenzymes remains to clarified. The transport of molybdenum from the environment across cell membranes, its storage in the cell, and its incorporation into molybdenum enzymes are problems that have scarcely been considered. Clearly, model studies, in cooperation with biochemical investigations, will be of great value in providing answers to many of these intriguing questions.

ACKNOWLEDGEMENT

This research was supported by AMAX Foundation, Climax Molybdenum Co., and NIH Grant GM08347. Thanks are gratefully expressed for this support and for the efforts of many colleagues and students who made this work possible.

REFERENCES

1. R. C. Bray, "Molybdenum Iron-Sulfur Flavin Hydroxylases and Related Enzymes", in "The Enzymes," 3rd ed., Vol XII, P. D. Boyer, Ed., Academic Press, N.Y., N.Y., 1975, p. 299.
2. E. I. Stiefel, "The Coordination and Bioinorganic Chemistry of Molybdenum", Prog. Inorg. Chem., 22, 1 (1977).
3. T. D. Tullius, D. M. Kurtz, Jr., S. D. Conradson and K. O. Hodgson, "The Molybdenum Site of Xanthine Oxidase, Structural Evidence from X-ray Absorption Spectroscopy", J. Am. Chem. Soc., 101, 2777 (1979).
4. S. Gutteridge, S. J. Tanner and R. C. Bray, "Comparison of the Molybdenum Centres of Native and Desulpho Xanthine Oxidase", Biochem. J., 175, 887 (1978).
5. R. D. Taylor, J. P. Street, M. Minelli and J. T. Spence, "Electrochemistry of Monomeric Molybdenum(V) - Oxo Complexes in Dimethylformamide", Inorg. Chem., 17, 3207 (1978).
6. A. F. Isbell and D. T. Sawyer, "The Electrochemistry of Molybdenum(VI,V)-8-Quinolinol Complexes in Dimethylformamide", Inorg. Chem., 10, 2449 (1971).
7. L. G. Ljungdahl, "Tungsten, a Biologically Active Metal", Trends in Biochem. Sci., 1, 63 (1976).
8. R. C. Bray, "Molybdenum Containing Enzymes", J. Less-Common Metals, 54, 527 (1977).
9. N. Pariyadath, W. E. Newton and E. I. Stiefel, "Monomeric Molybdenum(V) Complexes Showing Hydrogen-1, Hydrogen-2 and Nitrogen-14 Superhyperfine Splitting in Their Electron Paramagnetic Resonance Spectra. Implications for Molybdenum Enzymes", J. Am. Chem. Soc., 98, 5388 (1976).
10. K. Yamanouchi and J. Enemark, "Monomeric Molybdenum(V) Complexes. 1. Structure of Bis (2-aminobenzenethiolato)(diethyldithiocarbamato)molybdenum(V), Mo(NHC$_6$H$_4$S)$_2$(S$_2$CN(C$_2$H$_5$)$_2$)", Inorg. Chem., 17, 1981 (1978).
11. P. M. Boorman, C. D. Garner, F. E. Mabbs and T. J. King, "Oxotrichlorotriphenylphosphidesulphide-molybdenum(V), a Mononuclear Five-Co-ordinate Oxomolybdenum(V) Complex", J. Chem. Soc., Chem. Commun., 663 (1974).
12. I. W. Boyd, I. G. Dance, K. S. Murray and A. G. Wedd, "Mononuclear Oxo Thiolato Compounds of Molybdenum(V)", Aust. J. Chem., 31, 279 (1978).
13. R. D. Taylor, P. G. Todd, N. D. Chasteen and J. T. Spence, "Reduction of Nitrate by Monomeric Molybdenum(V) Complexes in Dimethylformamide", Inorg. Chem., 18, 45 (1979).

14. I. R. Marov, V. K. Belyaeva, Y. N. Dubrov and A. N. Ermakov, "EPR Study of Thio-oxinato-complexes of Molybdenum(V), Tungsten(V) and Copper(II)", Russ. J. Inorg. Chem., 14, 1515 (1972).

15. J. T. Spence, M. Minelli, P. Kroneck, M. T. Scullane and N. D. Chasteen, "Monomeric Molybdenum(V) Oxo Complexes with Tetradentate Aminoethanethiols", J. Am. Chem. Soc., 100, 8002 (1978).

16. J. S. Olsen, D. P. Ballou, G. Palmer, and V. Massey, "The Mechanism of Action of Xanthine Oxidase", J. Biol. Chem., 249, 4363 (1974).

17. R. Cammack, M. J. Barber and R. C. Bray, "Oxidation-Reduction Potentials of Molybdenum, Flavin and Iron-Sulfur Centres in Milk Xanthine Oxidase", Biochem. J., 157, 469 (1976).

18. M. J. Barber, R. C. Bray, R. Cammack and M. P. Coughlan, "Oxidation-Reduction Potentials of Turkey Liver Xanthine Dehydrogenase and the Origins of Oxidase and Dehydrogenase Behavior in Molybdenum-Containing Hydroxylases", Biochem. J. 163, 279 (1977).

19. P. W. Schneider, D. C. Bravard, J. W. McDonald and W. E. Newton, "Reactions of Oxo-bis(N,N-dialkyldithiocarbamato) Molybdenum(IV) with Unsaturated Organic Molecules and Their Biochemical Implications", J. Am. Chem. Soc., 94, 8640 (1972).

20. C. D. Garner, M. R. Hyde, F. E. Mabbs, and V. I. Routledge, "Kinetics and Mechanism of Oxidation of Trichloro-oxobis-(triphenylphosphineoxide) Molybdenum(V) by Nitrate in Dichloromethane", J. Chem. Soc., Dalton Trans., 1180 (1977).

21. W. H. Orme-Johnson, G. S. Jacob, M. T. Henzyl, and B. A. Averill, "Molybdenum in Enzymes", Adv. Chem. Ser., 162, 389 (1977).

Mo(VI) COMPLEXES OF N,S-DONOR LIGANDS: RELEVANCE TO MOLYBDENUM ENZYMES

Edward I. Stiefel, Kenneth F. Miller, Alice E. Bruce,
Narayanakutty Pariyadath, Jay Heinecke and James L. Corbin

Charles F. Kettering Research Laboratory
Yellow Springs, OH 45387 USA

Jeremy M. Berg and Keith O. Hodgson

Department of Chemistry
Stanford University
Stanford, CA 94305 USA

INTRODUCTION

The molybdenum enzymes other than nitrogenase have a common molybdenum cofactor[1,2] and spectroscopic studies of their molybdenum sites reveal these to be similar although not identical.[3] EPR* studies on xanthine oxidase in the Mo(V) oxidation state,[4] together with results from model compounds,[5] led to the suggestion of sulfur as a donor atom to molybdenum. However, in these original studies and in subsequent investigations,[3,6] the inorganic compounds used in the comparison were not structurally and, at times, not even stoichiometrically defined. In order to provide a comprehensive set of structurally defined oxo-molybdenum complexes containing sulfur-donor ligands for comparison with the enzymes by EPR*, EXAFS* and other spectroscopic probes, we have embarked on an extensive synthetic and isolation program to obtain relevant compounds in the Mo(IV), Mo(V) and Mo(VI) oxidation states. Recently, EXAFS studies on xanthine oxidase[7] and sulfite oxidase[8] have confirmed the presence of sulfur and terminal oxo ligands in the coordination sphere of molybdenum and have revealed distinct similarities to some of the model compounds discussed here.[9]

---

*EPR = electron paramagnetic resonance. EXAFS = extended X-ray absorption fine structure.

This paper summarizes some of our results on Mo(VI) complexes. In particular, we discuss complexes containing the MoO_2^{2+} structural core and bidentate or tetradentate ligands. Some of these complexes have proven useful as calibrants and test cases for EXAFS studies.[9,10] Structural studies reveal preferences of the Mo(VI) coordination sphere but somewhat surprisingly also show this sphere to be highly deformable from the standard octahedral structure and to be potentially fluxional in nature. Further, one structure reveals a partial disulfide bond in the "Mo(VI)" coordination sphere. The experimental findings and their implications for molybdoenzymes are discussed.

The Ligands

The synthesis of the new ligands used in this study will be reported elsewhere.[11] The ligands fall into two general categories, bidentate and tetradentate. The former series can all be viewed as derivatives of cysteamine formed by the substitution of the NH_2 or CH_2 hydrogens by methyl groups. Typical bidentate ligands discussed include I(a-c). Although the variation may appear minor,

it is sufficient to cause marked differences in the static and dynamic structural behavior of their Mo(VI) complexes.

The tetradentate ligands prepared include the new tripodal tetradentate ligands, II(a-c), and the known[12,13] linear

tetradentate ligands, III(a-c). Additional related bi-, tri- and tetradentate ligands will be reported elsewhere.[11]

IIIa   IIIb   IIIc

The Complexes

The complexes are of the form MoO_2L_2 or MoO_2L for the bi- and tetradentate ligands, respectively. They are formed by reaction of the ligand with $MoO_2(acac)_2$ (acac = $C_5H_7O_2^-$ = acetylacetonate) in CH_3OH (equations (1) and (2)) and are characterized by elemental

$$MoO_2(acac)_2 + 2LH \rightarrow MoO_2L_2 + 2acacH \qquad (1)$$

$$MoO_2(acac)_2 + LH_2 \rightarrow MoO_2L + 2acacH \qquad (2)$$

analysis, infrared and nmr spectroscopy. Complexes of the linear tetradentate ligands have also been reported recently by Spence[14] and Zubieta[15] and their respective coworkers. Infrared spectra of the complexes display the two-band pattern characteristic of a cis-dioxo molybdenum(VI) center. Proton nmr spectroscopy provides information about the placement of the bidentate and tetradentate ligands about molybdenum but does not in all cases (vide infra) unequivocally delineate the positions of the S and N donors. Furthermore, the bidentate ligands have been designed with the idea of forcing structural changes in the molybdenum coordination sphere by sterically hindering certain ligand arrangements which might otherwise have been favored. For this reason and for the testing of EXAFS distance predictions,[10] we have performed x-ray structural determinations on three complexes in this series.

STRUCTURAL STUDIES

The complexes $MoO_2((CH_3)_2NCH_2CH_2N(CH_2CH_2S)_2)$ and $MoO_2(CH_3SCH_2CH_2N(CH_2CH_2S)_2)$ were subjected to EXAFS analysis as part of a program aimed at assessing the results obtained by this new technique.[10] The x-ray crystallographically determined structures[9] are displayed in Figures 1 and 2. Both complexes adopt octahedral structures with cis oxo groups and trans thiolate donors. The deviations from 90 or 180° angles about the molybdenum atoms are attributable to the combined effects of interligand repulsion

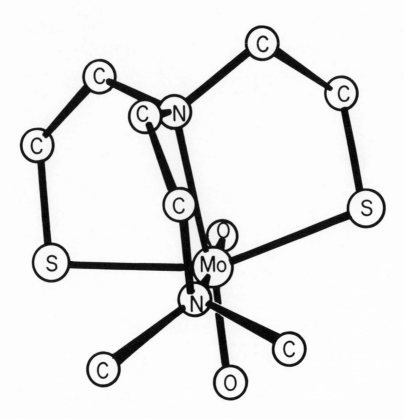

Fig. 1.   The Structure of $MoO_2((CH_3)_2NCH_2CH_2N(CH_2CH_2S)_2)$[9]

(especially involving the terminal $Mo-O_t$ linkages) and the constraints of the ligand systems.   Some key distances and angles are
given in Table 1.

    Comparison of the EXAFS and crystallographic results are shown
in Table 2 for these two complexes and for $Mo(NHSC_6H_4)_3$.[10,16]   The
comparison reveals that distance determinations by EXAFS and
crystallography on the average differ by less than 0.01 Å.   A
notable prediction by EXAFS was the long (2.8 Å) distance for
the thioether ligand.   The x-ray crystallographic confirmation of
this distance reveals the source of the length as the positioning
of the already weak thioether donor in a position <u>trans</u> to an oxo
group where it is subject to the well known <u>trans</u> bond-lengthening
influence of that ligand.   As discussed elsewhere,[9] the EXAFS for
the Mo(VI) complex of the thioether tripod ligand (Fig. 1) and for
the Mo(VI) state of sulfite oxidase are extremely similar.   Based
on this detailed structural comparison, it has been suggested[9] that
sulfite oxidase in the Mo(VI) state has two <u>cis</u> oxo groups, two

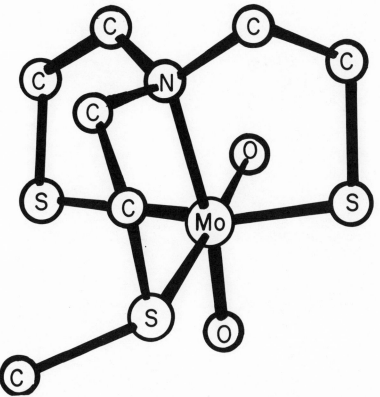

Fig. 2.   The Structure of $MoO_2(CH_3SCH_2CH_2N(CH_2CH_2S)_2)$ [9]

mutually <u>trans</u> thiolates and one thioether in a position <u>trans</u> to
an oxo group. The long thioether ligand appears to be maintained
in the Mo(V) and Mo(IV) states of sulfite oxidase,[8] thus indicating
the likelihood that the position <u>trans</u> to one oxo is always occupied.
This finding is of interest because it has been noted previously[6,17]
that it is the coordination position <u>cis</u> to an oxo which must be
used to transfer electrons to or from an oxomolybdenum site.

The requirement for an open site <u>cis</u> to $Mo-O_t$ originally led
us to the use of tripodal tetradentate ligands to mimic the sites
in molybdenum-containing oxidases and in nitrate reductase. In an
octahedral structure, the tripodal ligand assures the presence of
two <u>cis</u> coordination positions which are not occupied by the
tetradentate ligand donor atoms. In the dioxo-Mo(VI) structure,
both nonpolydentate ligand sites are occupied by oxo groups.
However, in a (to date hypothetical) mono-oxo Mo(IV) (or Mo(V))
complex of this ligand, the open site will be <u>cis</u> to oxo. Numerous
electronic structural studies show that, in d^1 and d^2 oxo complexes,

Table 1. Selected Distances (in Å), Angles (in °) and ν(Mo–O) (in cm^{-1}) for MoO$_2{}^{2+}$ Complexes of NS Ligands

Complex	Mo–O	O–Mo–O	Mo–S	Mo–N	S–Mo–S	ν(Mo–O)	ref.
MoO$_2$ (S–S–N, (CH$_3$)$_2$N)	1.699(2)	107.9°(1)	2.420(1)	2.372(2)	154.3°(3)	921	9
	1.705(2)		2.409(1)	2.510(2)		893	
MoO$_2$ (S–S–N, CH$_3$S)	1.694(5)	108.7°(3)	2.401(2)	2.420(5)	151.2°(7)	922,917	9
	1.694(5)		2.411(2)			891	
MoO$_2$(thiooxine)$_2$	1.694(6)	106.3(3)	2.411(3)	2.382(9)	160.9°(1)	921	18
	1.712(7)		2.411(3)	2.374(9)		888	
MoO$_2$(CH$_3$NHCH$_2$C(CH$_3$)$_2$S)$_2$	1.72	122°	2.42	2.28		883	This work
	1.73		2.42		69.7°	856	

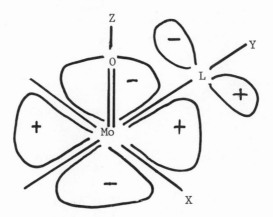

the d electrons reside in a d_{xy} orbital ($M-O_t$ being the z axis).
This orbital, as illustrated, is capable of π overlap with ligands
bound in the xy plane, i.e., <u>cis</u> to oxo. It is possible that in
oxidases and in nitrate reductase, the protein (or cofactor)
provides for molybdenum, a polydentate ligand environment which is
"designed" to provide just such an electronic and molecular
structural feature.

The last structure to be considered is that of the bidentate
ligand complex $MoO_2(CH_3NHCH_2C(CH_3)_2S)_2$. By comparison with the
structures of the tripodal N_2S_2 ligand (Fig. 1) and with
$MoO_2(thiooxine)_2$,[18] a <u>trans</u> thiolate arrangement might be predicted.
However, such an arrangement could, as shown in IV, lead to steric
interference between N-methyl and C-methyl substitutents of the
same or different ligands that cannot be totally removed by mere

IV

Table 2.   Comparison of Distances (in Å) Determined by
EXAFS and X-ray Crystallography[a]

Complex	Bond	EXAFS[b]	Crystallography[a]
$Mo(NHSC_6H_4)_3$	Mo–N	1.996	1.997[c]
	Mo–S	2.419	2.418[c]
$MoO_2((CH_3)_2NCH_2CH_2N(CH_2CH_2S)_2)$	Mo–O	1.694	1.702[d]
	Mo–S	2.424	2.415[d]
$MoO_2(CH_3SCH_2CH_2N(CH_2CH_2S)_2)$	Mo–O	1.693	1.694,1.706,1.697[e]
	Mo–S	2.401	2.406,2.405,1.401[e]
	Mo–S^1	2.803	2.709,2.809,2.781[e]

[a]The EXAFS numbers were reported[10] prior to the x-ray crystallographic
determinations.   [b]Predictions from ref. 10.   [c]Ref. 16.   [d]Ref. 9.
[e]Numbers from determinations of structures of two crystals, one of
which contained two molecules/unit cell (ref. 9 and additional data
to be published).

conformational change.   A structural change to a <u>cis</u> thiolate
arrangement, as shown in V or VI, would lead to partial or complete
elimination of the methyl interference, respectively.   However,
this change places one or two S-donor atoms <u>trans</u> to oxo in V and
in VI, respectively, and causes the strong σ-donor and π-donor oxo

and its <u>trans</u> thiolate ligand to interact with the same pσ orbital
and the same set of dπ orbitals.   This situation, although sterically
acceptable, is apparently energetically unfavorable as is vividly
illustrated by the crystallographic structure determination (Fig. 3)
with key distances and angles listed in Table 1.

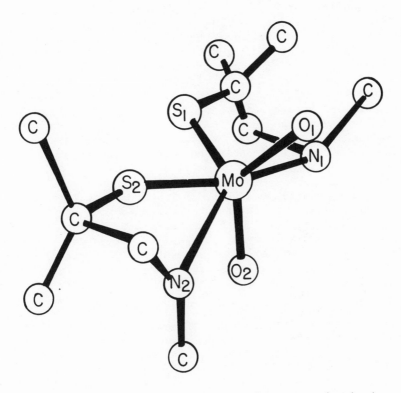

Fig. 3.   The Structure of $MoO_2(CH_3NHCH_2C(CH_3)_2S)_2$

The structure found for $MoO_2(CH_3NHCH_2C(CH_3)_2S)_2$ is in fact unlike IV, V, or VI, although it might be viewed as derived from VI.   The complex is six-coordinate and contains a <u>cis</u>-dioxo linkage but that is where its similarity to other Mo(VI) structures ends. The six-coordinate geometry is not close to octahedral by any of several criteria.   For example, the largest angles between donor atoms are 144.1° for N_1-Mo-N_2, 142.4 for S_1-Mo-N_2 and 141.0 for S_2-Mo-N_1.   Likewise, the <u>largest</u> "<u>trans</u>" angle about Mo-O_t is only 122° and is, in fact, the O_1-Mo-O_2 angle!   These are to be contrasted with the nominal 180° angles found in a regular octahedron.   Although the <u>trans</u> angles in the six-coordinate structures of Figures 1 and 2 are as small as 154°, they are as great as 168.5° and clearly show the closeness of the tripodal ligand structures to octahedral limits.   Moreover, the structure in Figure 3 is also not readily describable in terms of a trigonal-prismatic geometry.   Although triangles O_1-N_1-S_1 and O_2-N_2-S_2 are parallel and appear to be roughly eclipsed, the line connecting their centroids which passes through molybdenum is not perpendicular to these triangles.   It appears as if the triangles have slid apart from the trigonal-prismatic limit which is, in part, an artifact of the nonequilateral nature of the triangles, which is, in turn, a consequence of the vastly different Mo-O_t vs. Mo-S and Mo-N bond distances. This structure approaches closely the skew-trapezoidal bipyramid described by Kepert,[19] which until now had only been found in tin chemistry. In the structure shown in Figure 3, the S_2N_2 set of atoms forms the approximate trapezoid while the two O_t groups are at the apices of the bipyramid. The least-motion approach to an octahedral structure would appear to require the formation of the as yet unknown trans-MoO_2^{+2} isomer (not shown). Apparently, the placement of sulfurs trans to oxo is energetically unfavorable and the structure distorts such that the O-Mo-S angles range from 108.3 to 120.3°.

The above remarkable structural features notwithstanding, the most unusual feature of the structure is an S-Mo-S angle of 69.7° which occurs concomitantly with the short S-S distance of 2.76 Å for two donor atoms which would ordinarily not be considered as bound to each other. The 2.76 Å separation is roughly 1 Å shorter than the 3.7 Å van der Waals distance for two sulfur atoms.[20] It is 0.5 Å shorter than the 3.20 Å distance of nominally non-bonded adjacent sulfurs[21] in a metal coordination sphere. In the 8-membered ring of S_8^{+2}, there is a transannular sulfur-sulfur (S_3-S_7) distance of 2.83 Å which is considered, from both experimental[22] and theoretical[23] considerations, to be an important determinant of the stereochemistry of the ion. In S_5N_6, an S-S distance of 2.425 Å is clearly a bonding interaction of sufficient strength to cause the molecule to assume a shape described as a "molecular basket".[24] In bis and tris dithiolene complexes, inter ligand distances of ∿3.05 Å are suggested as possibly indicative of S-S bonding which might contribute to the stabilization of the square planar and trigonal prismatic geometries favored by these ligands.[25,26] Despite all of these distances being greater than the nominal S-S single bond distance of ∿2.1 Å, they appear to be of crucial import in determining the structural details of the particular compound. In $MoO_2(CH_3NHCH_2C(CH_3)_2S)_2$, the S-S interaction may likewise be crucial. It would appear to represent the partial formation of a disulfide bond which requires the concomitant transfer of charge to the molybdenum center. This transfer may be responsible for the opening of the O_t-Mo-O_t angle and possibly also for some lengthening of the Mo-O_t bond. A recent structural study of the Mo(V) complex $[Mo_2O_4(SC_6H_5)_4]^{2-}$ by Dance, Wedd and Boyd[21] showed one of the two molecules in the unit cell to have an inter-ligand S-S distance of 2.94 Å which was felt to be indicative of S-S overlap.

Taken together, our results and those of Dance et al.[21] illustrate a fascinating possibility for redox processes in bis(thiolate)-molybdenum systems. Thus, successive one-electron oxidations of Mo(IV) to (formally) Mo(V) and Mo(VI) may involve a substantial component of thiol-sulfide oxidation. If this S-S bond formation occurs in molybdoenzymes (and none of the extant data preclude such an occurrence), then it could explain, in part, the absence of identity between the spectroscopic, redox and catalytic properties of molybdoenzymes and their "model" systems. For example, the high g_z and low A_z ($Mo^{95,97}$) value of the Mo(V) epr signal[5] in xanthine oxidase may be attributable to strong delocalization of the unpaired electron into the S-S partial bond. Further, the great variability in redox potential for different molybdoenzymes (while also potentially explainable in other terms) could be due to different degrees of S-S bond formation. Recently, Musker and others[27-29] have shown that the presence of strategically juxtaposed sulfides in large ring multisulfur organic heterocyclic compounds is found to correlate with substantially lower potentials for one-electron oxidation and with increased reversibility in the electron-transfer process. It is possible that the <u>cis</u> positioning of molybdenum-bound thiolates brings these reactivity changes to

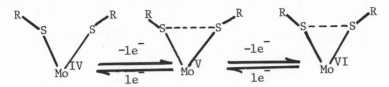

molybdenum sites in enzymes, such that a triatomic unit may be responsible for some of the unique spectroscopic and reactivity parameters of these sites.

Infrared spectral considerations show that the ν(Mo-O) vibrations of the $MoO_2(CH_3NHCH_2C(CH_3)S)_2$ complex are unusual. Table 1 reveals that ν(Mo-O) is substantially lower for this complex compared with other complexes with the same set of donor ligands. This lowering of ν(Mo-O) may be characteristic of the non-octahedral-partial disulfide structure. If this correlation holds up in additional cases currently under study, then this vibrational fingerprint might be useful for identifying such a structural arrangement in molybdoenzymes.

NMR AND DYNAMICAL CONSIDERATIONS

The background provided by the structure of $MoO_2(CH_3NHCH_2C(CH_3)_2S)_2$ allows us to look briefly at 220 MHz ^1H NMR data we have obtained for complexes in this class. The representative spectra shown in Fig. 4 can be used both for structural and dynamical evaluation of the complexes.

The spectrum of $MoO_2(NH_2C(CH_3)_2CH_2S)_2$ is discussed to illustrate the use of NMR to establish equivalence of the two bidentate ligands. The NMR spectrum of the free ligand, $NH_2C(CH_3)_2CH_2SH$, contains three

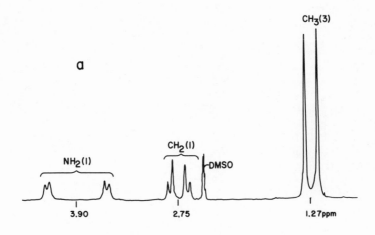

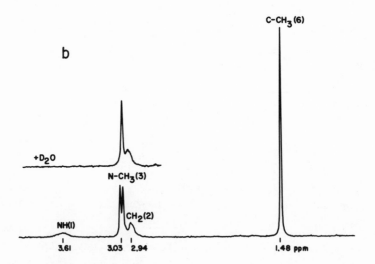

Fig. 4. ^{1}H NMR at 220 MHz and 17°C of a) $MoO_2(NH_2C(CH_3)_2CH_2S)_2$;
and b) $MoO_2(CH_3NHCH_2C(CH_3)_2S)_2$.

singlets of relative intensity 6:2:3, which are attributed respectively to the two equivalent methyl groups, the methylene group and
a common resonance for the two amino and one sulfhydryl proton.
Coordination to molybdenum eliminates the SH signal and removes the
equivalence of the methyl groups, the two methylene protons and the
two amino protons respectively. As displayed in Fig. 4a, the spectrum in d_6-dmso shows two methyl resonances centered at 1.27 ppm
and separated by 32 Hz. The inequivalence of the two methylene
protons and of the two amino protons are manifest by AB patterns

centered at 2.74 ppm (J_{AB} = 12 Hz) and 3.89 ppm (J_{AB} = 11 Hz), respectively. A structural isomer of type IV or VI (shown above with a different ligand) will account for the observed spectrum. However, isomer V, in the absence of exchange, would be expected to yield a spectrum containing four distinct methyl resonances, and a pair of AB patterns for each methylene and amino group.

In contrast to $MoO_2(NH_2C(CH_3)_2CH_2S)_2$, other MoO_2L_2 complexes of bidentate ligands display spectra where equivalent methyl substituents and equivalent methylene protons are observed at normal operating temperatures (17°C) and, in some cases, this equivalence is maintained at low temperatures. The spectrum of $MoO_2(CH_3NHCH_2C(CH_3)_2S)_2$ in $CDCl_3$ in Figure 4b consists of a singlet at 1.48 ppm, a broad peak at 2.94 ppm, a doublet at 3.03 ppm and a very broad singlet at 3.62 ppm. Consistent with the integration, these peaks are assigned to the two methyl groups on the carbon adjacent to sulfur, the methylene group, the amine-methyl group and the amine proton, respectively. The addition of D_2O results in the collapse of the doublet at 3.03 ppm to a singlet and the dis- appearance of the peak at 3.62 ppm which is the behavior anticipated upon exchanging the amine proton for a deuteron.

The crystal structure of $MoO_2(CH_3NHCH_2C(CH_3)_2S)_2$, discussed above and shown in Figure 3, shows that the methyl groups of the tertiary carbon adjacent to sulfur as well as the two methylene protons are inequivalent in the solid state. However, the spectrum of this complex in solution reveals that the solid state inequiva- lence of the two methyl groups and the two methylene protons have each been removed. In addition, any complexity which results from diasterioisomerism of the MoO_2L_2 complex, due to chiral centers associated with the nitrogen and molybdenum atoms, was not observed. Consequently, a molecular rearrangement process must exist to average the environments of both the methyl groups and the methylene protons.

The complexes $MoO_2(NH_2C(CH_3)_2CH_2S)_2$ and $MoO_2(CH_3NHCH_2C(CH_3)_2S)_2$ differ in the placement of the dimethyl substitution on the ethylene framework and by CH_3 substitution of the amine. Despite this seem- ingly minor change, the latter (Fig. 1b) is clearly fluxional on the NMR time scale at 17°C while the former displays a limiting ("static") structure under the same conditions. The range of structural fluxionality displayed for MoO_2L_2 complexes appears to be associated with the degree and position of methyl substitution. As the degree of methylation at nitrogen is increased, the degree of fluxionality increases in the series $H_2NCH_2C(CH_3)_2S^-$ < $CH_3NHCH_2C(CH_3)_2S^-$ < $N(CH_3)_2CH_2C(CH_3)_2S^-$.

CONCLUSIONS

The static and dynamic distortions of the Mo(VI) coordination sphere observed in these studies have potential impact on structural and mechanistic considerations for the molybdenum sites in enzymes. Heretofore, all suggestions for the structure of the molybdenum site have been based on octahedral geometries and have assumed structural rigidity. However, the results of our studies clearly show that relatively small, but sterically demanding, changes in the ligands can lead to severe distortions and relative flexibility in the molybdenum coordination sphere. Certainly, a set of donor ligands from a protein or cofactor could be equally demanding and, in attempting to model the molybdenum site, we must take account of this deformability in our choice or design of ligand systems.

ACKNOWLEDGEMENT

220 MHz spectra were obtained through the courtesy of Professor Rupert Wentworth and the NMR staff of Indiana University. We thank the National Science Foundation (Grants CHE 77-07026 to EIS and JLC and PCM 17105 to KOH) for support of this work. This paper constitutes Contribution No. 667 from the Charles F. Kettering Research Laboratory.

REFERENCES

1. S.-S. Pan, R. H. Erickson, K.-Y. Lee and A. Nason, Molybdenum-Containing Component Shared by the Molybdenum Enzymes as Indicated by the in vitro Assembly of Assimilatory Nitrate Reductase Using the Neurospora Mutant nit-1, in: "Proceedings of 1st International Symposium on Nitrogen Fixation," W. E. Newton and C. J. Nyman, eds., Washington State University Press, Pullman (1976), p. 293.
2. P. T. Pienkos, V.K. Shah, and W.J. Brill, Molybdenum Cofactors from Molybdenum Enzymes and in vitro Reconstitution of Nitrogenase and Nitrate Reductase, Proc. Nat. Acad. Sci. USA 74:5468 (1977).
3. R. C. Bray, Molybdenum-Containing Enzymes, J. Less-Common Metals 54:527 (1977).
4. R. C. Bray and L. S. Meriwether, Electron Spin Resonance of Xanthine Oxidase Substituted with Molybdenum-95, Nature 212:467 (1966).
5. L. S. Meriwether, W. F. Marzluff and W. G. Hodgson, Molybdenum-Thiol Complexes as Models for Molybdenum Bound in Enzymes, Nature 212:465 (1966).
6. E. I. Stiefel, The Coordination and Bioinorganic Chemistry of Molybdenum, Progr. Inorg. Chem. 22:1 (1977).

7. T. D. Tullius, D. M. Kurtz, Jr., S. D. Conradson, and K. O. Hodgson, The Molybdenum Site of Xanthine Oxidase. Structural Evidence from X-ray Absorption Spectroscopy, $\underline{J}$. $\underline{Am}$. $\underline{Chem}$. $\underline{Soc}$. 101:2776 (1979).

8. S. P. Cramer, H. B. Gray, and K. V. Rajagopalan, The Molybdenum Site of Sulfite Oxidase. Structural Information from X-ray Absorption Spectroscopy, $\underline{J}$. $\underline{Am}$. $\underline{Chem}$ $\underline{Soc}$. 101:2772 (1979).

9. J. M. Berg, K. O. Hodgson, S. P. Cramer, J. L. Corbin, A. Elsberry, N. Pariyadath, and E. I. Stiefel, Structural Results Relevant to the Molybdenum Sites in Xanthine Oxidase and Sulfite Oxidase. The Crystal Structures of MoO_2L, L = $(SCH_2CH_2)_2NCH_2CH_2X$ with X = SCH_3, $N(CH_3)_2$, $\underline{J}$. $\underline{Am}$. $\underline{Chem}$. $\underline{Soc}$. 101:2774 (1979).

10. S. P. Cramer, K. O. Hodgson, E. I. Stiefel, and W. E. Newton, A Systematic X-ray Absorption Study of Molybdenum Complexes. The Accuracy of Structural Information from Extended X-ray Absorption Fine Structure, $\underline{J}$. $\underline{Am}$. $\underline{Chem}$. $\underline{Soc}$. 100:2748 (1978).

11. J. L. Corbin, K. F. Miller, A. E. Bruce, J. Heinecke, N. Pariyadath and E. I. Stiefel, unpublished results.

12. J. L. Corbin and D. E. Work, 1-Alkyl- (or aryl-)amino-2-methylpropane-2-thiols. Some Bi- and Tetradentate Nitrogen-Sulfur Ligands from Schiff's Base Disulfides, $\underline{J}$. $\underline{Org}$. $\underline{Chem}$. 41:489 (1976).

13. K. D. Karlin and S. J. Lippard, Sulfur-Bridged Binuclear Iron (II) Complexes. Effect of Ligand Constraints on Their Physical Properties; Reactions with Carbon Monoxide and Alkyl Isocyanides, $\underline{J}$. $\underline{Am}$. $\underline{Chem}$. $\underline{Soc}$. 98:6951 (1976).

14. J. T. Spence, M. Minelli, P. Kroneck, M. I. Scullane and N. D. Chastien, Monomeric Molybdenum(V) Oxo Complexes with Tetradentate Aminoethanethiols, $\underline{J}$. $\underline{Am}$. $\underline{Chem}$. $\underline{Soc}$. 100:8002 (1978).

15. N. Kim, S. Kim, P. A. Vella and J. Zubieta, Synthesis and Characterization of the Complexes of Mo(VI) and Mo(V) with the Ligands $HSCH_2CH_2N(CH_3)(CH_2)_nN(CH_3)CH_2CH_2SH$, n = 2,3. The Crystal Structure of $H_2[Mo_2O_4(SCH_2CH_2N(CH_3)(CH_2)_3N-(CH_3)CH_2CH_2S)_2]$, $\underline{Inorg}$. $\underline{Nucl}$. $\underline{Chem}$. $\underline{Lett}$. 14:457 (1978).

16. K. Yamanouchi and J. H. Enemark, Structure of Tris [2-aminobenzenethiolato(2-)-N,S]Molybdenum(VI), $Mo(NHC_6H_4S)_3$, $\underline{Inorg}$. $\underline{Chem}$. 17:2911 (1978).

17. R. Durant, C. D. Garner, M. R. Hyde, F. E. Mabbs, J. R. Parsons and D. Richens, Oxygen Transfer Reactions Involving Certain Oxomolybdenum Complexes, $\underline{J}$. $\underline{Less-Common}$ $\underline{Metals}$ 54:459 (1977).

18. K. Yamanouchi and J. H. Enemark, Monomeric Molybdenum(V) Complexes. 2. Comparison of the Structure of $\underline{cis}$-Oxochlorobis(8-mercaptoquinolinato)molybdenum($\underline{V}$) and $\underline{cis}$-Dioxo-$\underline{bis}$(8-mercaptoquinolinato)molybdenum(VI), $\underline{Inorg}$. $\underline{Chem}$. 18:1626 (1979).

19. D. L. Kepert, Aspects of the Stereochemistry of Six-
    Coordination, Prog. Inorg. Chem. 23:1 (1977).
20. L. Pauling, The Nature of the Chemical Bond, Cornell University
    Press, Ithaca, N.Y., (1960), p. 260.
21. I. G. Dance, A. G. Wedd, and I. W. Boyd, The Formation and
    Molecular Structure of the Di-μ-oxo-di[di(benzene-
    thiolato)oxomolybdate(V)] Dianion, Aust. J. Chem.
    31:519 (1978).
22. R. J. Gillespie, J. Passmore, P. K. Ummat and O. C. Vaidya,
    Polyatomic Cations of Sulfur. 1. Preparation and Proper-
    ties of S_{16}^{2+}, S_8^{2+}, and S_4^{2+}, Inorg. Chem. 10:1327 (1971).
23. D. R. Salahub, A. E. Foti and V. H. Smith, Jr., Molecular
    Orbital Study of Structural Changes on Oxidation and
    Reduction of S_3, S_4, S_6, and S_8, J. Amer. Chem. Soc.
    100:7847 (1978).
24. T. Chivers and J. Proctor, Preparation and Crystal Structure
    of a New Sulphur Nitride, S_5N_6; a Molecular Basket,
    J. Chem. Soc. Chem. Comm. 642 (1978).
25. E. I. Stiefel, R. Eisenberg, R. C. Rosenberg and H. B. Gray,
    Characterization and Electronic Structures of Six-
    Coordinate Trigonal-Prismatic Complexes, J. Am. Chem.
    Soc. 88:2956 (1966).
26. R. Eisenberg, Structural Systematics of 1,1- and 1,2-Dithiolate
    Chelates, Progr. Inorg. Chem. 12:295 (1970).
27. W. K. Musker and T. L. Wolford, Long-Lived Radical Cations
    from Mesocyclic Dithioethers, J. Am. Chem. Soc.
    98:3055 (1976).
28. W. K. Musker and P. B. Roush, Preparation and Reactivity
    of Dithioether Dications [RS^+SR_2]: A New Functional
    Group, J. Am. Chem. Soc. 98:6745 (1976).
29. G. S. Wilson, D. D. Swanson, J. T. Klug, R. G. Glass,
    M. D. Ryan and W. K. Musker, Electrochemical Oxidation
    of Some Mesocyclic Dithioethers and Related Compounds,
    J. Am. Chem. Soc. 101:1040 (1979).

NEW MOLYBDENUM COMPLEXES WITH SULFUR DONOR LIGANDS

J.A. Broomhead and J. Budge

Department of Chemistry, Faculty of Science
Australian National University
Canberra, A.C.T., Australia, 2600

INTRODUCTION

Sulfur donors are prominent even in crude chemical descriptions of nitrogenase and are likely to be of importance in other molybdoenzymes. Recently, extended X-ray absorption fine structure studies have confirmed that molybdenum is indeed coordinated to sulfur in the Mo-Fe protein derived from Clostridium pasteurianum.[1] A study of the fundamental chemistry of Mo-S donor complexes is a logical accompaniment to understanding the more complex enzyme systems. For this reason, we are investigating molybdenum complexes in which the predominant donor is sulfur and where the coligands are those having significance in nitrogen fixation research. For example, one suggested pathway for reduction of N_2 to ammonia involves N_2 bridging between two molybdenum atoms to produce, in turn, bridged diazene and hydrazine complex intermediates and finally, a labile ammonia complex.[2]

DITHIOCARBAMATO COMPLEXES

Since dinitrogen is isoelectronic with both CO and NO^+, it seems reasonable that reactions of cis-$Mo(NO)_2(S_2CNR_2)_2$ might give rise to analogous dinitrogen complexes as well as complexes with other ligands which are substrates for nitrogenase. Although no dinitrogen complexes are obtained, the reaction of cis-$Mo(NO)_2$-$(S_2CNR_2)_2$ with azide or cyanate leads to the isolation of complexes of the type $MoA(S_2CNR_2)_2(Me_2SO)(NO)$ (A = N_3^-, NCO^-).[3] Isotopic labelling experiments and kinetic studies have established a mechanism for these somewhat complex reactions.[4]

The product complexes, as represented by $Mo(NCO)(S_2CNEt_2)_2$-$(Me_2SO)(NO)$, display a variety of substitution reactions of Me_2SO ligands.[5] The new mononitrosyl complexes are all yellow, diamagnetic, air-stable materials formulated as seven-coordinate complexes on the basis of analyses, ^{1}H nmr and infrared measurements. Two idealized geometries may be considered, the pentagonal bipyramid and the monocapped trigonal prism and, at present, structural information favors the first of these. Of the ten possible geometrical isomers (Figure 1), structure (a) has been confirmed for $Mo(NCO)(S_2CNEt_2)_2$-$(Me_2SO)(NO)$ $(L_1=Me_2SO,\ L_2=NCO)$[3] while (b) is preferred for the anion $[Mo(NCO)_2(S_2CNEt_2)_2(NO)]^-$. Evidence for structure (b) comes from the equivalence of the dithiocarbamate ligands in the ^{1}H nmr spectrum and the presence of only one infrared band attributable to NCO^- (vCN at $2220cm^{-1}$) both in solution and KBr disks. Upon replacement of Me_2SO by hydrazine, a product analyzing for $[Mo(NCO)(S_2CNEt_2)_2(NO)]_2N_2H_4$ is obtained. Two NH proton singlet resonances (δ 4.30, 4.59 ppm with integration 1.5:2.5) are observed and the dithiocarbamate proton resonances are broad. This complex probably has a hydrazine-bridged structure analogous to that found for $[Mo(S_2CNEt_2)_2(CO)_2]_2N_2H_4 \cdot CH_2Cl_2$.[6] In the latter compound, only one NH resonance (at $\delta 5.07$ ppm) is found and the dithiocarbamate proton resonances are sharp. These observations suggest that various isomers are present in the nitrosyl compound.

The ability of cis-$Mo(CO)_2(S_2CNEt_2)_2$ to add a further molecule of CO was first reported by Colton, Scollary and Tomkins.[7] Later, McDonald, Newton and others[8] described reactions with various acetylenes and with the diazene, diethyl azodicarboxylate, in which the CO ligands were displaced. The acetylene complexes and the diethyl azodicarboxylate analogues are specially interesting in that they could be reduced to ethylene or hydrazines respectively. We now report reactions of other N and P donor ligands with cis-$Mo(CO)_2(S_2CNEt_2)_2$ to give seven-coordinate products, $MoY(CO)_2(S_2CNEt_2)_2$ $(Y = NH_3, Me_2NNH_2, PhCONHNH_2, pyridine, Ph_3P)$. If the group Y possesses suitable additional donor atoms, then bridged complexes are obtained analogous to the hydrazine compound, viz. $[Mo(CO)_2(S_2CNEt_2)_2]_2B$ (B = ethylenediamine, N-methylhydrazine, pyrazine). These new complexes are orange or red and have been characterized by analysis, ^{1}H nmr and infrared measurements (Tables 1 and 2).

All these reactions reflect the tendency of the six-coordinate Mo(II)-sulfur center to increase its electron configuration from 16 to 18 by an increase in coordination number. The N sigma-donor ligands, in particular, are only weakly bound to the metal. For example, bright orange $Mo(NH_3)(CO)_2(S_2CNEt_2)_2$ loses ammonia upon recrystallization from $CH_2Cl_2/MeOH$. A long Mo-N bond (2.39Å) is found in the bridged hydrazine complex and similar long bond lengths are likely for the other N sigma-donor ligands. The

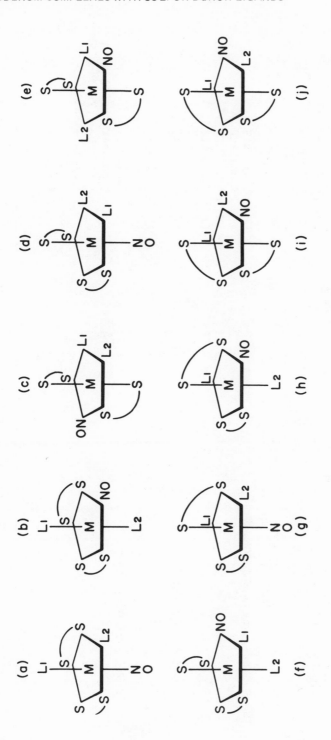

Geometrical Isomers for Pentagonal Bipyramidal Complexes

TABLE 1 : ^{1}H NMR SPECTRA

Compound	Chemical Shift [1]	Assignment
$[Mo(S_2CNEt_2)_2(CO)_2]_2\text{-}\mu\text{-}N_2H_3CH_3$	t 1.25	CH_3 of S_2CNEt_2
	s 2.31	CH_3 of $N_2H_3CH_3$
	q 3.74	CH_2 of S_2CNEt_2
	s 4.64	NH of $N_2H_3CH_3$
$[Mo(S_2CNEt_2)_2(CO)_2\{NH_2N(CH_3)_2\}]$	t 1.23	CH_3 of S_2CNEt_2
	s 2.50	CH_3 of $NH_2N(CH_3)_2$
	s 3.21	NH_2 of $NH_2N(CH_3)_2$
	q 3.77	CH_2 of S_2CNEt_2
$[Mo(S_2CNEt_2)_2(CO)_2NH_3]$	t 1.25	CH_3 of S_2CNEt_2
	s 2.42	NH_3
	q 3.72	CH_2 of S_2CNEt_2
$[Mo(S_2CNEt_2)_2(CO)_2]_2\text{-}\mu\text{-}en.CH_2Cl_2$ ^{2}	t 1.16	CH_3 of S_2CNEt_2
	m 2.64	CH_2 of en overlapping with residual proton peak of $(CD_3)_2SO$
	s,br 3.20	NH_2 of en
	q 3.66	CH_2 of S_2CNEt_2
	s 5.75	CH_2Cl_2
		Contd.

[1] Reference TMS in $CDCl_3$; $\delta \pm 0.02$ ppm, correct proton counts were obtained in each case. The abbreviations are t, triplet; d, doublet; s, singlet; q, quartet; m, multiplet; p, pair of.

2 in $(CD_3)_2SO$.

TABLE 1 Contd.

Compound	Chemical Shift[1]	Assignment
[Mo(S_2CNEt$_2$)$_2$(CO)$_2$(NH$_2$NHSO$_2$C$_6$H$_4$Me)]	t 1.25	CH$_3$ of S$_2$CNEt$_2^-$
	s 2.47	CH$_3$ of p tolyl
	q 3.73	CH$_2$ of S$_2$CNEt$_2^-$
	s 4.22(br)	NH$_2$ of hydrazine
	s 6.22(br)	NH of hydrazine
	7.36, 7.44, 7.77, 7.86	aromatic protons
[Mo(S_2CNEt$_2$)$_2$(CO)$_2$(N$_2$H$_3$CH$_3$)]	t 1.26	CH$_3$ of S$_2$CNEt$_2^-$
	s 2.72	CH$_3$ of N$_2$H$_3$CH$_3$
	q 3.74	CH$_2$ of S$_2$CNEt$_2^-$
	s 4.08	NH of N$_2$H$_3$CH$_3$
[Mo(S_2CNEt$_2$)$_2$(CO)$_2$(NH$_2$NHCOΦ)]	t 1.13	CH$_3$ of S$_2$CNEt$_2^-$
	q 3.64	CH$_2$ of S$_2$CNEt$_2^-$
	ps 5.28	overlapping CH$_2$Cl$_2$ and NH$_2$
	m 7.4 – 7.84	aromatic protons
	s 8.38	NH
[Mo(S_2CNEt$_2$)$_2$(CO)$_2$Py]	t 1.13	CH$_3$ of S$_2$CNEt$_2^-$
	q 3.64	CH$_2$ of S$_2$CNEt$_2^-$
	m 7.34	3,5 protons of pyridine
	m 7.72	4 proton of pyridine
	m 9.00	2,6 protons of pyridine

Contd.

1 Reference TMS in CDCl$_3$; $\delta \pm 0.02$ ppm, correct proton counts were obtained in each case. The abbreviations are t, triplet; d, doublet; s, singlet; q, quartet; c, complex; m, multiplet; p, pair of.

TABLE 1   Contd.

Compound	Chemical Shift [1]	Assignment
$[Mo(S_2CNEt_2)_2(CO)_2]-\mu-pyz$	t 1.23	CH_3 of $S_2CN\overline{Et_2}$
	q 3.73	CH_2 of $S_2CN\overline{Et_2}$
	s 8.80	pyz protons
$[Mo(S_2CNEt_2)_2(CO)_2PPh_3]$	m 0.94	CH_3 of hexane
	t 1.17	CH_3 of $S_2CN\overline{Et_2}$
	q 3.69	CH_2 of $S_2CN\overline{Et_2}$
	s 5.28	CH_2Cl_2
	m 7.34	PPh_3
$[Mo(S_2CNEt_2)_2CO(dppe)]$	t 1.01	CH_3 of $S_2CN\overline{Et_2}$
	m 2.51	CH_2 of dppe
	c 3.53	CH_2 of $S_2CN\overline{Et_2}$
	s 5.28	CH_2Cl_2
	7.1-7.8	aromatic protons
$[Mo\ NCO(S_2CNEt_2)_2(NO)(NH_2NHSO_2C_6H_4Me)]$	t 1.32	CH_3 of $S_2CN\overline{Et_2}$
	s 2.45	CH_3 of p-tolyl
	dq 3.76	CH_2 of $S_2CN\overline{Et_2}$
	d 4.13	NH_2 of hydrazine
	s 6.56	NH of hydrazine
	7.52 (four)	aromatic protons

[1] Reference TMS in $CDCl_3$; $\delta \pm 0.02$ ppm, correct proton counts were obtained in each case. The abbreivations are t, triplet; d, doublet; s, singlet; q, quartet; c, complex; m, multiple; p, pair of.

pyridine and pyrazine ligands are more firmly bound. Interestingly, when the bidentate phosphine, dppe ($Ph_2PCH_2CH_2PPh_2$), is used, a seven-coordinate bridged dicarbonyl complex analogous to the ethylenediamine case is <u>not</u> obtained. Instead, chelation occurs to form the mono-carbonyl complex, $Mo(CO)(S_2CNEt_2)_2(dppe)$. The infrared data (Table 2) show the characteristic modes of the N-N-diethyldithiocarbamato group and the coligands. Except for the phosphine coligand complexes, they all display three rather than two carbonyl absorptions. The highest frequency band varies in intensity, even in separate preparations of the same complex, while the remaining carbonyl bands are invariably of strong intensity. In CH_2Cl_2 solutions, three bands are still observed so that solid state effects are not responsible. The previously reported structure of $[Mo(S_2CNEt_2)_2(CO)_2]_2N_2H_4 \cdot CH_2Cl_2$ [6] shows that the carbonyl groups are <u>cis</u> and consequently, only two carbonyl absorptions are predicted and are found for crystals grown slowly from CH_2Cl_2/MeOH. However, material deposited more rapidly exhibits three carbonyl bands (Table 2). The above evidence is consistent with the presence of various isomeric forms in solution and, together with the nmr data, implies a degree of fluxionality about the Mo(II) center in these systems.

ATTEMPTED REACTIONS OF COORDINATED LIGANDS

The p-toluenesulfonylhydrazide complexes are of particular interest since this ligand is known to generate diazene by reaction with methoxide.[9] Reaction of $Mo(CO)_2(S_2CNEt_2)_2(NH_2NHSO_2C_6H_4CH_3)$ with methanol is, however, rapid and if any diazene complex is formed, it could not be detected by nmr. The parent dicarbonyl complex is isolated. With $Mo(NCO)(S_2CNEt_2)_2(NO)(NH_2NHSO_2C_6H_4CH_3)$, a slow reaction occurs but again no diazene containing products could be detected. Diazene complexes of chromium and manganese have been prepared by Sellman[10] using a copper(II)-H_2O_2 oxidation of bridged hydrazine complexes. However, a chemical parallel is not found for the bridged molybdenum hydrazine complex which decomposes during oxidation. Also, the ammonia complex could not be obtained by reduction of the bridged hydrazine complex, using H_2/Pt black or dithionite. In this respect, Sellman also finds that bridged hydrazine is resistant to reduction although terminal hydrazine complexes could be reduced.

CONCLUSIONS

Diazene and dinitrogen complexes of Mo(II) sulfur centers are likely to be unstable whereas both bridging hydrazine and terminal substituted hydrazine complexes, as well as the ammonia complex, have now been characterized. Further, the ammonia complex is labile and the complexes as a group show fluxional behavior. These properties fit the basic requirements of any catalytic system which is to involve molybdenum. The presence of CO and NO^+ ligands is likely to be

TABLE 2 : INFRARED SPECTRAL DATA (cm^{-1})

Complex	ν(CO)	ν(CN)[1]	ν(CS$_2$)		ν(NH)[2]
[Mo(S$_2$CNEt$_2$)$_2$(CO)$_2$][3]	2026w,1935s,1848s	1514s	1005w	1150m	–
[Mo(S$_2$CNEt$_2$)$_2$(CO)$_2$]$_2$-μ-N$_2$H$_4$ [4]	2022w,1927s,1843s	1503s	1003w	1148m	3264w,3211,3126vw
[Mo(S$_2$CNEt$_2$)$_2$(CO)$_2$]$_2$-μ-N$_2$H$_3$CH$_3$	2022w,1930s,1844s	1502s	1003w	1148m	–
[Mo(S$_2$CNEt$_2$)$_2$(CO)$_2${NH$_2$N(CH$_3$)$_2$}]	2020m,1927s,1839s	1502s	1004w	1148m	3265w
[Mo(S$_2$CNEt$_2$)$_2$(CO)$_2$NH$_3$]	2024w,1930sh,1910s, 1835s	1503s	1002w	1151m	3380w,3275w,3160wbr
[Mo(S$_2$CNEt$_2$)$_2$(CO)$_2$]$_2$-μ-en [5]	2024w,1925s,1847s	1495s	1009w	1146m	3313m,3249w
[Mo(S$_2$CNEt$_2$)$_2$(CO)$_2$(NH$_2$NHCOPh)]	2024m,1934s,1840s, 1667[6]	1503s	1008w	1151m	–
[Mo(S$_2$CNEt$_2$)$_2$(CO)$_2$(Py)]	2024m,1918s,1838s	1498s	1005w	1147m	–
[Mo(S$_2$CNEt$_2$)$_2$(CO)$_2$]$_2$-μ-Pyz	2024s,1929vs,1851s	1505s, 1509s	1004w	1150m	–
[Mo(S$_2$CNEt$_2$)$_2$(CO)$_2$PPh$_3$]	1934s,1851s	1518s	1001w	1149m	–
[Mo(S$_2$CNEt$_2$)$_2$CO(dppe)] [7]	1816s,1797sh	1487s	1008w	1145m	–
Mo(S$_2$CNEt$_2$)$_2$(CO)$_2$(NH$_2$NHSO$_2$C$_6$H$_4$Me)					

1 of dithiocarbamate   2 hexachlorobutadiene mull   3 Ref. 7   4 Ref. 6

5 en = ethylenediamine   6 of NH$_2$NHCOPh   7 Ref. 7   dppe = Ph$_2$PCH$_2$CH$_2$PPh$_2$

inimical to the coordination of N_2 and diazene, both of which make demands on the metal for pi back-bonding capability. We have, therefore, explored reactions designed to remove CO groups from individual complexes to be replaced by N_2. Although cis-Mo(CO)$_2$(S$_2$CNEt$_2$)$_2$ and Mo(CO)(S$_2$CNEt$_2$)$_2$(dppe) will react with Rh(Cl(PPh$_3$)$_3$ and with Me$_3$NO to lose the carbonyls, the Mo-S fragments generated decompose rather than form corresponding dinitrogen complexes. It is interesting that there is only one example of a molybdenum-dinitrogen complex which involves S donors; the very unstable deep-purple cis-Mo(N$_2$)$_2$(PMe$_2$Ph)$_2$(PhSCH$_2$CH$_2$SPh) reported by Aresta and Sacco.[11]

EXPERIMENTAL SECTION

Materials: Complete descriptions of the source materials for this work are given elsewhere.[3-5] All preparations use deoxygenated solvents in a dinitrogen atmosphere.

Dicarbonylbis(N,N-diethyldithiocarbamato)molybdenum(II). The directions given in reference 6 were followed until precipitation of the mixture of di- and tricarbonyl complexes. At this stage, the mixture was either: (a) filtered, washed with methanol and dried under vacuum which gives [Mo(S$_2$CNEt$_2$)$_2$(CO)$_2$]; or (b) re-dissolved by adding dichloromethane (20 cc) and the resulting solution (S) used in the following preparations.

μ-Methylhydrazinebis[dicarbonylbis(N,N-diethyldithiocarbamato)molyb-denum(II)]

Methylhydrazine (0.4 g, 8.68×10^{-3} mol) was added to S and the solution became dark-red. N_2 was passed through the solution for 0.5 h and the dark-red precipitate was then filtered off and washed with methanol. The product was dissolved in CH_2Cl_2 (10 cc) and the solution was filtered. On addition of n-hexane, small dark-red needles were obtained and these were filtered off and washed with CH_2Cl_2/nC$_6$H$_{14}$ (1:10). The product was dried at 0°C, under vacuum, for 9 h. The yield was 0.93 g (52%). (Found: C, 31.50; H, 5.04; N, 8.88; S, 27.02. $C_{25}H_{46}Mo_2N_6S_8O_4$ requires: C, 31.84; H, 4.92; N, 8.91; S, 27.20%).

Dicarbonylbis(N,N-diethyldithiocarbamato)(N,N-dimethylhydrazine)-molybdenum(II)

[Mo(S$_2$CNEt$_2$)$_2$(CO)$_2$] (0.52 g, 1.16×10^{-3} mol) was dissolved in CH_2Cl_2 (15 cc) and N,N-dimethylhydrazine (2 g, 3.3×10^{-2} mol) added. The solution was filtered and n-hexane containing methylhydrazine (2 g NH$_2$NMe$_2$ in 150 cc) was added gradually to precipitate an orange crystalline product. The crystals were filtered off and washed with the methylhydrazine-hexane solution and dried under vacuum at 0°C for 22 h. (Found: C, 33.45; H, 5.87; N, 11.56; S, 24.35.

$C_{14}H_{28}MoN_4S_4O_2$ requires C, 33.09; H, 5.55; N, 11.03; S, 25.24%).

Amminedicarbonylbis(N,N-diethyldithiocarbamato)molybdenum(II).

Ammonia was bubbled through S to give an orange precipitate, which was filtered off and washed with methanol. The product was dissolved in CH_2Cl_2 (10 cc) saturated with NH_3 and the solution was filtered. The addition of methanol (50 cc) gave the product as orange needles. These were filtered off, washed with methanol and dried under vacuum at 0°C for 9 h. The yield was 0.98 g (56%). (Found: C, 30.74; H, 4.88; N, 9.11; S, 27.44. $C_{12}H_{23}MoN_3S_4O_2$ requires C, 30.96; H, 4.98; N, 9.03; S, 27.55%). Recrystallization requires CH_2Cl_2 saturated with NH_3 otherwise the complex obtained loses NH_3 on vacuum storage.

μ-Ethylenediaminebis[dicarbonylbis(N,N-diethyldithiocarbamato)molybdenum(II)]

To a solution of $[Mo(S_2CNEt_2)_2(CO)_2]$ (0.53 g, 1.18×10^{-3} mol) in CH_2Cl_2 (20 cc) was added ethylenediamine (0.04 g, 6.7×10^{-4} mol) in CH_2Cl_2 (2 cc). After filtering the solution, methanol was added to deposit an orange crystalline product, which was filtered off, washed with methanol and dried as above. (Found: C, 30.78; H, 4.79; N, 8.10; S, 24.70; Cl, 6.36. $C_{27}H_{50}Mo_2N_6S_8O_4Cl_2$ requires C, 31.12; H, 4.84; N, 8.07; S, 24.61; Cl, 6.80%). The analysis was calculated for one molecule of dichloromethane of crystallization.

Benzoylhydrazinedicarbonylbis(N,N-diethyldithiocarbamato)molybdenum(II)

Benzoylhydrazine (0.51 g, 3.75×10^{-3}) in CH_2Cl_2 (10 cc) was added to S and the solution became dark-red. The solvent was evaporated in an N_2 stream and the residue was taken up in CH_2Cl_2 (15 cc). n-Hexane was added to the filtered solution to deposit a red crystalline product which was filtered off and washed with CH_2Cl_2/ n-hexane (1:3). The compound was dried as above to yield 1.3 g of product (2.10×10^{-3} mol, 53%). (Found: C, 37.82; H, 4.73; N, 9.37; S, 19.39; Cl, 4.66. $C_{19.4}H_{28.8}MoN_4S_4O_3Cl_{0.8}$ requires C, 37.66; H, 4.69; N, 9.06; S, 20.73; Cl, 4.58%). The analysis was calculated for 0.4 molecules of CH_2Cl_2.

Dicarbonylbis(N,N-diethyldithiocarbamato)pyridinemolybdenum(II)

$[Mo(S_2CNEt_2)_2(CO)_2]$ (0.42 g, 9.3×10^{-4} mol) was dissolved in CH_2Cl_2 (10 cc) and pyridine (0.1 g, 1.26×10^{-3} mol) in CH_2Cl_2 (2 cc) was added. The solution went deep red and after addition of n-hexane, the orange crystalline product was filtered off, washed and dried as above. (Found: C, 39.02; H, 4.95; Mo, 18.09; N, 7.94; S, 24.01. $C_{17}H_{25}MoN_3S_4O_2$ requires C, 38.70; H, 4.78; Mo, 18.19; N, 7.97; S, 24.31%).

μ-Pyrazinebis[dicarbonylbis(N,N-diethyldithiocarbamato)molybdenum(II)]

[Mo(S$_2$CNEt$_2$)$_2$(CO)$_2$] (0.40 g, 8.92x10^{-4} mol) was dissolved in CH$_2$Cl$_2$ (15 cc) and pyrazine (0.04 g, 5x10^{-4} mol) in CH$_2$Cl$_2$ (1 cc) added slowly. The solution became dark red. After filtration, n-hexane was added to precipitate a deep red crystalline material which was filtered off, washed and dried as above. (Found: C, 35.54; H, 4.79; Mo, 19.46; N, 8.26; S, 26.16. C$_{28}$H$_{44}$Mo$_2$N$_6$S$_8$O$_4$ requires C 34.42; H, 4.54; Mo, 19.65; N, 8.60; S, 26.25%).

Dicarbonylbis(N,N-diethyldithiocarbamato)triphenylphosphinemolybdenum(II)

Triphenylphosphine (0.99 g, 3.78x10^{-3} mol) was added to S and the solution became dark red. The solvent was evaporated in N$_2$ stream and the residue dissolved in CH$_2$Cl$_2$ (20 cc) and filtered. After the addition of n-hexane, a bright-red crystalline product was filtered off and washed with CH$_2$Cl$_2$/n-hexane (1:5) and recrystallized as above. The yield was 1.8 g. (Found: C, 52.73; H, 5.69; N, 3.85; S, 17.45; P, 3.99; Cl, 0.33. C$_{33.64}$H$_{43.48}$MoN$_2$S$_4$O$_2$PCl$_{0.08}$ requires C, 52.75; H, 5.72; N, 3.66; S, 16.75; P, 4.04; Cl, 0.37%). The analysis has been calculated for 0.6 molecules of C$_6$H$_{14}$ and 0.04 molecules of CH$_2$Cl$_2$ of crystallisation. The ^{1}H nmr spectrum of the initial product suggested that approximately 0.25 molecules of CH$_2$Cl$_2$ was present. Evidently, the complex may crystallize with a variable amount of solvent.

Carbonylbis(N,N-diethyldithiocarbamato)(1,2-bis{diphenylphosphino}ethane)molybdenum(II)

1,2-Bis(diphenylphosphino)ethane (1.51 g, 3.78x10^{-3} mol) in CH$_2$Cl$_2$ (10 cc) was added to S. Gas evolution was evident and the solution became dark red. The solvent was evaporated off in N$_2$ stream and the residue dissolved in CH$_2$Cl$_2$ (10 cc). After filtration, n-hexane was added to deposit a dark red crystalline compound, which was filtered off, washed and recrystallized as above. (Found: C, 54.41; H, 5.75; N, 3.47; S, 15.12; P, 6.94. C$_{37}$H$_{44}$MoN$_2$S$_4$P$_2$O requires C, 54.29; H, 5.42; N, 3.42; S, 15.66; P, 7.56%). The ^{1}H nmr spectrum of the initial product indicated about 0.15 molecules CH$_2$Cl$_2$. However, the analysis is consistent with no solvent and again, it appears that a variable amount of CH$_2$Cl$_2$ may be occluded.

Dicarbonylbis(N,N-diethyldithiocarbamato)p-toluenesulfonylhydrazinemolybdenum(II)

[Mo(S$_2$CNEt$_2$)$_2$(CO)$_2$] (0.89 g, 2.1x10^{-3} mol) was dissolved in CH$_2$Cl$_2$ (12 cc) and p-toluenesulfonylhydrazine (0.31 g, 2.1x10^{-3} mol) added. The solution was stirred for 5 min, filtered and n-hexane added to the filtrate to give a red crystalline product. This was filtered off and washed with CH$_2$Cl$_2$/n-hexane (1:3, 3x10 cc). The yield was

0.73 g. The product was recrystallized from these solvents and dried at 0°C under vacuum for 9 hr. (Found: C, 35.8; H, 5.0; N, 8.3; S, 24.4. $C_{19}H_{31}MoN_4S_4O_4$ requires C, 35.9; H, 4.9; N, 8.8; S, 25.2%).

Cyanotobis(N,N-diethyldithiocarbamato)nitrosyl-p-toluenesulfonyl-hydrazinemolybdenum

A solution of $MoNCO(S_2CNEt_2)_2(Me_2SO)(NO)$ (0.1 g, 0.184×10^{-3} mol prepared as in reference 3) and $NH_2NHSO_2C_6H_4Me$ (0.034 g, 0.184×10^{-3} mol) in CH_2Cl_2 (5 cc) was stirred for 5 min and filtered. Carbon tetrachloride was added to give yellow crystals which were washed with dichloromethane/carbon tetrachloride (1:9). The yield was 0.06 g. (for ^{1}H nmr see Table 1).

Physical Measurements: The infrared spectral data in Table 2 were recorded on a Unicam SP200G instrument for CH_2Cl_2 solutions and KBr discs unless otherwise indicated. [1]H nmr spectra were obtained with a Jeol MH-100 spectrometer.

REFERENCES

1. S.P. Cramer, K.O. Hodgson, W.O. Gillum and L.E. Mortenson, The Molybdenum Site of Nitrogenase. Preliminary Structural Evidence from X-Ray Absorption Spectroscopy, J.Amer.Chem. Soc., 100:3398 (1978).

2. J. Chatt and R.L. Richards, Dinitrogen Complexes and Nitrogen Fixation, in "The Chemistry and Biochemistry of Nitrogen Fixation", J.R. Postgate, ed., Plenum Press, London (1971) p.57.

3. J.A. Broomhead, J.R. Budge, W. Grumley, T. Norman and M. Sterns, The Synthesis and Characterization of Azido and Cyanato Dithiocarbamato Nitrosyl Complexes of Molybdenum: Reduction of Coordinated Azide in Azidobis(N,N-diethyldithiocarbamato)(dimethylsulfoxide)nitrosylmolybdenum, Aust.J. Chem., 29:275 (1976).

4. J.A. Broomhead and J.R. Budge, Mechanism of the Reaction of cis-Bis(N,N-diethyldithiocarbamato)dinitrosylmolybdenum with Azide or Cyanate and Related Reactions, Inorg.Chem., 17:2414 (1978).

5. J.A. Broomhead and J.R. Budge, Molybdenum and Tungsten Nitrosyl Complexes with Dithiocarbamate Ligands, Aust.J.Chem., 32:000 (1979).

6. J.A. Broomhead, J.R. Budge, J.H. Enemark, R.D. Feltham, J.I. Gelder and P.L. Johnson, Preparation and X-Ray Structure of a Hydrazine-Bridged Dinuclear Molybdenum Complex $[Mo(S_2CN(C_2H_5)_2(CO)_2]_2N_2H_4 \cdot CH_2Cl_2$, Adv.Chem.Ser., 162:421 (1977).

7. R. Colton, G.R. Scollary and I.B. Tomkins, Carbonyl Halides of
    the Group VI Transition Metals. V. Carbon Monoxide Carriers
    and some Sulfur and Nitrogen Derivatives of Molybdenum
    Halocarbonyls, Aust.J.Chem., 21:15 (1968).

8. J.W. McDonald, W.E. Newton, C.T.C. Creedy and J.L. Corbin,
    Binding and Activation of Enzymic Substrates by Metal
    Complexes. III. Reactions of $Mo(CO)_2[S_2CN(C_2H_5)_2]_2$,
    J.Organometal Chem., 92:C25 (1975).

9. T.A. Geissman, "Principles of Organic Chemistry", 3rd ed.,
    W.H. Freeman, San Francisco (1968) p.227.

10. D. Sellman, Syntheses, Properties and Reactions of Dinitrogen,
    Diimine, Hydrazine and Ammonia Complexes of Transition
    Metals, in "Recent Developments in Nitrogen Fixation",
    W. Newton, J.R. Postgate and C. Rodriguez-Barrueco, ed.,
    Academic Press, London (1977) p.52.

11. M. Aresta and A. Sacco, Nitrogen Fixation III. Dinitrogen-,
    carbonyl-, and hydrido, complexes of molybdenum, Gazz.Chim.
    Italiana, 102:755 (1972).

STRUCTURAL STUDIES OF MOLYBDENUM COMPLEXES

K. Yamanouchi, J. T. Huneke and John H. Enemark

Department of Chemistry
University of Arizona
Tucson, Arizona 85721 USA

INTRODUCTION

Molybdenum-containing enzymes are vitally important in
nitrogen uptake, nitrogen metabolism and several biochemical
reactions.[1,2] The detailed coordination environment about the
molybdenum atoms of such enzymes is still unknown because no
molybdoenzyme has had its structure determined by x-ray
crystallography. However, the molybdenum centers of several
enzymes have been probed directly by electron paramagnetic
resonance (EPR) spectroscopy[1] and by x-ray absorption
spectroscopy (EXAFS).[3-5] Comparison of the spectral data from
enzymes with spectral data from coordination compounds suggests
that sulfur atoms are coordinated to molybdenum in the enzymes.
EXAFS data from xanthine oxidase[4] and sulfite oxidase[5] in their
oxidized forms also indicate the presence of terminal oxo groups
attached to the molybdenum. In order to effectively interpret
the spectral results from enzymes, it is important to have
available a series of compounds whose stoichiometry and
stereochemistry have been definitively established by x-ray
structure determination. Of particular interest are compounds
of sulfur donor ligands and oxomolybdenum(V) species.

SULFUR RICH COMPLEXES

At the time we initiated this work, there were relatively
few structural studies of complexes in which the molybdenum atoms
were ligated exclusively by sulfur atoms.[2] Definitive
stereochemical information for Mo-S distances on this class of
compounds is of interest in view of recent EXAFS data for

nitrogenase which provide convincing evidence for sulfur ligation of molybdenum and which rule out the presence of Mo=O groups.[3]

Figure 1 shows two isomeric forms of the binuclear complex $[Mo_2S_4(dme)_2]^{2-}$, which contains three different kinds of Mo-S distances.[6] The $Mo_2S_4(dme)_2^{2-}$ anion is the first example of syn and anti isomers of the $Mo_2S_4^{2+}$ core being isolated with the same ligand system. The structure of syn-$Mo_2S_4(S_2CNEt_2)_2$ has also been determined.[7] These structure determinations establish the Mo-S_t distance for a terminal sulfido group is $\sim$2.10 Å. Thus, the bonding can be written as Mo=S, analogous to Mo=O.[8]

Sulfur-rich molybdenum complexes are known to result from the reaction of dithiolene ligands with Mo(VI).[9,10] Recently, this reaction has been reinvestigaged by Bravard and Newton. One of the products is $Mo_2(S_2)(S_2C_2Ph_2)_4$, and a stereoview of the structure[11] is shown in Figure 2. The two Mo atoms are bridged by four S atoms. Two of the bridging S atoms are supplied by the two bridging dithiolene ligands, and two of the bridging atoms are supplied by the bridging S_2 unit. The coordination about each molybdenum is completed by the two S atoms from a chelating dithiolene and one S atom from a bridging dithiolene ligand. Thus, each Mo atom is coordinated by seven S atoms. The approximate stereochemistry about each Mo atom is a monocapped trigonal prism. The two prisms share a rectangular face formed by the four bridging S atoms. The Mo...Mo distance is 2.778(1) Å. The approximate symmetry of $Mo_2(S_2)(S_2C_2Ph_2)_4$ is C_2. Selected distances are shown in Table I. The S(9)-S(10) distance is 2.044(3) Å. Two other examples of binuclear molybdenum complexes bridged by four S atoms have recently been reported.[12,13]

Table I. Selected Distances in $Mo_2(S_2)(S_2C_2Ph_2)_4$

Mo(1)-Mo(2)	2.778(1)	S(9)-S(10)	2.044(3)
Mo(1)-S(1)	2.454(2)	Mo(2)-S(3)	2.438(2)
Mo(1)-S(2)	2.442(2)	Mo(2)-S(4)	2.454(2)
Mo(1)-S(3)	2.468(2)	Mo(2)-S(1)	2.499(2)
Mo(1)-S(5)	2.371(2)	Mo(2)-S(7)	2.353(2)
Mo(1)-S(6)	2.375(2)	Mo(2)-S(8)	2.382(2)
Mo(1)-S(9)	2.503(2)	Mo(2)-S(10)	2.485(2)
Mo(1)-S(10)	2.428(2)	Mo(2)-S(9)	2.421(2)
S(1)-S(3)	2.742(2)		

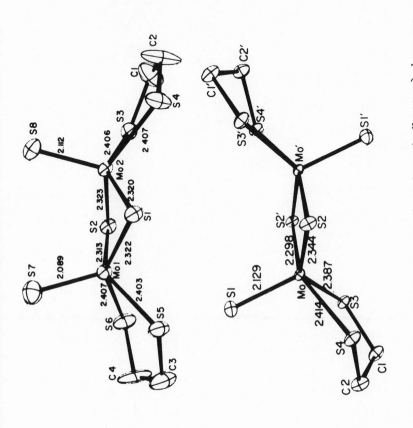

Fig. 1. Perspective view of the syn (top) and anti (bottom) isomers of the $Mo_2S_4(dme)_2^{2-}$ anion showing selected bond lengths (dme is the dianion of 1,2-dimercaptoethane). Reproduced from ref. 6.

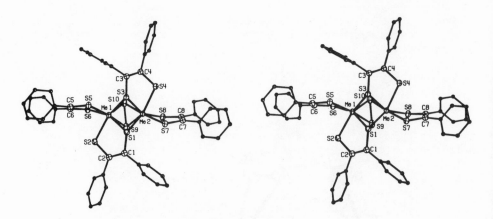

Fig. 2. Stereoview of the $Mo_2(S_2)(S_2C_2Ph_2)_4$ molecule parallel
to the approximate C_2 axis.

MONOMERIC OXO-MOLYBDENUM(V) COMPLEXES

There are relatively few structural studies of monomeric oxo-
molybdenum(V) complexes, especially of complexes which also
contain sulfur as a donor atom.[2] There are two reports of the
structures of five-coordinate complexes, $MoOCl_3(SPPh_3)$ and
$[MoO(SPh)_4]^-$, which contain sulfur-donor atoms.[14,15] Recently,
we have determined the structures of several six-coordinate
oxomolybdenum(V) complexes[16-19] including $[MoOCl_4(H_2O)]^-$,
$MoOCl_3(ox)$, $[MoOCl_2(salphO)]^-$ and $MoOCl(tox)_2$, (ox = 8-hydroxy-
quinolinate, salphO = N-2-hydroxyphenylsalicylideneiminate and
tox = 8-mercaptoquinolinate).

The sterochemistry of $MoOCl(tox)_2$ is of particular interest
because it is a six-coordinate oxomolybdenum(V) compound with a
sulfur donor ligand whose solution EPR spectrum mimics some of
the features of xanthine oxidase.[20] A complex of this
stoichiometry having trans stereochemistry was first proposed
from solution EPR studies of Marov et al.[21] More recently, pure
compounds of this stoichiometry have been isolated and the
kinetics of their oxidation by nitrate investigated.[20] We have
succeeded in obtaining single crystals of $MoOCl(tox)_2$ and its
structure is shown in Figure 3.[19] Note that the oxygen and
chlorine atoms are cis to each other, and not trans as originally
proposed from solution studies.[21] The sulfur atoms are mutually
trans to one another and cis to the terminal oxo group. One
nitrogen atom is trans to the chlorine atom and the other is
trans to the oxo group. The Mo-N distance trans to the oxo group
is ~0.2 Å longer than the Mo-N distance trans to Cl. Exposure

of solutions of MoOCl(tox)$_2$ to air results in oxidation of the complex to MoO$_2$(tox)$_2$, a Mo(VI) complex.[19] The structure of MoO$_2$(tox)$_2$ is shown in Figure 4. The two oxo groups are cis to one another; the two sulfur atoms are mutually trans to one another and cis to the oxo groups. The two nitrogen atoms are trans to the oxo groups with Mo-N distances of 2.382(9) and 2.374(9) Å. The stereochemistries of the molybdenum(V) and molybdenum(VI) complexes are remarkably similar (Tables II, III), the major difference being the Mo-N(1) distance when the atom is trans to chloride versus trans to oxygen. The bond angles in both structures deviate substantially from those in an ideal octahedron. For example, in MoOCl(tox)$_2$ the O-Mo-S(1) angle of 106.3(2)° differs substantially from the O-Mo-S(2) angle of 89.2(2)°. Similar differences in O-Mo-S angles occur in MoO$_2$(tox)$_2$. In fact, irregular coordination geometries are a general feature of oxomolybdenum complexes containing chelate ligands.[22]

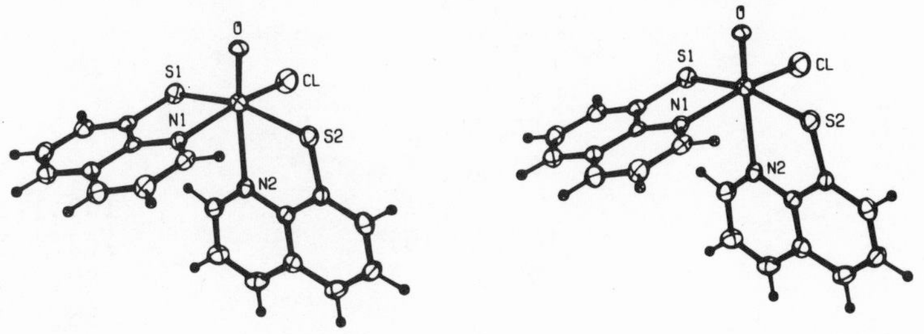

Fig. 3.   Stereoview of the MoOCl(tox)$_2$ molecule. The hydrogen atoms have been assigned arbitrarily small thermal parameters for clarity. Reproduced from ref. 19.

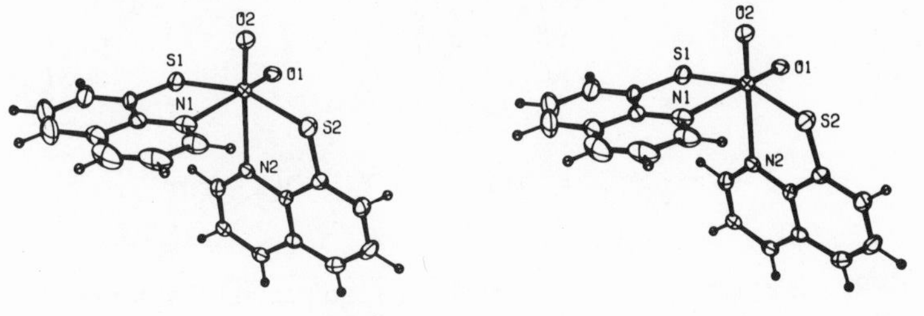

Fig. 4.   Stereoview of the MoO$_2$(tox)$_2$ molecule. The hydrogen atoms have been assigned arbitrarily small thermal parameters for clarity. Reproduced from ref. 19.

Table II. Interatomic Distances (Å) in MoOCl(tox)$_2$, $\underset{\sim}{1}$, and MoO$_2$-(tox)$_2$, $\underset{\sim}{2}$.

	($\underset{\sim}{1}$)	($\underset{\sim}{2}$)		($\underset{\sim}{1}$)	($\underset{\sim}{2}$)
Mo-X[a]	2.342(2)	1.694(6)	Mo-Y[a]	1.716(4)	1.712(7)
Mo-S(1)	2.403(2)	2.411(3)	Mo-S(2)	2.414(2)	2.411(3)
Mo-N(1)	2.210(6)	2.382(9)	Mo-N(2)	2.408(6)	2.374(9)

[a]X = Cl and Y = O for $\underset{\sim}{1}$. X = O(1) and Y = O(2) for $\underset{\sim}{2}$.

Table III. Interatomic Angles (deg) in MoOCl(tox)$_2$, $\underset{\sim}{1}$, and MoO$_2$-(tox)$_2$, $\underset{\sim}{2}$.

	($\underset{\sim}{1}$)	($\underset{\sim}{2}$)
S(1)-Mo-N(1)[a]	80.21(6)	76.7(3)
X-Mo-S(1)	86.26(8)	88.1(3)
X-Mo-S(2)	92.67(8)	102.1(3)
X-Mo-N(1)	161.28(15)	162.4(3)
X-Mo-N(2)	88.57(15)	89.7(3)
S(1)-Mo-N(2)	88.62(14)	88.1(2)
S(1)-Mo-S(2)	164.30(7)	160.9(1)
X-Mo-Y	102.96(17)	106.3(3)
S(2)-Mo-N(2)	75.69(14)	75.9(2)
Y-Mo-S(2)	89.24(17)	88.0(3)
Y-Mo-S(1)	106.28(18)	104.8(3)
Y-Mo-N(2)	161.53(21)	159.5(3)
Y-Mo-N(1)	93.27(21)	86.3(3)
S(2)-Mo-N(1)	96.86(16)	90.3(3)
N(1)-Mo-N(2)	78.29(19)	81.2(3)

[a]X = Cl and Y = O for $\underset{\sim}{1}$. X = O(1) and Y = O(2) for $\underset{\sim}{2}$.

The EPR spectrum of MoOCl(tox)$_2$ in solution is nearly axial.[20] However, in the solid state, the complex has Cl and θ atoms cis to one another and irregular geometry (Figure 3). Clearly, a much larger catalog of spectral and structural data for mononuclear molybdenum(V) complexes is needed before the EPR data from enzymes can be correlated with a specific molybdenum environment. The present study also reemphasizes the danger of attempting to deduce the stereochemistry about a transition metal center solely from EPR data.

The $MoOCl(tox)_2$ and $MoO_2(tox)_2$ complexes are the first pair of molybdenum(V) and molybdenum(VI) complexes to be structurally characterized. The similarity of their structures provides some support for one proposed pathway for the reaction of certain molybdoenzymes with substrates (reaction 1). In the forward reaction of (1), an oxomolybdenum(IV) center reduces the substrate by abstracting an oxygen atom to form a cis-dioxomolybdenum(VI) center. The reverse reaction, substrate oxidation, involves transfer of an oxygen atom from a cis-dioxomolybdenum(VI) center to the substrate leaving an oxomolybdenum(IV) center. Such oxo-transfer reactions have been observed for the oxidation of phosphines by molybdenum(VI)[23,24] but have not yet been generally observed in model systems. The cis-stereochemistry of $MoOCl(tox)_2$ and $MoO_2(tox)_2$ and the facile reductions of nitrate by a number of cis-oxochloromolybdenum(V) complexes[20,25] suggest that such a pathway is feasible for molybdoenzymes. However, additional studies on the stereochemistry and reactions of monomeric molybdenum(IV, V, VI) complexes of biologically relevant ligands are needed to provide further insight concerning this and other possible mechanisms for the action of molybdoenzymes.

$$RO + Mo^{IV} \overset{O}{\underset{}{\overset{\|}{\Longleftrightarrow}}} \quad R + O=Mo^{VI} \overset{\overset{O}{\|}}{} \tag{1}$$

ACKNOWLEDGEMENTS

We thank Drs. W. E. Newton, D. T. Sawyer, and J. T. Spence for samples and for helpful discussions. We also thank Drs. D. Chasteen, S. P. Cramer, K. O. Hodgson, E. I. Stiefel and A. G. Wedd for preprints of results and for helpful discussions. We are grateful to the University of Arizona Computing Center for a generous allocation of computer time, the National Institutes of Environmental Health Sciences (Grant ES 00966) for financial support, and Climax Molybdenum Company for gifts of molybdenum compounds. We acknowledge the National Science Foundation for funds for the diffractometer (Grant CHE-76-05481).

REFERENCES

1. R. C. Bray, Molybdenum Iron-Sulfur Flavin Hydroxylases and Related Enzymes, in "The Enzymes", P. D. Boyer, ed., Academic Press, New York (1975) p. 299.
2. E. I. Stiefel, The Coordination and Bioinorganic Chemistry of Molybdenum, Prog. Inorg. Chem., 22:1 (1977).

3.   S. P. Cramer, K. O. Hodgson, W. O. Gillum, and L. W.
     Mortenson, The Molybdenum Site of Nitrogenase.
     Preliminary Structural Evidence from X-Ray Absorption
     Spectroscopy, J. Am. Chem. Soc., 100:3398 (1978).

4.   T. D. Tulius, D. M. Kurtz, Jr., S. D. Conradson and K. O.
     Hodgson, The Molybdenum Site of Xanthine Oxidase.
     Structural Evidence from X-Ray Absorption Spectroscopy,
     J. Am. Chem. Soc., 101:2776 (1979).

5.   S. P. Cramer, H. B. Gray and K. V. Ragagopalan, The
     Molybdenum Site of Sulfite Oxidase. Structural
     Information from X-Ray Absorption Spectroscopy,
     J. Am. Chem. Soc., 101:2772 (1979).

6.   G. Bunzey and J. H. Enemark, Molybdenum Complexes of
     Aliphatic Thiols. The Structures of the Syn and Anti
     Isomers of the Di-μ-sulfido-bis(sulfido-1,2-
     dimercaptoethanatomolybdate(V)) Anion,
     $[Mo_2S_4(S_2C_2H_2)_2]^{2-}$ Inorg. Chem. 17:682 (1978;
     G. Bunzey, J. H. Enemark, J. K. Howie, and D. T.
     Sawyer, Molybdenum Complexes of Aliphatic Thiols. The
     Isolation and Characterization of Two Isomeric Forms
     of the Redox Active Binuclear Mo(V) Anion,
     $[Mo_2S_4(S_2C_2H_2)_2]^{2-}$ J. Am. Chem. Soc., 99:4168 (1977).

7.   J. T. Huneke and J. H. Enemark, The Mo=S Bond Distance in
     Di-μ-sulfidobis(sulfido-N,N-diethyldithiocarbamato-
     molybdenum(V)), Inorg. Chem., 17:3698 (1978).

8.   B. Spivack, Z. Dori, and E. I. Stiefel, The Crystal and
     Molecular Structure of a Mo(V) Complex Having a
     Multiple Bonded Terminal Sulfur Atom, Inorg. Nucl.
     Chem. Lett., 11:501 (1975).

9.   G. N. Schrauzer, V. P. Mayweg and W. Heinrich, Coordination
     Compounds with Delocalized Ground States. α-Dithio-
     diketone-Substituted Group VI Metal Carbonyls and
     Related Compounds, J. Am. Chem. Soc., 88:5174 (1966).

10. J. A. McCleverty, J. Locke, B. Ratcliffe and E. J. Wharton,
     Transition Metal Dithiolene Complexes. X. Oxy-Metal
     Bis-1,2-dithiolenes, Inorg. Chim. Acta, 3:283(1969).

11. D. C. Bravard, W. E. Newton, J. T. Huneke, K. Y. Yamanouchi,
     and J. H. Enemark, unpublished data.

12. A. Muller, W. O. Noter and B. Krebs, $[(S_2)_2Mo(S_2)_2Mo(S_2)_2]^{2-}$
     a Novel Complex Containing Only S_2^{2-} Ligands and a Mo-Mo
     Bond, Angew. Chem, Int. Ed. Engl., 17:279 (1978).

13. W. E. Silverthorn, C. Couldwell and K. Prout, Synthesis,
     X-Ray Crystal Structure, and Chemistry of the Cation
     $[(\eta\text{-MePh})\text{-Mo-}(\mu\text{-SMe})_4Mo(\eta\text{-MePh})]^{2+}$; Easy D_2O-NaOD
     Catalyzed H-D Exchange of the Methyl Hydrogens of a
     η-Bonded Toulene Ligand, Chem. Commun., 1009 (1978).

14. P. M. Boorman, C. D. Garner, F. E. Mabbs, and T. J. King,
     Oxotrichlorotriphenylphosphinesulphidemolybdenum(V)
     $[MoOCl_3(SPPh_3)]$, A Mononuclear Five-Coordinate Oxo-
     molybdenum(V) Complex, Chem. Commun., 663 (1974).

15. J. R. Bradbury, M. F. Mackay and A. G. Wedd, The Crystal and
     Molecular Structure of Tetraphenylarsonium Oxo-tetra-
     (benzenethiolato)molybdate(V), Aust. J. Chem.,
     31:2423 (1978).
16. G. Bunzey, J. H. Enemark, J. E. Gelder, K. Yamanouchi and
     W. E. Newton, Preparation and Structure of a Compound
     Containing a Triply Bridged Binuclear Mo(V) Cation and
     the [MoOCl$_4$(H$_2$O)]$^-$ Anion, J. Less-Common Metals,
     54:101 (1977).
17. K. Yamanouchi, J. T. Huneke, J. H. Enemark, R. D. Taylor,
     and J. T. Spence, Monomeric Molybdenum(V) Complexes.
     3. The Structure of 8-Hydroxyquinolinium Oxotrichloro-
     8-hydroxyquinolinatomolybdate(V),
     [C$_9$H$_8$NO][MoOCl$_3$(C$_9$H$_6$NO)], Acta Crystallogr., Sect. B,
     in press.
18. K. Yamanouchi, S. Yamada, J. H. Enemark, unpublished data.
19. K. Yamanouchi and J. H. Enemark, Monomeric Molybdenum(V)
     Complexes. 2. Comparison of the Structures of Cis-
     Oxochlorobis-(8-mercaptoquinolinato)molybdenum(V) and
     cis-Dioxobis-(8-mercaptoquinolinato)molybdenum(VI),
     Inorg. Chem., 18:1626 (1979).
20. R. D. Taylor, P. G. Todd, N. D. Chasteen, and J. T. Spence,
     Reduction of Nitrate by Monomeric Molybdenum(V)
     Complexes in Dimethylformamide, Inorg. Chem., 18:44
     (1979); J. T. Spence, M. Minelli, P. Kroneck, M. I.
     Scullane, N. D. Chasteen, Monomeric Molybdenum(V)
     Oxo Complexes with Tetradentate Aminoethanethiols,
     J. Am. Chem. Soc., 100:8003 (1978).
21. I. N. Marov, V. K. Belyaeva, Y. N. Dubrov, A. N. Ermakov,
     EPR Study of Thio-oxinato-complexes of Molybdenum(V),
     Tungsten(V), and Copper(II); Russ. J. Inorg. Chem.,
     17:515 (1972); I. N. Marov, E. M. Reznik, V. K.
     Belyaeva, Y. N. Dubrov, Mixed Thio-oxinatomolybdenum(V)
     Complexes, ibid, 17:700 (1972).
22. K. Yamanouchi and J. H. Enemark, unpublished data.
23. P. W. Schneider, D. C. Bravard, J. W. McDonald, and W. E.
     Newton, Reactions of Oxobis(N,N-dialkyldithiocarbamato)
     molybdenum(IV) with Unsaturated Organic Compounds and
     their Biochemical Implications, J. Am. Chem. Soc.,
     94:8640 (1972); G. J.-J. Chen, J. W. McDonald, and
     W. E. Newton, Synthesis of Mo(IV) and Mo(V) Complexes
     Using Oxo Abstraction by Phosphines, Mechanistic
     Implications, Inorg. Chem., 15:2612 (1976).
24. R. Barral, C. Bocard, I. Seree de Roth, and L. Sajus,
     Activation of Molecular Oxygen in a Homogeneous Liquid
     Phase by Molybdenum Complexes, Kinet. Catal., 14:130
     (1973).

25. C. D. Garner, M. R. Hyde, F. E. Mabbs, V. I. Routledge,
      Kinetics and Mechansim of Oxidation of Trichloro-oxobis-
      (triphenylphosphine oxide)molybdenum(V) by Nitrate in
      Dichloromethane, J. Chem. Soc., Dalton 1180 (1975);
      Kinetics and Mechansim of Substitution Reactions of
      Trichlorooxobis[tris(dimethylamino)phosphine-oxido]-
      molybdenum(V) in Dichloromethane solution, J. Chem.
      Soc., Dalton 1198 (1977).

SOME OXYGEN-BRIDGED COMPLEXES OF MOLYBDENUM(VI)

C. Knobler, A.J. Matheson and C.J. Wilkins

Chemistry Department
University of Canterbury
Christchurch, New Zealand

Features of oxygen bridging in molybdenum(VI) complexes are discussed with particular reference to three newly-established oxygen-bridged structures. A pinacol (H_2pin) complex, typical of derivatives from highly-substituted vicinal diols, is constituted $[MoO(Hpin)(pin)]_2O$, with an Mo_2O_3 core. There is intramolecular hydrogen bonding across the oxygen bridge. An Mo_2O_5 core exists in the complex $[Mo_2O_5(Hnta)_2]^{2-}$ formed by nitrilotriacetic acid (H_3nta). As compared with core-bridged structures, a compound $[MoO_2(Hpin)(CH_3O)]_2 \cdot 2CH_3OH$ has methoxy ligand bridges. For an Mo_2O_3 core a bridge vibration gives a band close to 750 cm^{-1} and for the Mo_2O_5 core in $[Mo_2O_5(Hnta)]^{2-}$ the band is at 770-800 cm^{-1}. Compounds having a bridge through a ligand oxygen show a band in the range 625-690 cm^{-1}.

INTRODUCTION

The facility with which molybdenum(VI) develops six-coordination towards oxo-ligands, including carboxylate anions and anions from di- or poly-hydric alcohols is well known,[1] but the varied ways in which combinations of terminal oxygen atoms, O_t, core-bridging oxygen, O_b, and ligand oxygen, O_ℓ, are used continues to attract attention. We discuss bridged structures having Mo_2O_3 and Mo_2O_5 cores, and structures having bridging by ligand oxygen, $O_{\ell b}$. The Mo_2O_3 core in vicinal diol complexes is of interest in showing reversible protonation through reaction with water,[2,3] and the Mo_2O_5 core occurs in the complex formed by mannitol[4,5] (as an example of a polyhydroxy ligand). It is a possibility that

319

polyhydroxy ligands play some part in the uptake and mobilization of molybdenum in biological systems.

Under aqueous conditions the core used by a ligand can be influenced by pH,[1] but the core used in any particular complex can also depend upon the match between ligand conformation and possible alternative core structures. At pH 6-7, nitrilotriacetic acid, $N(CH_2COOH)_3$ (H_3nta), uses an MoO_3 core,[6,7] but at pH 2 it coordinates with an Mo_2O_5 core[8] to give the anion $[Mo_2O_5(Hnta)_2]$.[2-] At pH 2 the malate ligand (trianion, mal^{3-}) uses an Mo_4O_{11} chain in the anion $[Mo_4O_{11}(mal)_2]^{4-}$ in which each malate spans three successive metal centers.[9]

In $\underline{simple}$ cases where structures of reference compounds are known, the type of core present in a compound of undetermined structure can be inferred from the infrared spectrum. This approach is necessarily dependent upon identification of the (usually strong) $Mo-O_t$ and $Mo-O_b$ bands among the ligand bands, and to this end Newton and McDonald[10] have made careful assignments of molybdenum-oxygen frequencies in various complexes containing thio-ligands. We remark on infrared criteria in relation to classes of oxygen-bridged compounds for which there are recent structure determinations. Features of these structures will be considered in turn.

STRUCTURES HAVING THE Mo_2O_3 CORE

Simple vicinal diols, H_2L, produce complexes of the type $MoO_2(HL)_2$, with intermolecular hydrogen bonding in the crystal lattice.[11] But when the carbon atoms carrying the hydroxyl groups are sufficiently substituted (e.g. one tertiary carbon, or two secondary carbon atoms), an alternative type of yellow complex containing an Mo_2O_3 core is formed.[2,3] A recent determination of the crystal structure of the yellow pinacol (H_2pin) complex, $[Mo_2O_3(Hpin)_2(pin)_2]$ (I) has shown the molecular structure to be as in Fig. 1 with an $O_tMoO_bMoO_t$ core.[12] In a strict sense the two ligands coordinated with each molybdenum differ, one being singly deprotonated and the other doubly deprotonated. The formal distinction is, however, reduced through development of hydrogen-bonding between the hydroxylic proton of one ligand (Hpin) and a ligand (pin) of the second kind attached to the other molybdenum center.

This yellow pinacolate (I) is reversibly convertible[13] to a colorless derivative $MoO_2(Hpin)_2$ (II) according to the scheme:

$$[Mo_2O_3(Hpin)_2(pin)_2] + H_2O \underset{\text{azeotropic distillation}}{\overset{\text{moist ethanol}}{\rightleftarrows}} 2MoO_2(Hpin)_2$$

(I)   (II)

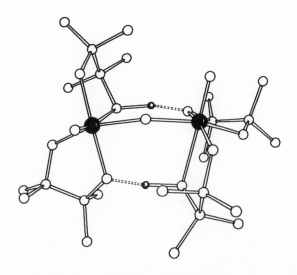

Fig. 1.   The coordination environment in the yellow pinacolate
           [Mo$_2$O$_3$(Hpin)$_2$(pin)$_2$] showing the oxygen bridging and the
           intramolecular hydrogen bonding. All skeletal atoms of
           the pinacolate ligands are shown as open circles. (From
           Ref. 13.)

The ready reversibility of this process would indicate the total
Mo-O bond energies associated with (I) and (II) to be rather closely
similar (since the number of O-H bonds remains unchanged). For the
molybdenum-oxygen core in (I), the bond energy for one Mo-O$_t$ and one
Mo-O$_b$ linkage will be less than that in (II), with two Mo-O$_t$ bonds
per molybdenum. However, compensation towards equalisation will be
provided by the shortening and strengthening of an Mo-O$_\ell$ bond in
(I) [from ca. 2.3Å for a bond trans to O$_t$ in such compounds[11] as
(II), to 1.89Å trans to O$_b$ in (I)].

    It seems reasonable to assume that the structure of the yellow
pinacolate is representative of those of all the yellow complexes
which are formed by substituted vicinal diols and give an infrared
band near 750 cm^{-1} attributable to the MoO$_b$Mo vibration.[2,3] Since
the formation of the yellow compounds is promoted by a wide variety
of substituents irrespective of their polarity or electronic properties, it is probable that the effect of the substituents is to
hinder the molecular packing required for inter-molecular hydrogen
bonding. This leads in effect to a relative stabilization of the
alternative structure having intra-molecular hydrogen bonding.

STRUCTURES HAVING THE Mo_2O_5 CORE

The Mo_2O_5 core, $(O_t)_2MoO_bMo(O_t)_2$, provides linear O_b bridging in the aquo-oxalato anion of the salt[14] $K_2[Mo_2O_5(C_2O_4)_2(H_2O)_2]$ and in the dimethylformamide complex[15] $[Mo_2O_5(Cl)_2(dmf)_4]$. The core may also have a bent O_b bridge when ligand bridging, $O_{\ell b}$, is superimposed.[4,5]

Determination of the crystal structure of the salt $Na_2[Mo_2O_5(Hnta)_2] \cdot 8H_2O$ (III) showed the tridentate ligand to be coordinated through the nitrogen and an oxygen of each of two carboxylate groups (Fig. 2). The Mo_2O_5 core has a linear bridge and is of closely similar geometry to that in the aquo-oxalato complex. The sodium, potassium and ammonium salts of the $[Mo_2O_5(Hnta)_2]^{2-}$ anion provide reference spectra characteristic of the linear-bridged Mo_2O_5 core,[13] with strong $\nu(Mo-O_t)$ bands at 910-930 cm^{-1} and $\nu(Mo-O_b)$ at 775-790 cm^{-1} (Table 1). In other compounds[14,16] having this core the spectra contain one or more bands in the intermediate 830-860 cm^{-1} range. The reason for this variability has not been established, but a force-constant analysis[17] for the cis-trioxo group $Mo(O_t)_2O_b$ leads to the expectation of coupling between the $\nu(Mo-O_t)$ and $\nu(Mo-O_b)$ modes.

LIGAND-BRIDGED STRUCTURES

Ligand bridging may be anticipated when the normal ligancy of the coordinated groups would not provide for completion of six-coordination. The structure of the dinuclear neopentylglycol (dianion, npg) complex[18] $[MoO_2(npg)(H_2O)]_2$ (IV) affords an example of bridging by one end of a chelating ligand. On the other hand, in the compound $[MoO_2(CH_3O)(Hpin)]_2 \cdot 2CH_3OH$ (V) obtained by reaction of the yellow pinacolate (I) with dry methanol,[13] the monofunctional

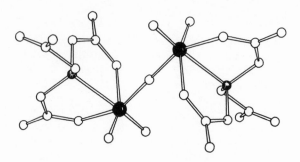

Fig. 2.   Coordination around the molybdenum centers in the complex anion $[Mo_2O_5(nta)_2]^{2-}$. Two of the three acetato groups attached to the nitrogen (shaded) are coordinated, each through a single oxygen. (From Ref. 13.)

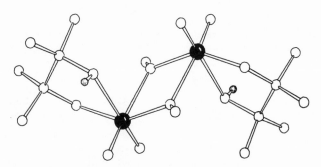

Fig. 3. The coordination environment around the molybdenum centres
in the compound [MoO$_2$(Hpin)(CH$_3$O)]$_2$.2CH$_3$OH, showing the
double bridging by the two methoxy groups. The lattice
methanol is not shown. (From Ref. 13.)

methoxy ligand provides bridging (Fig. 3). In the crystal there is
extended intermolecular hydrogen bonding from the hydroxyl groups
of the pinacolate ligands to the lattice methanol.

No structures with bridging by an open-chain vicinal diol have
yet been reported. The lack of such examples suggests that a five-
membered ring is best suited to simple chelation with an individual
molybdenum center, but that lengthening of a potentially chelating
chain, as in the neopentylglycol compound (IV), enables an increased
angle of bite at a first metal centre, leaving an O$_\ell$ oxygen more
favorably placed to develop bridging to a second metal center.
[The angle is 149° in (IV), but 71 - 76° in (I) and (V).]

In the ligand-bridged compounds the Mo-O$_{\ell b}$ distances [(IV),
2.00, 2.22Å; (V), 2.01, 2.22Å] are greater than Mo-O$_b$ distances

Table 1. Infrared Frequency Ranges

Core	ν(Mo-O$_t$)		ν(Mo-O$_b$)	
MoO$_2$	a	885-950		-
MoO$_3$	b	840-900		-
Mo$_2$O$_3$	a	920-950	a	750-760s
Mo$_2$O$_5$	a,c	900-930	a,c	750-790s
Mo$_4$O$_{11}$	d	885-935	d	850-870m ⎫
				790-800m ⎬
				~720s ⎭

[a] For compounds cited in Ref. 13. [b] For compounds
cited in Ref. 7. [c] Other Mo$_2$O$_5$ complexes give
bands in the range 830-860 cm^{-1} as well. [d] For
[Mo$_4$O$_{11}$(mal)$_2$]$^{4-}$ (in the NH$_4^+$ and Cs$^+$ salts).

[(I), 1.90; (III), 1.88Å]. In consequence, the $MoO_{\ell b}Mo$ infrared bands in the ligand-bridged compounds lie within a lower frequency range[13] (625 - 690 cm^{-1}) than do the MoO_bMo bands in the Mo_2O_3 and Mo_2O_5 core-bridged structures which show no bands below 750 cm^{-1} (Table 1).

OTHER REMARKS ON INFRARED SPECTRA

Infrared spectra can serve to identify the simple MoO_2 and MoO_3 cores and can provide corroborative evidence on the Mo_2O_3 core, but cannot be used confidently for Mo_2O_5 since this core can show variability in its bands. As regards more complicated structures[9] the Mo_4O_{11} chain of the malic acid complex $[Mo_4O_{11}(mal)_2]^{4-}$ gives several groups of bands (Table 1), some in the same ranges as for the Mo_2O_5 core. The mannitol complex $[Mo_2O_5(man)]^-$ has a bent O_b bridge with two $O_{\ell b}$ bridges. Its salts[4,5] show bands close to 770 and 612 cm^{-1}, in positions such as would correspond to $\nu(MoO_bMo)$ and $\nu(MoO_{\ell b}Mo)$ respectively. Thus for structures which may have the complexity of the mannitol and malic acid derivatives, spectra can provide some indication of broad features, but the need for X-ray structure determinations will continue.

ACKNOWLEDGEMENTS

We are grateful to Professor B.R. Penfold and Dr W.T. Robinson for their help and interest which made possible the determination of the structures of compounds (I), (III), and (V). We also thank Dr Brit Hedman, Umea University, and Dr Joyce Waters, University of Auckland, for providing infrared spectra of the mannitol complex.

REFERENCES

1.   P.C.H. Mitchell, Oxo-species of molybdenum-(V) and -(VI), Quart. Rev. Chem. Soc., 20:103 (1966).
2.   Snam Progetti S.P.A., British Patent, 1,226,937 (1971).
3.   R.A. Sheldon, Molybdenum-catalyzed epoxidation of olefins with alkyl hydroperoxides.II. Isolation and structure of the catalyst, Rec. Trav. Chim., 92:367 (1973).
4.   J.E. Godfrey and J.M. Waters, Ammonium μ-oxo-μ-mannitolatotetra-oxodimolybdate monohydrate, Cryst. Struct. Comm., 4:5 (1975).
5.   B. Hedman, Multicomponent polyanious, 15. The molecular and crystal structures of $Na[Mo_2O_5\{O_3(OH)C_6H_8(OH)_2\}]\cdot2H_2O$, a protonized mannitodimolybdate complex, Acta Cryst., B 33:3077 (1977).
6.   R.J. Butcher, H.K.J. Powell, C.J. Wilkins and S.H. Yong, New amino-acid complexes of molybdenum-(V) and -(VI), J.C.S. Dalton Trans., 356 (1976).

7.   R.J. Butcher and B.R. Penfold, Structural studies of molyb-
     denum(VI) coordination chemistry. Crystal and molecular
     structure of trispotassium [cis-trioxo(nitrilotriacetato)molyb-
     date(VI)] monohydrate, J. Cryst. Mol. Structure 6:13 (1976).

8.   C. Knobler, B.R. Penfold and W.T. Robinson. Unpublished data.

9.   M.A. Porai-Koshits, L.A. Aslanov, G.V. Ivanova and T.N.
     Polynova, X-ray diffraction of ammonium dimolybdomalate, Zhur.
     Strukt. Khim., 9:475 (1968).

10. W.E. Newton and J.W. McDonald, Synthesis and infrared spectra
     of ^{16}O- and ^{18}O-substituted oxomolybdenum complexes, J. Less-
     Common Metals, 54:51 (1977).

11. F.A. Schroeder, J. Scherle and R.G. Hazell, Chemistry of molyb-
     denum and tungsten. XV. Structure of cis-dioxobis(2-hydroxy-
     ethyl-1-oxo)molybdenum(VI), Acta Cryst. B 31:531 (1975).

12. A.J. Matheson and B.R. Penfold. Unpublished data.

13. C. Knobler, B.R. Penfold, W.T. Robinson, C.J. Wilkins and
     S.H. Yong, J.C.S. Dalton Trans. In press.

14. F.A. Cotton, S.A. Morehouse and J.S. Wood, The identification
     and characterization by X-ray diffraction of a new binuclear
     molybdenum(VI) oxalate complex, Inorg. Chem., 3:1603 (1964).

15. L.O. Atovmyan, Yu. A. Sokolova and V.V. Tkachev, Crystal
     structures of the dioxy compounds of Mo(VI). $[MoO_2Cl_2(DMF)]_2O$
     and $K_2[MoO_2(C_6H_4O_2)_2]2H_2O$, Dokl. Akad. Nauk. S.S.S.R., 195:1355
     (1970).

16. M. Cousins and M.L.H. Green, Some oxo- and oxochloro-cyclo-
     pentadienylmolybdenum complexes, J. Chem. Soc., 1567 (1964).

17. F.A. Cotton and R.M. Wing, Properties of metal-to-oxygen
     multiple bonds, especially molybdenum-to-oxygen bonds, Inorg.
     Chem., 4:867 (1967).

18. C.K. Chew and B.R. Penfold, Structural studies of molybdenum
     (VI) coordination chemistry: Crystal and molecular structure
     of $Mo_2O_4(2,2$-dimethylpropane-$1,3$-diolate$)_2 \cdot 2H_2O$ containing a
     μ-dioxo bridge, J. Cryst. Mol. Structure, 5:413 (1976).

ELECTROCHEMICAL AND SPECTROSCOPIC STUDIES OF MOLYBDENUM-CATECHOL COMPLEXES: MODELS FOR MOLYBDOENZYMES AND BIOLOGICAL TRANSPORT

John P. Wilshire, Luis Leon, Paula Bosserman, and Donald T. Sawyer

Department of Chemistry
University of California
Riverside, California 92521, U.S.A.

INTRODUCTION

The chemistry of molybdenum is important because of its involvement in a variety of biological processes. Although present in trace quantities, molybdenum is crucial for the activity of at least five enzymes: xanthine oxidase,[1] aldehyde oxidase,[2] nitrate reductase,[3] sulphite oxidase,[4] and nitrogenase.[5] All of these enzymes catalyze redox-type chemical reactions which involve the transfer of two or more electrons per substrate molecule. Electron paramagnetic resonance (EPR) studies[1-3,6-8] indicate that, in all but nitrogenase, the molybdenum undergoes oxidation state changes during the catalytic process. Also, the resting state of these enzymes appears to have molybdenum in the +6 oxidation state and molybdenum oxidation state changes are observed only when reducing agents are added to the enzyme.[1] All of the enzymes contain two atoms of molybdenum, and in the cases of xanthine oxidase, aldehyde oxidase, and nitrate reductase, two molecules of flavin as well.[4] Some researchers believe[9] that molybdenum alternates between the +6 and +5 oxidation states during enzyme activity. However, there is evidence[10] that the +4 oxidation state also may be involved.

In spite of the importance of these molybdoenzymes and the considerable effort expended in their study, the structure and function of the active site and the exact role of the molybdenum in the enzyme are still not

understood. For this reason, the molybdenum enzymes are
well suited for model compound studies. Numerous
molybdenum-containing model complexes have been prepared
by our group[11-16] and by others.[17] To date, only two
complexes appear to be viable reaction models for the
molybdenum enzymes. The dithiocarbamate complexes, cis-
$Mo^{VI}O_2(S_2CNR_2)_2$ (R = Et, Pr^n or Bu^i), catalyze the oxidation of a tertiary phosphine to the corresponding
phosphine oxide by oxygen atom transfer from the Mo(VI)
complex.[18,19] Recent studies[20] of the tetradentate amino-
ethanethiol complexes of molybdenum(V) yield EPR parameters similar to those observed for molybdenum in xanthine oxidase. None of these model complexes appears to
catalyze the oxidation of xanthine nor to effect the
reduction of molecular oxygen to superoxide.

To be a viable model for the role of molybdenum in
xanthine oxidase, the ligand employed must be able to
stabilize the metal over a range of oxidation states.
In addition, the complex formed must have oxidation-
reduction potentials in the range to act as a good oxidant for xanthine and a good reductant for dioxygen.
As catechol complexes of transition metals exhibit unique
coordination chemistry and redox properties,[21-24] studies
of such complexes have been undertaken[21,22,25] to determine their viability as models for biological processes.
We report here the formation and the properties of two
complexes of molybdenum that result from the reaction of
molybdenum(VI) with 3,5-di-tert-butylcatechol. Their
spectroscopic and electrochemical properties are summarized, and their viability as reaction models for
xanthine oxidase is discussed.

EXPERIMENTAL SECTION

Equipment. Cyclic voltammetric measurements were
made with a three-electrode potentiostat-amperostat[26]
constructed with solid-state operational amplifiers.
The voltammograms were recorded on a Houston Instrument
Omnigraphic Model 2000 X-Y recorder. A Princeton Applied
Research Corporation Model 173 Potentiostat-Galvanostat
with a Model 179 Digital Coulometer was used for controlled potential coulometry.

The electrochemical cell consisted of Brinkman
Model EA875-20 cell, with corresponding plastic cell top.
The cell top supported the electrodes and the auxilary
compartment (a Pyrex tube with a medium-porosity frit on
the end). The top also had access holes for the addition and withdrawal of samples.

The working electrode for cyclic voltammetry was a Beckman platinum-inlay electrode. For controlled-potential electrolysis, a cylindrical platinum mesh electrode was used. The auxilary electrode was a 1-cm^2 foil of platinum. The reference electrode consisted of a silver wire coated with AgCl in a Pyrex tube having a small soft-glass cracked-bead tip. The electrode was filled with a solution of aqueous tetramethylammonium chloride (Matheson, Coleman and Bell) with the concentration adjusted such that the electrode potential was 0.000 V vs. SCE. The reference electrode was placed inside a luggin capillary in the cell assembly.

Infrared spectra of the complexes were recorded as nujol mulls on NaCl plates, or as KBr pellets with a Perkin-Elmer Model 283 infrared spectrophotometer. The UV-visible spectra for solutions of the complexes were recorded with a Cary Model 17-D spectrophotometer and a Cary Model 219 spectrophotometer. Proton NMR spectra were obtained on a Varian Model E17-390 spectrometer. ESR spectra were obtained on a Varian Model V4500 spectrometer. Vapor pressure osmometric measurements were performed with a Mechrolob Inc. Model 301A. Magnetic susceptibility measurements were made by the Gouy method, or by the NMR method developed by Evans.[27] The measurements were corrected for diamagnetism.[28]

Reagents. Acetonitrile, dimethylsulfoxide (DMSO), N,N-dimethylformamide (DMF), and pyridine (py) (all Burdick and Jackson Laboratories, "Distilled in Glass") were used as purchased. Benzene (Mallinckrodt AR), toluene (Mallinckrodt AR), ethyl ether (Mallinckrodt AR) and ethanol (Pharmco) also were used as purchased.

Tetraethylammonium perchlorate (TEAP) was used as the supporting electrolyte. TEAP was prepared by the stoichiometric combination of reagent grade perchloric acid and reagent grade tetraethylammonium bromide. The crystalline product was collected, washed, and recrystallized from cold water. Tetraethylammonium hydroxide (TEAOH) was obtained from Eastman Kodak Co. as a 25% solution in methanol, and was determined by titration to contain 1.42 $\underline{M}$ OH$^-$.

Imidazole and 3,5-di-tert-butylcatechol were obtained from the Alrich Chemical Co. Bis(acetylacetonato)dioxomolybdenum(VI), which was used for the syntheses, was synthesized and purified as previously published.[29]

Preparation of Complexes. 1. Bis[bis(3,5-di-tert-butylcatecholato)oxomolybdenum(VI)] (1). A 20-ml solution of $Mo^{VI}O_2(acac)_2$ (0.33g, 0.001 mol) in methanol was combined with a 20-ml solution of 3,5-di-tert-butylcatechol (0.44g, 0.002 mol) in methanol and stirred. After two hours, a dark red-purple solution was obtained, which was filtered and then evaporated to a small volume to yield a dark blue-purple precipitate. The crystals were filtered and air dried, and then recrystallized from a hot toluene solution.

Addition of water to the filtered methanol solution caused a flocculent green precipitate (2) to be formed. The precipitate was filtered and air dried, and then recrystallized from an ethanol-water solution.

When ethanol (absolute) or acetone were used for the synthesis the same purple complex (1) was isolated. If CH_3CN, DMSO or DMF were used, a greenish-brown solution resulted which yielded a precipitate of 2.

Anal. Calcd for $Mo_2C_{56}H_{80}O_{10}$: C, 60.85; H, 7.30; Mo, 17.36. Found:[30] C, 61.09; H, 7.50; Mo, 17.17. The molecular weight of 1 in toluene solution was determined by vapor-pressure osmometry to be 1130 g mol^{-1}, which is consistent with a binuclear complex $[MoO(cat)_2]_2$. (Calcd mol. wt., 1128 g mol^{-1}).

RESULTS

Although reactions of molybdenum with various catechols have been studied previously,[31-36] most of these studies have been in aqueous media and, with the exception of the molybdenum carbonyls, have not included the isolation and characterization of stable complexes. The first complex isolated from the reaction of molybdate ion and 3,5-di-tert-butylcatechol is the purple binuclear species, bis[bis(3,5-di-tert-butylcatecholato)-oxomolybdenum(VI)], (1). In polar solvents such as dimethylsulfoxide and dimethylformamide, 1 dissociates to a monomeric species (2). The formation of 1 and 2 have been described previously,[37] and the crystal structure of 1 has been determined.[38]

Spectroscopy. The UV-visible spectra for 1 in CH_2Cl_2 solution and for 2 and its derivatives in DMF solutions are illustrated by Figure 1. When 1 or 2 is dissolved in DMF, a dark olive-green solution results. Similar olive-green solutions are obtained with CH_3CN,

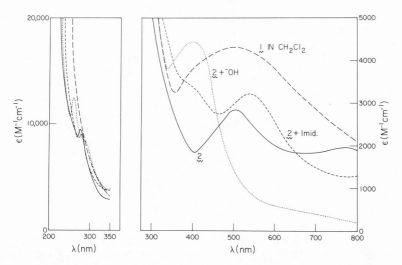

Figure 1. Absorption spectra of 1 mM bis(3,5-di-tert-butylcatecholato)oxomolybdenum(VI) (2): in DMF (——); DMF solution plus one equivalent of tetraethylammonium hydroxide (····); DMF solution plus excess imidazole (----); and of 0.5 mM bis[bis(3,5-di-tert-butylcatecholato)-oxomolybdenum(VI)] (1) in CH_2Cl_2 (— — —). Molar absorptivities, ϵ, are based on a mononuclear formulation.

DMSO, DMA or any moderately strong coordinating solvent. When dissolved in MeOH, EtOH, $(CH_3)_2CO$, CH_2Cl_2, toluene or benzene, species 1 or 2 yields an intense purple colored solution. Furthermore, the colors of the solutions are interconvertable by the addition of the appropriate solvent. If a strongly coordinating ligand (imidazole, N-methylimidazole, or pyridine) is added to either the olive-green or purple solutions, an intense violet solution results. The addition of less strongly coordinating ligands (triphenylphosphine) did not effect any change in solution color. There, thus, appears to be a considerable solvent effect on 1 and 2.

Spectrophotometric titrations have confirmed that both the olive-green and purple complexes contain two catechols per molybdenum.[37] Molecular weight determinations by vapor pressure osmometry establish that the purple solutions contain a chromophore with a molecular weight of 1130 g mol^{-1}, and that the olive-green solutions contain a species with a molecular weight of 570 g mol^{-1}. The changes in color and molecular weight

are indicative of a monomer-dimer equilibrium (equation
1) which is affected by solvent. Dilution experiments
of 1 in CH_2Cl_2 indicate that the solution remains purple down to a concentration of 4.7 x 10^{-6}M, at which
point the solution becomes green. The value of the
equilibrium constant, K_{eq}, for formation of the dimer
in CH_2Cl_2 is 44 M^{-1}.

$$2 \; MoO(cat)_2 \; \xrightleftharpoons{K_{eq}} \; [MoO(cat)_2]_2 \qquad\qquad (1)$$

olive-green, 2   purple, 1

Thus, in the presence of coordinating solvents, 2 is
dominant, while in non-coordinating media, the dimeric
form (1) is favored.

The absorption spectra are indicative of the oxidation state of the molybdenum. On the basis of the
molar absorptivities for 3,5-di-tert-butylcatechol
(λ, 279 nm; ϵ, 2130), its semiquinone anion (λ, 338 nm;
ϵ, 1240; λ, 377 nm; ϵ, 1010), and its quinone (λ, 400
nm; ϵ, 1550) in CH_3CN, the ligands bound to molybdenum
in 1 and 2 have the characteristics of catecholate
dianions. Hence, for these two neutral bis(catecholato)-
oxomolybdenum complexes (1 and 2), the molybdenum must
be present as Mo(VI). When one equivalent of tetraethylammonium hydroxide (TEAOH) per molybdenum is added
to 1, an orange solution results (3).

The infrared spectrum of 1 and 2 (as KBr discs,
nujol mulls, and CCl_4 solutions) is dominated by bands
due to the catechol ligands. Additional strong bands
are observed at 975 cm^{-1} and 966 cm^{-1}. The 975 cm^{-1}
band can be assigned to the C-O bonds of the ligands
and the 966 cm^{-1} band to the Mo-axial oxygen bond. The
strong terminal axial oxygen band and the absence of
bands for a Mo-O-Mo vibration in 1 confirm that the
terminal oxygen is not involved in bridging for this
dimeric complex.[39,40]

Species 1 and 2 do not exhibit any EPR activity
and magnetic susceptibility measurements indicate that
both are diamagnetic.

The proton nmr spectrum of 2 in deuterated dimethylsulfoxide exhibits a single set of resonances for the
3,5-di-tert-butyl groups of the ligand (mean resonances
at 1.34 ppm and 1.23 ppm vs TMS), which are consistent
with the catechol form of the ligand.[41] The peaks for

the ring protons of the ligand have mean values of
6.65 ppm (6.67 and 6.62).

Electrochemistry. The cyclic voltammograms for 2
and its derivatives in DMF are presented in Figure 2.~
When 2 is scanned negatively, three irreversible reduction ~peaks are observed at -0.4 V, -0.80 V, and at
-1.70 V vs SCE; the reverse scan yields an irreversible
oxidation at -0.75 V. A positive scan of 2 yields a
reversible couple at +0.25 V. If the positive scan is

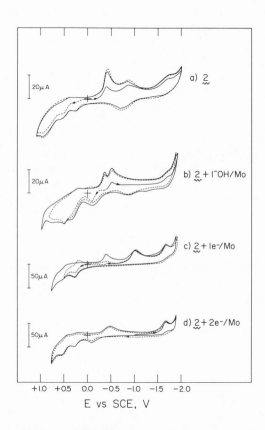

Figure 2. Cyclic voltammograms in DMF
(0.1 M tetraethylammonium perchlorate (TEAP)) of:
(a) 1 mM 2; (b) 1 mM 2 plus one equivalent TEAOH;
(c) 1 mM 2 after reduction at -0.5 V (one electron transferred per Mo); and (d) 1 mM 2 after
further reduction at -1.2 V (one additional
electron transferred per Mo). Measurements
were made at a platinum electrode (area 0.23 cm^2)
at a scan rate of 0.1 V sec^{-1}; temperature 23°C.

taken to +1.0 V prior to reversal of the scan, the
reduction peak at -0.4 V increases in intensity. The
latter peak is due to the reduction of o-quinone, which
is generated when the voltammogram is scanned to potentials sufficiently positive to oxidize the catecholate
ligand (greater than +0.75 V). Similar voltammograms
for 2 are observed when DMSO is the solvent.

     Cyclic voltammograms for 1 in CH_2Cl_2 do not exhibit
the reduction peak at -0.8 V; instead a peak at -1.2 V
is observed. When CH_3CN is used as the solvent, reduction peaks at -0.4 V, -0.8 V, -1.2 V, and -1.7 V are
observed. Furthermore, the relative peak currents at
-0.8 V and -1.2 V are concentration dependent. For low
concentrations (<1 mM), the peak at -0.8 V is dominant,
while at higher concentrations (>1 mM), the -1.2 V peak
increases at the expense of the -0.8 V peak. The solution is olive-green at low concentrations and tends
toward purple at high concentrations. These observations
are consistent with the monomer-dimer of equation 1 with
the monomeric form (2) predominating at low concentrations and being reduced at -0.8 V. The reduction peak
at -1.2 V is due to the dimer (1). In non-coordinating
solvents, such as CH_2Cl_2, 1 is the dominant form and the
peak at -0.8 V is not observed. In DMSO or DMF, the
monomer is dominant and the reduction peak at -1.2 V is
not observed. The addition of strongly coordinating
ligands (imidazole, 1-methylimidazole, CN^-) to CH_2Cl_2
solutions of 1 cause a total loss of the -1.2 V peak
and the appearance of a peak at -0.8 V. Hence, these
ligands add to the axial site on the molybdenum and
cause the dimer to dissociate.

     Controlled potential coulometry of 2 in DMF at
-0.5 V results in the transfer of one electron per
molybdenum and the formation of a lime-green solution
of species 4. The cyclic voltammogram for 2 after it
has been reduced at -0.5 V (species 4) exhibits reduction waves at -1.0 V and -1.7 V (Figure 3a). In addition, a reversible oxidation is observed at +0.25 V,
which corresponds to the oxidation of bound catechol
to bound semiquinone.

     Analysis of the absorption spectrum of 4 indicates
that the ligands are still present as catecholate dianions, and, hence, that the Mo(VI) has been reduced.
This result is confirmed by the EPR spectrum for the
reduced solution, which exhibits a resonance at g = 1.95
for Mo(V) (Figure 4a). The EPR spectrum of the reduced

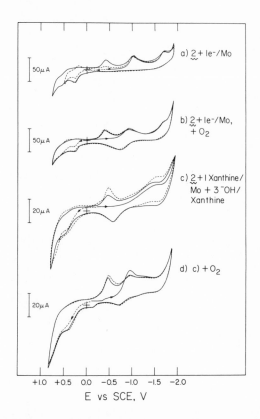

Figure 3.   Cyclic voltammograms of DMF
(0.1 M TEAP) solutions of: (a) 1 mM 2 after
reduction at -0.5 V (one electron transferred
per Mo); (b) solution (a) plus dioxygen;
(c) 1 mM 2 plus one equivalent of xanthine
and three equivalents of TEAOH per Mo, plus
three equivalents TEAOH per xanthine; and
(d) solution (c) plus dioxygen.

species (4) at 77°K (Figure 4b) has considerable anisotropy, which is consisent with a square pyrimidal coordination geometry around molybdenum.

Upon addition of dioxygen to a solution of 4 (2
electrolyzed at -0.5 V), the ESR signal at g = 1.95
disappears and a signal at g = 2.00 appears (Figure 4c).
In addition, a shoulder is observed at 263 nm in the
absorption spectrum.   These two observations are consistent with the formation of semiquinone.   Apparently,
dioxygen is reduced by 4 to give $O_2^{\cdot-}$ and/or SQ$^{\cdot}$ as well
as Mo(VI).

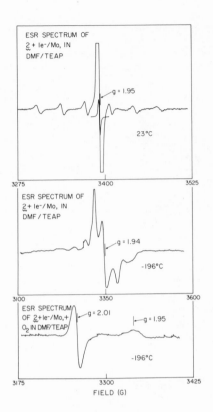

Figure 4. EPR spectra of DMF solutions
of: (a) 1 mM 2 after reduction at -0.5 V
(one electron transferred) at 23°C; (b)
frozen solution spectrum (-196°C) of (a);
and (c) frozen solution spectrum of (a)
plus dioxygen. Spectra referenced to
DPPH (g = 2.0037).

A controlled potential reduction of 4 at -1.2 V
results in the transfer of one electron per molybdenum
to give an orange-brown solution which is EPR silent.
An initial negative scan after reduction yields a cyclic
voltammogram (Figure 2d) with a reduction wave at -1.6 V
and oxidation waves at -0.1, +0.25, and +0.5 V for the
reverse scan. Unlike the one-electron reduction species
(4), the two-electron reduction product is air stable.
The addition of H_2O_2 to the solution has no significant
effect on the reduced solution either.

When one equivalent of xanthine per molybdenum
(plus three equivalents of OH⁻ to effect solution) is
added to 2 in DMF, reduction again occurs to yield a
lime-green species. Initial negative scans of the product solution exhibit a single reduction peak at -1.5 V
(Figure 3c). An ESR resonance at g = 1.94 is observed
for the solution and its absorption spectrum is similar
to that for 4 (2 plus one electron). These results
indicate that 2 is reduced to $[Mo^VO(cat)_2]^-$ by xanthine
in the presence of base. The Mo(V) EPR signal is a
maximum when the xanthine-to-molybdenum stoichiometry
is 1:1.

The addition of dioxygen to the xanthine-reduced
solution of 2 eliminates the Mo(V) EPR signal and yields
a species with a single resonance at g = 2.00. A
shoulder at 263 nm again is noted in the absorption
spectrum. Apparently, O_2^- and/or SQ^- is produced from
the reduced molybdenum complex. The cyclic voltammogram for the product solution (Figure 3d) is similar to
that observed when O_2 is added to $[Mo^V(O)(cat)_2]^-$ (4).

DISCUSSION AND CONCLUSIONS

The spectroscopic, magnetic, and electrochemical
results for the bis(catecholato)oxomolybdenum(VI) system
indicate that a dimer species (1) is formed in CH_2Cl_2
and related non-complexing solvents. Addition of donor
solvents (DMSO or DMF) or of strong ligands (imidazole
or CN⁻) causes the dimer to dissociate to a mononuclear
species (2). These and related reactions for the molyb-
denum-catechol system are summarized in Table I.

The X-ray crystal structure[38] of the dimeric purple
species (1) shows that one oxygen of one catechol ligand
per molybdenum is bidentate and bound to the second
molybdenum atom. This arrangement provides a di-oxo
bridge without involving the terminal Mo=O. This result
is in line with the normal Mo=O stretching frequency
observed in the IR spectrum, which would have been
shifted had the axial oxygen been involved in bridging.

Although most Mo(VI) complexes are colorless, 1
and 2 are intensely colored due to a ligand-to-metal
charge band between 400 nm and 600 nm. When the axial
ligand is a weak σ donor, the charge transfer band is
observed near 400 nm and the complex has a green color.
When the ligand is a strong σ donor, the charge transfer
is facile and the band occurs at lower energy (500 nm -
600 nm). Thus, with ligands such as imidazole, the

Table I. Summary of Reactions for the Molybdenum–Catechol System.

(a) $[Mo^{VI}O(Cat)_2]_2 \xrightarrow{\text{solvent}} 2Mo^{VI}O(Cat)_2(sol)$

$\underset{\sim}{1}$ $\underset{\sim}{2}$

Purple; stable in   Green; stable in DMSO,
CH_2Cl_2, toluene.   DMF.

(b) $\underset{\sim}{1} + 2H_2O \rightarrow 2\ Mo^{VI}O(Cat)_2(H_2O)$
                     Green

(c) $\underset{\sim}{2} + imidazole \rightarrow Mo^{VI}O(Cat)_2(imidazole) + sol$
                              Purple

(d) $\underset{\sim}{1} + 2\ OH^- \xrightarrow[H_2O]{O_2} [Mo_2^{VI}O_5(Cat)_2]^{2-} + 2H_2Cat$
                              $\underset{\sim}{3}$
                         Orange

(e) $\underset{\sim}{2} + e^- \xrightarrow[\text{DMF, } CH_3CN]{\text{DMSO}} [Mo^VO(Cat)_2(sol)]^-$
                              $\underset{\sim}{4}$

                        $E_{p,c} = -0.5$ V vs SCE

(f) $\underset{\sim}{4} + O_2 \xrightleftharpoons{\text{DMSO}} \underset{\sim}{2} + O_2^{\cdot-} \xrightleftharpoons{H_2O} \frac{1}{2}\underset{\sim}{3} + SQ^{\cdot-}$

(g) $\underset{\sim}{4} + e^- \xrightarrow[\text{DMF, } CH_3CN]{\text{DMSO}} [Mo^{IV}O(Cat)_2]_{tr}^{2-} \xrightarrow{\text{DMSO}}$
                              $\underset{\sim}{5}$

         $[Mo^{VI}O(Cat)_2]_{tr}^{2-} + DMS \xrightarrow{H_2O}$

         $\frac{1}{2}\underset{\sim}{3} + OH^- + H_2Cat$ $E_{p,c} = -1.2$ V

(h) $\underset{\sim}{5} \xrightarrow[H_2O]{\text{DMF, } CH_3CN} \frac{1}{2}\underset{\sim}{3} + H_2 + OH^- + H_2Cat$

charge transfer is observed at 510 nm and the solution
is purple. The purple color of the dimeric species (1)
indicates that the bidentate catechol oxygen which is~
involved in bridging forms a strong σ donor bond to
molybdenum. The charge transfer absorption for 1 is at
540 nm.

Electrochemical reduction of 2 by one electron
yields $[Mo^VO(cat)_2]^-$ (4), which exhibits an EPR resonance at g = 1.94. When dioxygen is added to 4, the EPR
resonance at g = 1.94 decreases and a resonance at g =
2.0 is observed. This resonance is characteristic of
O_2^- or semiquinone in DMF solution and indicates that 4
undergoes reoxidation by electron transfer to the oxygen.
This activation of oxygen by metal ions has been observed
previously.[42] Addition of a second electron to 2 yields
an orange-yellow, EPR-silent species which is air-stable
and unreactive with H_2O_2. The cyclic voltammogram and
absorption spectrum for this species are similar to those
for 2 plus one hydroxide per molybdenum or 3 plus hydroxide.~ The stability of the species to O_2 and H_2O_2 indicates that the metal is in a higher oxidation state and
that the reduced species has been oxidized by solvent or
H_2O. A reasonable reaction scheme is outlined by Reaction (g) in Table I, with a Mo(IV) intermediate reducing
DMSO to dimethylsulfide (DMS).

Finally, the reaction of 2 with xanthine to yield
a Mo(V) species, followed by the reduction of dixoygen
when it is added, mimics the reactions of xanthine oxidase.[1] As with 2, when xanthine is added to xanthine
oxidase, an EPR resonance due to a Mo(V) center appears.
The addition of dioxygen to the reduced enzyme results
in the formation of superoxide ion. Hence, the
$Mo^{VI}O(cat)_2$ complex (2) exhibits characteristics that
make it worthy of consideration as a model for the
enzyme. Xanthine is a two-electron reductant and equi-
molar mixtures of xanthine and 2 should cause reduction
of 2 to Mo(IV). This reaction may occur, but the
appearance of an EPR resonance for Mo(V) indicates that
any Mo(IV) produced by xanthine oxidation is either
dismutated or oxidized by O_2 or solvent to Mo(V).

Complexation of Mo(VI) by catechol yields species
that exhibit unique redox chemistry as well as reduction
to a mononuclear Mo(V) complex. The reactions of these
species are summarized in Table I and an interconversion
scheme is presented in Figure 5.

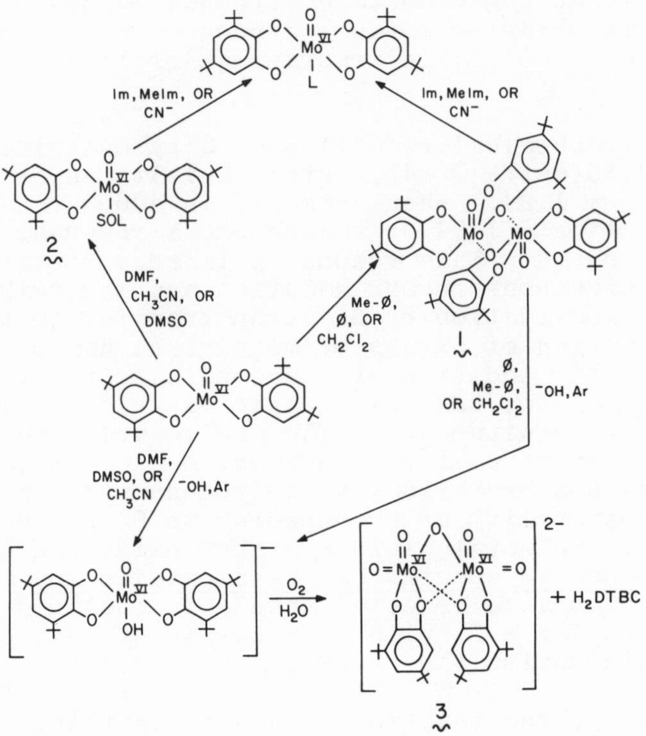

Figure 5. Reaction scheme for $Mo^{VI}(0)(cat)_2(sol)$ (2).

ACKNOWLEDGMENT

    This work was supported by the National Science Foundation under Grant No. CHE76-24555.

REFERENCES

1. R. C. Bray, P. F. Knowles, and L. S. Meriwether, "ESR and the Role of Molybdenum in Enzymic Catalysis by Milk Xanthine Oxidase," in: "Magnetic Resonance in Biological Systems," A. Ehrenberg, B. G. Malmstrom, and T. Vanngard, eds., Pergamon Press, Oxford (1967), p. 249.

2. H. Beinert and W. H. Orme-Johnson, "Electron Spin Relaxation as a Probe for Active Centers of Paramagnetic Enzyme Species," in ref. 1, p. 221.

3. A. Nason, "Nitrate Reductases," in: "The Enzymes,"
    P. D. Boyer, H. Lardy, and K. Myrback, eds.,
    Vol. 7, Academic Press, New York, N.Y. (1963),
    p. 587.
4. R. C. Bray and J. C. Swann, "Molybdenum-Containing
    Enzymes," Struct. and Bonding, 11:107 (1972).
5. K. B. Taylor, "Enzymology of Nitrogent Fixation in
    Cell-free Extracts of Clostridium Pasteurianum,"
    J. Biol. Chem., 244:171 (1969).
6. G. Palmer and V. Massey, "Electron Paramagnetic
    Resonance and Circular Dichroism Studies on
    Milk Xanthine Oxidase," J. Biol. Chem.,
    244:2614 (1969).
7. C. H. Fewson and D. J. D. Nicholas, "Nitrate Reduc-
    tase from Pseudomonas Aeruginosa," Biochim.
    Biophys. Acta, 49:335 (1961).
8. D. J. D. Nicholas, P. W. Wilson, W. Heinen, G.
    Palmer, and H. Beinert, "Use of Electron Para-
    magnetic Resonance Spectroscopy in Investiga-
    tions of Functional Metal Components in Micro-
    organisms," Nature, 196:433 (1962).
9. D. J. D. Nicholas and P. J. Wilson, "A Dissimila-
    tory Nitrate Reductase from Neurospora Crassa,"
    Biochim. Biophys. Acta, 86:466 (1964).
10. V. Massey, H. Komai, G. Palmer, and G. B. Elion,
    "Mechanism of Inactivation of Xanthine Oxidase
    by Allopurinol and other Pyrazolo[3,4-d]pyri-
    midines," J. Biol. Chem., 245:2837 (1970).
11. A. F. Isbell, Jr., and D. T. Sawyer, "The Electro-
    chemistry of Molybdenum(VI,V)-8-Quinolinol
    Complexes in Dimethyl Sulfoxide," Inorg. Chem.,
    10:2449 (1971).
12. D. T. Sawyer, J. N. Gerber, L. W. Amos, and L. J.
    DeHayes, "Electrochemical and NMR Studies of
    Molybdenum-Flavin Model Compounds in Aprotic
    Solvents," J. Less-Common Metals, 36:487 (1974).
13. T. L. Riechel and D. T. Sawyer, "Electrochemical
    Studies of Vanadium(III), -(IV), and -(V) Com-
    plexes of 8-Quinolinol in Acetonitrile. Forma-
    tion of a Binuclear Mixed-valence (IV,V) Com-
    plex," Inorg. Chem., 14:1869 (1975).
14. J. K. Howie and D. T. Sawyer, "Electrochemical
    Studies of Oxo- and Sulfido-bridged Binuclear
    Molybdenum(V) Complexes in Aprotic Media,"
    Inorg. Chem., 15:1892 (1976).
15. D. T. Sawyer, J. K. Howie, and W. H. Doub, "Oxida-
    tion-reduction Chemistry of Oxo- and Sulfido-
    bridged Binuclear Molybdenum(V) Complexes in
    Aprotic Media," J. Less-Common Metals, 54:425
    (1977).

16. D. T. Sawyer and W. H. Doub, Jr., "Formation of Model Compounds of Biological Molybdenum(IV)," Inorg. Chem., 14:1736 (1975).

17. For recent work on molybdenum model compounds, see E. I. Stiefel, "The Coordination and Bioinorganic Chemistry of Molybdenum," in: "Progress in Inorganic Chemistry," Vol. 22, S. J. Lippard, ed., John Wiley and Sons, New York, N.Y. (1977), p. 1, and references therein.

18. R. Durant, C. F. Garner, M. R. Hyde, and F. E. Mabbs, "Kinetics and Mechanism of the Oxygen-transfer Reaction Between Bis(diethyldithio-carbamato)dioxomolybdenum(VI) and Triphenyl-phosphine," J. Chem. Soc. Dalton Trans., 955 (1977).

19. C. D. Garner, R. Durant, and F. E. Mabbs, "Molybdenum(VI) Complexes as Oxidants," Inorg. Chim. Acta, 24:L29 (1977).

20. J. T. Spence, M. Minelli, P. Kroneck, M. I. Scullane, and N. D. Chasteen, "Monomeric Molybdenum(V) Oxo Complexes with Tetradentate Aminoethane-thiols," J. Am. Chem. Soc., 100:8002 (1978).

21. K. D. Magers, C. G. Smith, and D. T. Sawyer, "Reversible Binding of Dioxygen by Tris(3,5-di-tert-butylcatecholato)manganese(III) in Dimethyl Sulfoxide," J. Am. Chem. Soc., 100:989 (1978).

22. K. D. Magers, C. G. Smith, and D. T. Sawyer, "Polarographic and Spectroscopic Studies of the Manganese(II), -(III), and -(IV) Complexes Formed by Polyhydroxy Ligands," Inorg. Chem., 17:515 (1978).

23. P. A. Wicklund and D. G. Brown, "Synthesis and Characterization of Some Cobalt(III) Catechol Complexes," Inorg. Chem., 15:396 (1976).

24. P. A. Wicklund, L. S. Beckmann, and D. G. Brown, "Preparation and Properties of a Stable Semi-quinone Complex," Inorg. Chem., 15:1996 (1976).

25. J. P. Wilshire and D. T. Sawyer, "Reversible Binding of Dioxygen, Nitric Oxide, and Carbon Monoxide by Bis(3,5-di-tert-butylcatecholato)-vanadium(IV)," J. Am. Chem. Soc., 100:3972 (1978).

26. A. D. Goolsby and D. T. Sawyer, "Verratile Solid-State Potentiostat and Amperostat," Anal. Chem., 39:411 (1967).

27. D. F. Evans, "The Determination of the Paramag-
     netic Susceptibility of Substances in Solu-
     tion by Nuclear Magnetic Resonance," J. Chem.
     Soc., 2003 (1959).
28. B. N. Figgis and J. Lewis, "The Magnetochemistry
     of Complex Compounds," in: "Modern Coordina-
     tion Chemistry, Principles and Methods," J.
     Lewis and R. G. Wilkins, eds., Interscience,
     New York, N.Y. (1960), Ch. 6.
29. W. C. Fernelius, K. Terada, and B. E. Bryant,
     "Molybdenum(VI) Dioxyacetylacetonate," Inorg.
     Synth., 6:147 (1960).
30. Carbon and hydrogen analyses were performed by
     Galbraith Laboratories, Inc., Knoxville,
     Tenn., 37921. The molybdenum content was
     determined thermogravimetrically.
31. G. P. Haight, Jr., and V. Paragamian, "Color Com-
     plexes of Catechol with Molybdate," Anal.
     Chem., 32:642 (1960).
32. D. H. Brown and J. D. McCollum, "Complexes of
     Tungstate and Molybdate Ions Containing Two
     Organic Ligands," J. Inorg. Nucl. Chem.,
     25:1483 (1963).
33. R. N. Soni and M. Bartusek, "Molybdate Complexes
     with o-diphenols," J. Inorg. Nucl. Chem.,
     33:2557 (1971).
34. K. Kustin and S.-T. Liu, "Kinetics and Complex
     Formation of Molybdate with Catechol," J. Am.
     Chem. Soc., 95:2487 (1973).
35. K. Gilbert and K. Kustin, "Kinetics and Mechanism
     of Molybdate and Tungstate Complex Formation
     with Catechol Derivatives," J. Am. Chem. Soc.,
     98:5502 (1976).
36. C. G. Pierpont, H. H. Downs, and T. G. Rukavina,
     "Neutral Tris(o-benzoquinone) Complexes of
     Chromium, Molybdenum, and Tungsten," J. Am.
     Chem. Soc., 96:5573 (1974).
37. J. P. Wilshire, L. Leon, P. Bosserman, and D. T.
     Sawyer, "Formation of Bis[bis(3,5-di-tert-
     butylcatecholato)oxomolybdenum(VI)] and
     Related Stable Molybdenum-catechol Complexes,"
     J. Am. Chem. Soc., 101:3379 (1979).
38. R. M. Buchanan and C. G. Pierpont, "Synthesis,
     Structure, and Properties of the Oxygen
     Deficient Bis(3,5-di-tert-butyl)catecholato-
     oxomolybdenum(VI) Dimer, $[MoO(O_2C_6H_2-
     (t-But)_2)_2]_2$," Inorg. Chem., 18:1616 (1979).

39. R. M. Wing and K. P. Callaham, "The Characteriza-
      tion of Metal-oxygen Bridge Systems," Inorg.
      Chem., 8:871 (1969).
40. W. P. Griffith, "Oxy-complexes and Their Vibra-
      tional Spectra," J. Chem. Soc. (A), 211 (1969).
41. The mean values of the proton nmr resonances (vs
      TMS) for the 3,5-di-tert-butyl groups of the
      free catechol, its anion, and the related
      o-quinone are 1.33 ppm and 1.21 ppm, 1.29
      ppm and 1.17 ppm, and 1.18 ppm and 1.16 ppm,
      respectively. The mean resonance values
      (two doublets) for the two ring protons of
      free catechol, its anion and its related
      quinone are 6.69 ppm (6.72 and 6.66), 6.19
      ppm (6.31 and 6.08) and 6.28 (6.49 and 6.07),
      respectively.
42. D. T. Sawyer and E. J. Nanni, Jr., Unpublished
      results.

ELECTROCHEMISTRY OF MOLYBDENUM COMPLEXES.   EFFECT OF STRUCTURE AND

SOLUTION ENVIRONMENT ON REDOX PROPERTIES.

Franklin A. Schultz

Department of Chemistry
Florida Atlantic University
Boca Raton, Florida 33431

INTRODUCTION

   The occurrence of molybdenum in at least six metalloenzymes
having redox function (nitrogenase, nitrate reductase, xanthine
oxidase, aldehyde oxidase, sulfite oxidase, and formate
dehydrogenase)[1] has created a need for greater understanding of the
oxidation-reduction chemistry of this element.   This realization
prompted us to begin an investigation of electrochemical behavior
of suitable inorganic (i.e., model) molybdenum complexes with the
expectation that the results would be useful in interpreting the
redox behavior of this element in its various enzymes.   A number
of important questions can be addressed using contemporary electrochemical techniques such as cyclic voltammetry, potential step
electrolysis, and controlled potential coulometry.   These questions
include: (1) whether electron transfer into the molybdenum center
occurs by multiple steps involving a single electron or by a single
multi-electron step; (2) the redox potentials of the electron
transfer steps; (3) whether electron transfer leads to species with
unusual stability or reactivity; and (4) the extent to which
processes in (1)-(3) are influenced by molecular structure and
solution environment.   Trends have emerged with important
implications for the redox properties of molybdoenzymes.

   The principal compounds investigated are two series of binuclear
Mo(V) complexes of the type $Mo_2X_2Y_2(LL)_2$, where X = terminal O or S,
Y = $\mu-O_2$, $\mu-OS$, or $\mu-S_2$, and LL = N,N-diethyldithiocarbamate
(Et_2dtc), 1,1-dicyanoethylene-2,2-dithiolate (i-mnt), or ethyl
cysteinate (Etcys), and $Mo_2O_2Y_2(L-L)^{2-}$, where $\overline{Y}$ = $\mu-O_2$, $\mu-OS$, or
$\mu-S_2$ and (L-L) = EDTA or (cysteinate)$_2$.   Due to solubility properties,
the former series is examined in nonaqueous aprotic solvents and

the latter in aqueous buffers. Although recent x-ray absorption measurements[2-4] have shown that the molybdenum sites in nitrogenase, sulfite oxidase, and xanthine oxidase are probably mononuclear with respect to molybdenum, these two series of well-characterized binuclear complexes are still effective model compounds because they enable a systematic study to be made of the relationship between molecular structure and redox properties. This is accomplished by replacing bridging or terminal oxygen atoms on molybdenum by sulfur and by coordinating various ligands to the $Mo_2X_2Y_2^{2+}$ cores. For example, in the case of Et_2dtc, the entire family of compounds from $Mo_2O_4(Et_2dtc)_2$ through $Mo_2S_4(Et_2dtc)_2$ has been prepared and studied.[5] In certain cases, it also has been possible to compare the behavior of several different ligands attached to the same $Mo_2X_2Y_2^{2+}$ center. Coordination of EDTA or cysteine produces water-soluble complexes and enables a comparison to be made of electrochemical behavior in protic and aprotic solvents. Also reported are preliminary results for monomeric complexes with catechol (1,2-benzenediol) ligand in water and with 1,1-dithiolate ligands in non-aqueous media. Results from these several investigations reveal that the redox behavior of molybdenum complexes is particularly sensitive to substituents in the bridging and terminal positions of the coordination sphere, that the electrochemistries of mono- and binuclear complexes in a given solvent environment often have many similarities, and that the most significant differences in redox behavior occur upon changing from an aprotic to protic medium.

BINUCLEAR MOLYBDENUM(V) COMPLEXES: NON-AQUEOUS MEDIA

Data for electrochemical reduction of the $Mo_2X_2Y_2(Et_2dtc)_2$, $Mo_2O_2Y_2(Etcys)_2$, and $Mo_2O_2S_2(\underline{i}\text{-mnt})_2^{2-}$ complexes in dimethyl-sulfoxide (DMSO) are summarized in Table 1. Comparative results for other compounds of similar structure are taken from the recent literature.

A common expectation is that single-electron rather than multi-electron transfers will prevail in nonaqueous electrochemistry. The complex $Mo_2O_4(Et_2dtc)_2$ is reduced in an irreversible two-electron step at $\underline{ca}$. -1.46 V vs. SCE, as are most of the other di-μ-oxo compounds in Table 1. However, as sulfur is substituted for oxygen -- first in bridging and then terminal positions -- it becomes apparent that the general pattern of behavior for these compounds is sequential one-electron reductions to Mo(V)-Mo(IV) and Mo(IV)$_2$ species. Figure 1 shows that the remaining $Mo_2X_2Y_2(Et_2dtc)_2$ complexes display successive one-electron reduction waves. For the two sulfido-bridged complexes, $Mo_2O_3S(Et_2dtc)_2$ and $Mo_2O_2S_2(Et_2dtc)_2$, the first but not the second reduction wave becomes reversible at fast scan rates. Both reductions become reversible at fast scan rates for the two terminal sulfido compounds, $Mo_2OS_3(Et_2dtc)_2$ and $Mo_2S_4(Et_2dtc)_2$. The number of electrons transferred in each step

Table 1.   Electrochemical Data for Reduction of $Mo_2X_2Y_2(LL)_2$ Complexes[a]

Compound[b]	$(E_{1/2})_1$	$(E_{1/2})_2$	Electrode Process	Ref
$Mo_2O_4(Et_2dtc)_2$	-1.46(i)		$2e^-$	5
$Mo_2O_4(Etcys)_2$	-1.65	-2.00(i)	$1e^-,1e^-$	6
$Mo_2O_4(tbz)_2(py)_2$[d]	-1.52(i)		$2e^-$	7
$Mo_2O_4Q_2(py)_2$	-1.47(i)		$2e^-$	8
$Mo_2O_4(Prtxn)_2$[e]	-0.45(i)		$2e^-$	9
$Mo_2O_3S(Et_2dtc)_2$	-1.41	-1.74(i)	$1e^-,1e^-{\rightarrow}2e^-$	5
$Mo_2O_3S(Etcys)_2$	-1.52	-1.85(i)	$1e^-,1e^-$	6
$Mo_2O_3S(tbz)_2(py)_2$[d]	-1.31(i)		$2e^-$	7
$Mo_2O_3S(Bu_2dtc)_2$[f]	-1.35	-1.78(i)	$1e^-,1e^-$	10
$Mo_2O_2S_2(Et_2dtc)_2$	-1.30	-1.61(i)	$1e^-,1e^-{\rightarrow}2e^-$	5
$Mo_2O_2S_2(Etcys)_2$	-1.43	-1.59	$1e^-,1e^-$	6
$Mo_2O_2S_2(\underline{i}\text{-mnt})_2^{2-}$[f]	-1.42	-1.70(i)	$1e^-,1e^-$	5
$Mo_2O_2S_2(Bu_2dtc)_2$[f]	-1.28	-1.67(i)	$1e^-,1e^-$	10
$Mo_2OS_3(Et_2dtc)_2$	-0.97	-1.34	$1e^-,1e^-{\rightarrow}2e^-$	5
$Mo_2OS_3(Bu_2dtc)_2$f	-0.92	-1.75(i)	$1e^-,1e^-$	10
$Mo_2S_4(Et_2dtc)_2$	-0.84	-1.27	$1e^-,1e^-{\rightarrow}2e^-$	5
$Mo_2S_4(Bu_2dtc)_2$f	-0.86	-1.48	$1e^-,1e^-$	10
$Mo_2S_4(dme)_2^{2-}$	-1.87		$1e^-$	11

[a] Data recorded for ~1mM complexes in 0.1 M tetraethyl-ammonium perchlorate or tetraethylammonium tetrafluoro-borate in DMSO unless other wise noted.
[b] Ligand abbreviations not given previously are: tbz = thiobenzoate; py = pyridine; Q = 8-quinolinolate; Prtxn = iso-propylthioxanthate; Bu_2dtc = N,N-di-$\underline{n}$-butyldithiocarbamate; dme = dimercaptoethane dianion.
[c] For reversible processes, $(E_{1/2})_1$ and $(E_{1/2})_2$ are reported as the mean of cathodic and anodic peak potentials in V vs. SCE.   Irreversible processes are designated by (i), and the cathodic peak potential at 0.1 V s^{-1} is given.
[d] 0.1 M tetrabutylammonium perchlorate/pyridine
[e] 0.1 M tetrabutylammonium hexafluorophosphate/dichloromethane
[f] 0.1 M tetraethylammonium perchlorate/N,N-dimethylformamide

is verified by comparing voltammetric and chronoamperometric current parameters at short times to those of $Mo_2O_2S_2(\underline{i}\text{-mnt})_2^{2-}$, which undergoes a demonstrably reversible one-electron reduction under all conditions in DMSO.

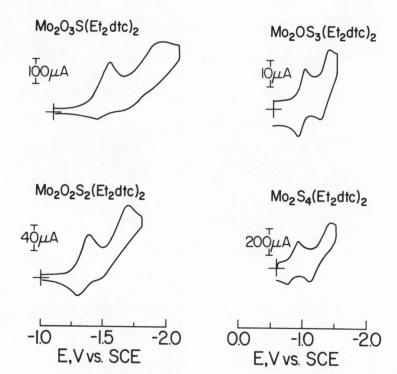

Fig. 1.   Cyclic voltammograms for reduction of oxo- and sulfido-
bridged Mo(V)-diethyldithiocarbamate complexes in DMSO.
Compound, concentration (mM), and scan rate (V s^{-1}) as
follows: $Mo_2O_3S(Et_2dtc)_2$, 1.13, 5; $Mo_2O_2S_2(Et_2dtc)_2$, 0.81,1;
$Mo_2OS_3(Et_2dtc)_2$, 1.00, 20; $Mo_2S_4(Et_2dtc)_2$, 0.60, 20.

At long times, the pair of one-electron waves for the
$Mo_2X_2Y_2(Et_2dtc)_2$ complexes collapses into a single irreversible
two-electron wave.   This behavior is the result of a chemical
reaction of the Mo(V)-Mo(IV) species producing a substance which is
further reducible by one electron at the potential of the first
wave.   Controlled potential coulometry indicates that two electrons
are transferred in the overall reduction of these compounds.   The
mechanism of the intervening chemical reaction is unknown, but it is
apparently a higher than first-order process because a concentration
dependence is observed in the increase in current parameter and
shift in peak potential of the first reduction wave at low scan
rates.   The rate of the reaction decreases with increasing sulfur
substitution.   The di-μ-oxo Et_2dtc complex apparently represents
the extreme case in which the intervening chemical reaction is so
rapid that only irreversible two-electron behavior is observed.

The behavior of the $Mo_2O_2Y_2(Etcys)_2$ complexes is revealing in
this regard. $Mo_2O_4(Etcys)_2$ is reduced by two irreversible one-

electron steps at slow scan rates in DMSO. The chemical reaction consuming the electrode product does not produce a new electroactive substance; thus, a transition from one-electron to two-electron behavior is not observed. A reverse peak following the first $Mo_2O_4(Etcys)_2$ reduction is observed by cyclic voltammetry at scan rates above 20 V s^{-1}. Therefore, the $Mo_2O_4^{2+}$ center can be reduced by a single electron in aprotic solvents. As sulfur is substituted for oxygen in the bridge, similar changes in redox potential and product stability occur as for the Et$_2$dtc complexes, with the result that two reversible one-electron reductions are observed for $Mo_2O_2S_2(Etcys)_2$ at scan rates greater than 2 V s^{-1}. The reaction of the $Mo_2O_2Y_2(Etcys)_2^-$ electrode products has been monitored by cyclic voltammetry and gives a good fit to a first-order EC mechanism. Rate constant data and schemes summarizing the electrochemical behavior of the Et$_2$dtc and Etcys complexes are presented in Figure 2. An interesting but unanswered question is why the intermediates of the Et$_2$dtc and Etcys complexes decay by different mechanisms. The answer may be related to the different coordination geometries around molybdenum in the two Mo(V)$_2$ complexes:[1] approximately square pyramidal for $Mo_2X_2Y_2(Et_2dtc)_2$ and distorted trigonal bipyramidal for $Mo_2O_2Y_2(Etcys)_2$. The more hindered geometry in the latter case may prevent proper alignment between two reactant molecules needed for a second-order reaction.

Replacement of oxygen by sulfur in bridging and terminal positions around molybdenum(V) has two significant effects on electrochemical behavior. The first is to increase the ease of reduction of the binuclear unit in the sequence. $Mo_2O_4 < Mo_2O_3S < Mo_2O_2S_2 < Mo_2OS_3 < Mo_2S_4$. The shift in half-wave potential is ~+100 mV per S atom except for the first substitution at a terminal site where the shift is ~+300 mV. The second effect is to increase the stability of the Mo(V)-Mo(IV) and Mo(IV)$_2$ electrode products in the same sequence. Both these results are understandable in terms of recent electronic state calculations[12,13] of $Mo_2O_4^{2+}$ and $Mo_2O_2S_2^{2+}$ centers. The $Mo_2X_2Y_2^{2+}$ cores are viewed as integral structural units which have as their lowest energy unfilled orbitals Mo-Mo bonding or antibonding molecular orbitals containing substantial contributions from the bridging and terminal atoms. Inclusion of formally empty sulfur d orbitals in calculations for the $Mo_2O_2S_2^{2+}$ center results in significant bonding interactions between terminal and bridging sites and a much more uniform charge distribution over the entire structure. Thus, the presence of sulfur atoms promotes electron delocalization over the binuclear framework. This delocalization would be expected to provide (i) orbitals of lower energy for occupancy by an other electron and (ii) stabilization of reduced species by more effective charge distribution, making them potentially less reactive in decomposition reactions. These expectations are clearly realized in our studies of the Et$_2$dtc and Etcys complexes. This picture also predicts that the formally mixed-valent Mo(V)-Mo(IV) species are best viewed as radical anions in which the unpaired electron is located

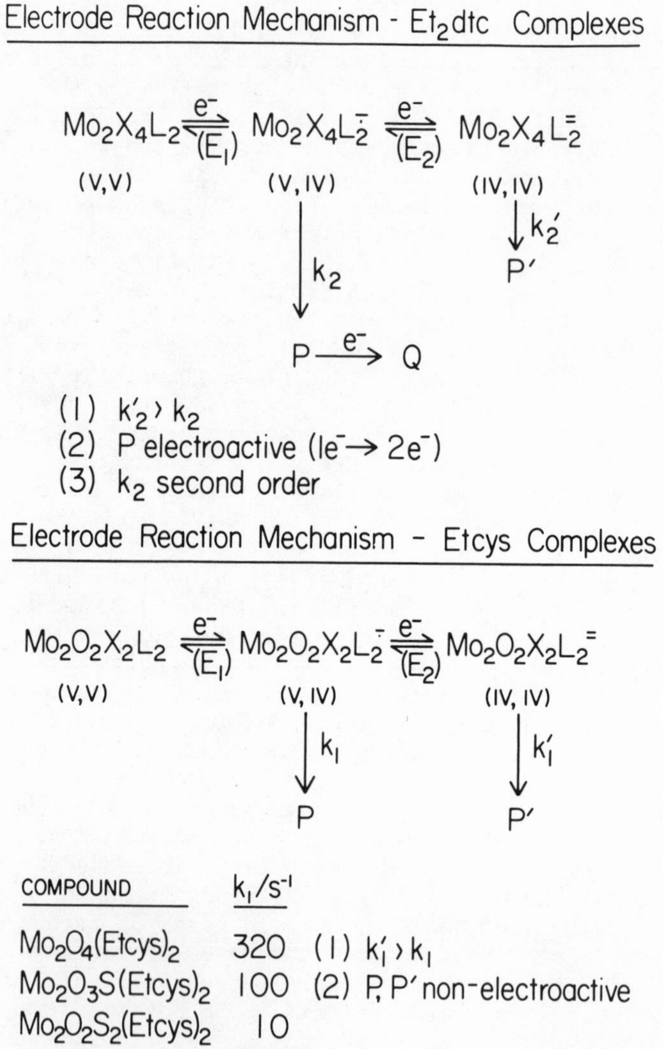

Electrode Reaction Mechanism - Et$_2$dtc Complexes

$$Mo_2X_4L_2 \underset{(E_1)}{\overset{e^-}{\rightleftharpoons}} Mo_2X_4L_2^{\bar{}} \underset{(E_2)}{\overset{e^-}{\rightleftharpoons}} Mo_2X_4L_2^{\bar{\bar{}}}$$

(V,V)   (V,IV)   (IV,IV)

$$\Big\downarrow k_2 \qquad\qquad \Big\downarrow k_2'$$

$$P \qquad\qquad P'$$

$$P \overset{e^-}{\longrightarrow} Q$$

(1) $k_2' > k_2$
(2) P electroactive ($1e^- \rightarrow 2e^-$)
(3) k_2 second order

Electrode Reaction Mechanism - Etcys Complexes

$$Mo_2O_2X_2L_2 \underset{(E_1)}{\overset{e^-}{\rightleftharpoons}} Mo_2O_2X_2L_2^{\bar{}} \underset{(E_2)}{\overset{e^-}{\rightleftharpoons}} Mo_2O_2X_2L_2^{=}$$

(V,V)   (V,IV)   (IV,IV)

$$\Big\downarrow k_1 \qquad\qquad \Big\downarrow k_1'$$

$$P \qquad\qquad P'$$

COMPOUND	k_1/s^{-1}	
Mo$_2$O$_4$(Etcys)$_2$	320	(1) $k_1' > k_1$
Mo$_2$O$_3$S(Etcys)$_2$	100	(2) P, P' non-electroactive
Mo$_2$O$_2$S$_2$(Etcys)$_2$	10	

Fig. 2. Electrode reaction mechanisms for Mo$_2$X$_4$(Et$_2$dtc)$_2$ and
Mo$_2$O$_2$X$_2$(Etcys)$_2$ complexes.

in a molecular orbital distributed over the entire Mo$_2$X$_2$Y$_2{}^{2+}$
framework.

The electronic state calculations also suggest that the
energetics of electron transfer into the binuclear unit should be
controlled more directly by the composition of the Mo$_2$X$_2$Y$_2$ core than
the ligands attached to it. The majority of the compounds in Table 1
fulfill this prediction. The most convincing argument is provided by
Mo$_2$O$_2$S$_2{}^{2+}$ complexes with Et$_2$dtc, Bu$_2$dtc, Etcys, and i-mnt ligands

which undergo reversible one-electron reduction within a range of
only 150 mV. However, one should be cognizant of possible exceptions
to this generalization. The dimercaptoethane complex, $Mo_2S_4(dme)_2{}^{2-}$,
undergoes reversible one-electron reduction a full one volt more
negative than the corresponding dithiocarbamate complex. While a
combination of saturated hydrocarbon framework in the ligand and
dinegative charge on the complex may explain the observed shift in
$(E_{1/2})_1$, neither of these factors operating singly appears to effect
a change in the half-wave potential of $Mo_2O_2S_2(Etcys)_2$ or
$Mo_2O_2S_2(\underline{i}\text{-mnt})_2{}^{2-}$ relative to $Mo_2O_2S_2(Et_2dtc)_2$. The thioxanthate
complex, $Mo_2O_4(S_2CS\text{-}\underline{i}\text{-Pr})_2$, undergoes an irreversible two-electron
reduction at -0.45 V, which is one volt more positive than the other
di-μ-oxo complexes. While comparisons based on irreversible processes have no thermodynamic foundation, the magnitude of the
difference demands an explanation. Zubieta[9,14] suggests that thioxanthate is a less effective electron-donating ligand than the other
1,1-dithiolates and thereby facilitates reduction of the molybdenum
center. There is precedence for this behavior among mononuclear
tetrakis(1,1-dithiolate) complexes (vide infra) where the thioxanthate complex exhibits a far more positive Mo(V/IV) redox potential
than the Et_2dtc and $\underline{i}$-mnt complexes. Additional studies are clearly
needed with this and other ligands in electrochemically reversible
complexes to determine the extent to which the ligand can influence
redox properties of the binuclear Mo(V) core.

BINUCLEAR MOLYBDENUM(V) COMPLEXES: AQUEOUS MEDIA

     Pronounced changes in electrochemical behavior occur when
compounds containing $Mo_2O_2Y_2{}^{2+}$ centers are examined in aqueous media.
All of the $Mo_2O_2Y_2(EDTA)^{2-}$ and $Mo_2O_2Y_2(cys)_2{}^{2-}$ complexes undergo reduction in a single four-electron step from Mo(V) to Mo(III) dimers
in aqueous buffers. The number of electrons involved is established
by comparative measurements of voltammetric and chronoamperometric
current parameters and by controlled potential coulometry. No
evidence of an intermediate oxidation state has been observed in the
reduction process. Electrochemical results for these compounds are
collected in Table 2.

     The four-electron reduction wave exhibits an unusual dependence
on buffer composition which has been studied extensively for the
di-μ-oxo complexes. The cathodic peak potential depends on the
concentration of the protonated buffer species, but not on pH or
concentration of unprotonated buffer. This has been interpreted as
evidence for a concerted electron-proton transfer mechanism in which
the protonated buffer species actually participates in the transition
state of the electrode reaction.[15,16] Proton-bearing species
apparently are required for conversion of terminal Mo=O to Mo-OH$_2$
groups upon reduction from Mo(V) to Mo(III). The overall process
requires four protons and four electrons and is noteworthily parallel

Table 2. Electrochemical and Catalytic Data for $Mo_2O_2Y_2(EDTA)^{2-}$ and $Mo_2O_2Y_2(cys)_2^{2-}$ Complexes in Aqueous Buffers

Compound	$E_{1/2}$[a]	k_1[b]	Catalytic Rate and Product Ratio[c]	Ref
		Acetate (pH 4.7)		
$Mo_2O_4(EDTA)^{2-}$	−0.88	0	0	6
$Mo_2O_3S(EDTA)^{2-}$	−0.70	0	NM	6
$Mo_2O_2S_2(EDTA)^{2-}$	−0.69	0	NM	6
		Borate (pH 9.2)		
$Mo_2O_4(EDTA)^{2-}$	−1.06	0	0	6,15,17
$Mo_2O_3S(EDTA)^{2-}$	−0.99	>0	0.05(1:5)	6,17
$Mo_2O_2S_2(EDTA)^{2-}$	−0.98	>0	0.33(1:3)	6,17
$Mo_2O_4(cys)_2^{2-}$	−1.05	0.004	0.30(4:1)	6,16−18
$Mo_2O_3S(cys)_2^{2-}$	−1.04	1	0.32(4:1)	6,17
$Mo_2O_2S_2(cys)_2^{2-}$	−1.14	100	0.32(5:1)	6,17
		Phosphate (pH 7.3−8.3)		
$Mo_2O_4(EDTA)^{2-}$	−0.99	0	0	15
$Mo_2O_4(cys)_2^{2-}$	−1.08	1.3	0.17(2.4:1)	16,17
		Ammonia (pH 9.3)		
$Mo_2O_4(EDTA)^{2-}$	−1.03	>0	NM	15
$Mo_2O_4(cys)_2^{2-}$	−0.92	0.4	0.11(3.1)	16,17

[a] Data recorded for ~1mM complexes at a hanging Hg drop electrode. $E_{1/2}$ is the mean of cathodic and anodic peak potentials in V vs. SCE.
[b] k_1 is the first-order rate constant in s^{-1} for decay of $Mo(III)_2$ electrode product (see Fig. 3).
[c] The first number cited is the rate of acetylene reduction expressed as µmoles C_2H_4 plus C_2H_6 produced per minute from controlled potential electrolysis at −1.40 V of a 1.7mM solution of the complex under 1 atm C_2H_2 at a 12 cm² Hg pool cathode. The $C_2H_4 : C_2H_6$ ratio is shown in parentheses.

to the coupled transfer to protons and electrons in multiples of two in the known reactions of molybdoenzymes.[19]

Buffer species appear to replace water molecules within the Mo(III) coordination sphere upon reduction of the Mo(V) dimers. Sykes et al.[20] isolated a binuclear complex in which an acetate ion in addition to EDTA and oxo groups was found to bridge the two Mo(III) atoms by coordination at sites formerly occupied by Mo=O groups. We

have observed[15] that the absorption spectrum of electrochemically
reduced $Mo_2O_4(EDTA)^{2-}$ is sensitive to buffer type and that the peak
potential for Mo(III) reoxidation shifts with both buffer type and
concentration. These results indicate coordination or bridging of
the Mo(III) centers by other buffer species (borate, phosphate, and
ammonia) as well. The basic process envisioned for reduction of the
aqueous $Mo_2O_2Y_2^{2+}$ complexes is presented in the first part of Fig.3.

    Substitution of sulfur into bridging positions of aqueous
$Mo_2O_2Y_2^{2+}$ complexes produces changes which both parallel and contrast
with the behavior observed in an aprotic environment. Sulfido bridging enhances both the ease of reduction and electrochemical reversibility of the binuclear centers. The effect is much more pronounced
for the EDTA complexes, and the major change occurs upon substitution
of the first S. Sulfido bridging atoms, however, dramatically <u>decrease</u>
the stability of Mo(III) dimers produced electrochemically in aqueous
media. This phenomenon is much more apparent for the cysteinate than
the EDTA complexes. In cyclic voltammetric experiments, the magnitude
of the Mo(III)$_2$ reoxidation peak at constant scan rate decreases in
the sequence $Mo_2O_4(cys)_2^{2-} > Mo_2O_3S(cys)_2^{2-} > Mo_2O_2S_2(cys)_2^{2-}$. The
first-order rate constant for the decomposition reaction increases
approximately two orders of magnitude with each S substitution.[17]
The decomposition reaction of the $Mo_2O_4(cys)_2^{2-}$ reduction product

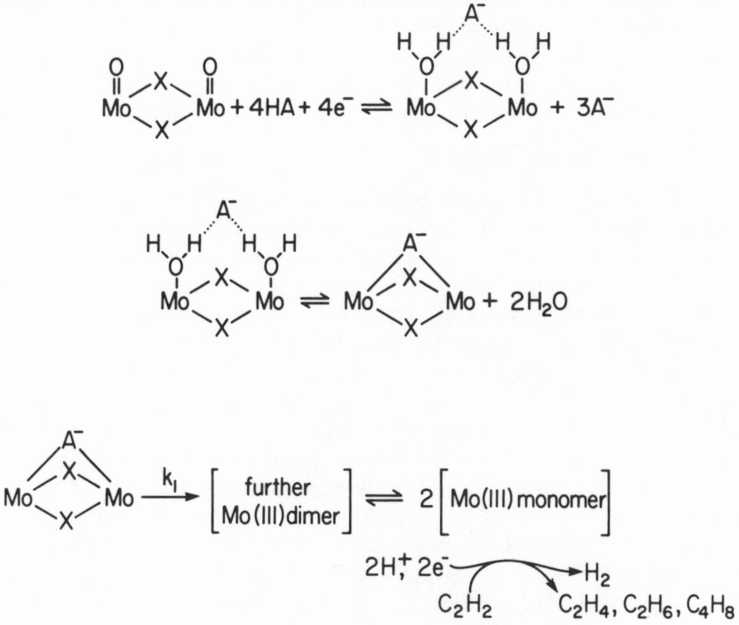

Fig. 3.   Proposed mechanism for electrochemical reduction of
        binuclear molybdenum(V) complexes and catalytic reduction
        of acetylene by Mo(III) complexes.

also has been investigated quantitatively in a number of media by double potential-step chronocoulometry.[16] The data suggest a first-order EC mechanism in which the rate determining chemical step is dependent on buffer type, but not on pH or buffer concentration. An intramolecular, buffer-specific decomposition reaction is consistent with coordination of these species immediately following the electron transfer step as shown in Figure 3. Bridging sulfido apparently exerts its effect by labilizing Mo(III)-S and possibly other bonds within the molybdenum coordination sphere. Similar labilizing effects have been noted for Cr(III)-thiol complexes[21,22] and for sulfido-bridged Mo(V) complexes.[23] Bridging of the two Mo centers by the EDTA backbone in $Mo_2O_2Y_2(EDTA)^{2-}$ greatly stabilizes the Mo(III)$_2$ derivatives of these species. Under most conditions, it is possible to electrochemically cycle $Mo_2O_2Y_2(EDTA)^{2-}$ complexes between the Mo(V)$_2$ and Mo(III)$_2$ states in a quantitative manner. Under other conditions, slow decomposition (detected by changes in uv-visible spectra) occurs following reduction of $Mo_2O_4(EDTA)^{2-}$ in ammonia buffer and of $Mo_2O_3S(EDTA)^{2-}$ and $Mo_2O_2S_2(EDTA)^{2-}$ in borate buffer. The nature of these reactions has not been investigated.

Electrocatalytic Acetylene Reduction

Complexes which undergo decomposition after reduction to the Mo(III)$_2$ state also display catalytic activity. When these compounds are reduced at a potential negative of the Mo(V)$_2$/Mo(III)$_2$ reduction wave in the presence of acetylene, catalytic reduction of C_2H_2 to ethylene, ethane, and small amounts of 1,3-butadiene occurs. Table 2 summarizes some of these results.

The rate of catalytic acetylene reduction increases linearly with increasing H^+ concentration, exponentially with negative electrode potential, and linearly with the square root of initial Mo(V) dimer concentration. The last fact in conjunction with the number of electrons required to reduce the Mo(V)$_2$ unit suggests that the catalytically active material is a Mo(III) monomer. Production of this substance from electrochemically generated Mo(III)$_2$ complexes is more complicated than represented in Figure 3, but the number and nature of intermediate steps have not been determined. It is certain that the initial Mo(III)$_2$ electrode products are not present in catalytic systems, since their characteristic voltammetric oxidation peaks and electronic absorption bands are absent. Dissociation of Mo(III) centers appears to be a necessary prerequisite for catalysis. When Mo(V) dimers such as $Mo_2O_4(EDTA)^{2-}$ in borate buffer are reduced to a binuclear Mo(III) product which remains structurally intact, no catalysis is observed. Another important feature of these systems is that catalysis ensues following reduction to the Mo(III) state only when potential is continuously applied to the electrode. If acetylene is added to a prereduced solution without applying potential, no reduction of C_2H_2 is observed. This result and the exponential

dependence of rate on electrode potential suggest that the catalyst
is adsorbed on the electrode surface. In this state, the dissociated
Mo(III) center can bind acetylene and mediate the transfer of
electrons to it from the electrode. The adsorbed material also
catalyzes the reduction of H^+ to H_2 at the electrode surface, which
occurs as a major current-consuming process concurrent with acetylene
reduction.

The electrocatalytic system is complicated, and a number of important factors regarding it need to be clarified. These include
characterization of the catalytically active species, the mechanism
of its production from the $Mo(III)_2$ electrode product, the mechanism
of acetylene reduction, and whether other nitrogenase substrates are
reduced. However, several important features have emerged which may
be significant relative to the chemistry of molybdoenzymes. These
include: (1) Mo(III) is viable as a catalytically active oxidation
state, (2) dissociation of molybdenum centers is a prerequisite to
catalytic activity, and (3) coordinated or bridging sulfido groups
facilitate this process as well as electron transfer into the
molybdenum center.

MONONUCLEAR MOLYBDENUM COMPLEXES: AQUEOUS AND NON-AQUEOUS MEDIA

Association of catalytic activity with monomeric species and
evidence for an active site in enzymes that is mononuclear with
respect to molybdenum have focused attention on these compounds.
Electrochemistry of tris(dithiolene)-molybdenum complexes was investigated several years ago.[24] However, the "non-innocent" nature of
these ligands leads to significant metal-ligand delocalization of
electrons and places the redox properties of these compounds in a
special category. Recently, several tetrakis complexes with more
nearly innocent 1,1-dithiolate ligands (MoS_8 coordination sphere)
have been investigated. These results are presented in Table 3.

The dithiocarbamate (Et_2dtc) and i-mnt complexes appear to
represent one class of behavior, and the thioxanthate (Prtxn) and
dithiobenzoate (dtb) complexes another. The Et_2dtc and i-mnt
complexes exhibit three sequential one-electron transfers encompass-
ing the Mo(VI) through Mo(III) oxidation states. The Prtxn and dtb
complexes display Mo(V/IV) and Mo(IV/III) couples. Both the mononuclear complexes in Table 3 and the sulfido-bridged binuclear
complexes in Table 1 undergo sequential single electron transfers,
behavior which is expected in aprotic, non-aqueous solvents. However,
the separation between successive half-wave potentials is much
greater for monomeric species, consistent with the greater difficulty
of transferring electrons sequentially into a monometallic site. The
most interesting result in Table 3 is the extremely positive value of
$E_{1/2}$ for the Mo(V/IV) couple of the Prtxn and dtb complexes relative
to the Et_2dtc and i-mnt complexes. A similar positive shift was

Table 3. Voltammetric Half-Wave Potentials of Molybdenum-
Tetrakis(1,1-dithiolate) Complexes[a]

Compound	VI/V	V/IV	IV/III	Ref.
$Mo(S_2CNEt_2)_4$	+1.16	-0.50	-1.22(i)	25
$Mo[S_2C_2(CN)_2]_4^{4-}$	+0.65	-0.71	-1.33(i)	26
$Mo(S_2CS-\underline{i}-Pr)_4$		+0.38	-1.14	14
$Mo(S_2CC_6H_5)_4$		+0.59	-0.79(i)	14

[a]Half-wave potentials in V vs. SCE in 0.1 M tetraethylammonium
tetrafluoroborate/acetonitrile for $Mo[S_2C_2(CN)_2]_4^{4-}$ and in
0.1 M tetrabutylammonium hexafluorophosphate/dichloromethane
for remaining compounds. Irreversible processes are designated
(i). The formula for the Mo(IV) oxidation state is listed.

noted in the peak potential for the irreversible two-electron reduction of $Mo_2O_4(Prtxn)_2$. The prevailing explanation of these results
is that Et_2dtc and $\underline{i}$-mnt are more effective electron-donating ligands
than Prtxn and dtb and make reduction of the Mo(V) center more difficult.[9,14,25] However, relationships between redox properties and
structural, magnetic, and spectroscopic properties of these compounds
remain to be fully explored. The extent to which the 1,1-dithiolate
ligands influence the potential of the Mo(V/IV) couple is remarkable,
and the observed dependence of redox potential on ligand could prove
to be relevant to the behavior of molybdenum in enzymatic systems.

Although binuclear and polynuclear species are prevalent in the
aqueous chemistry of molybdenum, at least one mononuclear system with
well-behaved electrochemistry exists. The cis-dioxo Mo(VI) complex,
$MoO_2L_2^{2-}$, where L^{2-} = catechol dianion, undergoes reversible electrochemical reduction to monomeric Mo(V) and Mo(III) species in aqueous
buffers[27,28] containing an excess of catechol to stabilize complex
formation. The two processes are observed as fully reversible one-
and two-electron reductions at moderate scan rates by cyclic voltammetry. Between pH 3 and 7, the half-wave potentials follow the
relationships $(E_{1/2})_1$ = +0.206 - 0.116 pH and $(E_{1/2})_2$ =-0.547 - 0.058
pH. The pH dependences are consistent with the uptake of two protons
in each electron transfer step and conversion of an oxo group to a
coordinated water molecule.

$$MoO_2L_2^{2-} + 2H^+ + e^- \rightleftarrows MoO(H_2O)L_2^- \qquad (E_{1/2})_1 \qquad (1)$$

$$MoO(H_2O)L_2^- + 2H^+ + 2e^- \rightleftarrows Mo(H_2O)L_2^- \qquad (E_{1/2})_2 \qquad (2)$$

If a complexing agent, such as pyridine, is present in addition to
excess catechol, this molecule can replace the H_2O formed at the
Mo=O site. Thus, the species $MoO(py)L_2^-$ and $Mo(py)_2L_2^-$ result upon

reduction to the Mo(V) and Mo(III) states, respectively, in excess
pyridine. From the shift in half-wave potentials with pyridine concentration, the following formation constants have been evaluated:

$$MoO(H_2O)L_2^- + py \rightleftarrows MoO(py)L_2^- + H_2O \qquad \beta_1^V = 40 \text{ M}^{-1} \qquad (3)$$

$$Mo(H_2O)_2L_2^- + 2py \rightleftarrows Mo(py)_2L_2^- + 2H_2O \qquad \beta_2^{III} = 6 \times 10^5 \text{ M}^{-2} \quad (4)$$

If no molecule capable of occupying the water-coordinated site is
present, the complex dimerizes after reduction to the Mo(V) state.
This reaction occurs even in the presence of excess pyridine, because
coordination of $MoO(H_2O)L_2^-$ by py is not complete, and equilibrium
(3) is sufficiently labile to shift in the direction of dimer formation. The rate of the dimerization reaction increases with decreasing pH, pyridine concentration, and catechol concentration. The
mechanism appears to be:

$$MoO(H_2O)L_2^- + 2H_2O \rightleftarrows MoO(OH)_2(H_2O)L^- + H_2L \qquad (5)$$

$$2 \text{ MoO(OH)}_2(H_2O)L^- + H^+ \rightarrow Mo_2O_4L_2^{2-} + 3H_2O + H_3O^+ \qquad (6)$$

$$2 \text{ MoO(OH)}_2(H_2O)L^- \longrightarrow Mo_2O_4L_2^{2-} + 4H_2O \qquad (7)$$

Reactions (5)-(7) involve formation of the reactive intermediate, a
monomeric 1:1 Mo(V)-catechol complex, and its dimerization via acid-
dependent and acid-independent pathways.

    In amine buffers (ammonia, "tris," and ethylenediamine) at pH >7
$MoO_2L_2^{2-}$ is reduced to red-orange Mo(V) and yellow Mo(III) monomers
which are stable on the time scale of controlled potential coulometry
experiments. The Mo(V) complex in ammonia buffer has a visible ab-
sorption band at 470nm with $\varepsilon \simeq 2{,}500$ M^{-1}cm^{-1} [29] and a room temperature epr spectrum with a single strong line and six satellite lines
due to ^{95,97} Mo hyperfine splitting.[26] The epr parameters, g=1.939
and A(^{95,97}Mo) = 49G, are similar to those reported by Trachevskii[30]
for monomeric Mo(V)-catechol complexes. The Mo(V) epr signal disappears when the complex is reduced to the Mo(III) state. Stabilization of the Mo(V) monomer is attributed to replacement of H$_2$O by a
non-labile ligand, X, in alkaline media to give MoO(X)L$_2^-$. Whether
X is an amine, hydroxide ion, or a third catechol molecule acting as
a monodentate ligand is unknown.

    The electrochemistry of the monomeric molybdenum-catechol
complexes is similar in many respects to that of binuclear Mo(V)
species in aqueous media. In both systems, Mo(V) to Mo(III) reduction
is accompanied by conversion of a terminal oxo to coordinated aquo
group, which can be substituted by a buffer molecule. It is intriguing that the Mo(IV) state is bypassed in these reductions, particularly as a number of aquomolybdenum(IV) species have been identified
recently.[31-34] A possible answer is that Mo(IV) has special

structural requirements that make it inaccessible from the Mo(V) structures studied in this work. However, we have noted[29] that, if $MoO_2L_2^{2-}$ is exhaustively reduced to the Mo(III) state in alkaline media and reoxidized by one electron, a stable, red Mo(IV) complex is produced. The composition and means of formation of this compound are under continuing investigation in our laboratory.

ACKNOWLEDGMENT

The support of this research by the National Science Foundation under Grant No. CHE 76-18703 is gratefully acknowledged. The author also wishes to thank his coworkers, whose efforts are cited in the references, and Dr. W. E. Newton and his staff of the Charles F. Kettering Research Laboratory for preparing many of the compounds discussed in this work and for many thoughtful discussions on the chemistry of molybdenum.

REFERENCES

1. E. I. Stiefel, The coordination and bioinorganic chemistry of molybdenum, Progr. Inorg. Chem., 22:1 (1977).
2. S. P. Cramer, L. O. Hodgson, W. O. Gillum, and L. E. Mortenson, The molybdenum site of nitrogenase. Preliminary structural information from x-ray absorption spectroscopy, J. Amer. Chem. Soc., 100:3398 (1978).
3. S. P. Cramer, H. B. Gray, and K. V. Rajagopalan, The molybdenum site of sulfite oxidase. Structural information from x-ray absorption spectroscopy, J. Amer. Chem. Soc., 101:2772 (1979).
4. T. D. Tullius, D. M. Kurtz, Jr., S. D. Conradson, and K. O. Hodgson, The molybdenum site of xanthine oxidase. Structural evidence from x-ray absorption spectroscopy, J. Amer. Chem. Soc., 101:2776 (1979).
5. F. A. Schultz, V. R. Ott, D. S. Rolison, D. C. Bravard, J. W. McDonald, and W. E. Newton, Synthesis and electrochemistry of oxo- and sulfido-bridged molybdenum(V) complexes with 1,1-dithiolate ligands, Inorg. Chem., 17:1758 (1978).
6. F. A. Schultz, V. R. Ott, and D. S. Swieter, Di-μ-oxo, μ-oxo-μ-sulfido, and di-μ-sulfido complexes of molybdenum(V) with EDTA, cysteine, and cysteine ester ligands. Preparation and electrochemical and spectral properties, Inorg. Chem., 16:2538 (1977).
7. M. R. DuBois, Synthesis and characterization of a series of thiobenzoate complexes of molybdenum(V), Inorg. Chem., 17:2405 (1978).
8. J. K. Howie and D. T. Sawyer, Electrochemical studies of oxo- and sulfido-bridged binuclear molybdenum(V) complexes in aprotic media, Inorg. Chem., 15:1892 (1976).

9. J. Hyde, K. Venkatasubramanian, and J. Zubieta, Synthesis and chemical and structural investigation of molybdenum(IV) thioxanthate complexes. An unusual example of η^3 coordination for a dithio acid ligand, Inorg. Chem., 17:414 (1978).

10. H. C. Faulkner III, Electrochemical studies of oxo- and sulfido-bridged molybdenum(V) dimers, J. Electrochem. Soc., 125:287C (1978).

11. G. Bunzey, J. H. Enemark, J. K. Howie, and D. T. Sawyer, Molybdenum complexes of aliphatic thiols. Isolation and characterization of two isomeric forms of the redox active binuclear Mo(V) anion, $[Mo_2S_4(S_2C_2H_4)_2]^{2-}$, J. Amer. Chem. Soc., 99:4168 (1977).

12. D. H. Brown, P. G. Perkins, and J. J. Stewart, The electronic states of the dioxo-di-μ-oxo-dimolybdate(V) group, J. Chem. Soc., Dalton Trans., 1105 (1972).

13. D. H. Brown and P. G. Perkins, The electronic structure and spectra of the di-μ-sulfido-bis[oxo-(L-cysteinato)molybdate-(V)]$^{2-}$ ion, Rev. Roum. Chem., 20:515 (1975).

14. J. Hyde and J. Zubieta, Preparation and characterization of tetrakis(thioxanthato)molybdenum(IV) complexes, J. Inorg. Nucl. Chem., 39:289 (1977).

15. V. R. Ott and F. A. Schultz, Electrochemical reduction of a binuclear dioxo-bridged molybdenum(V) complex with ethyl-enediaminetetraacetate, J. Electroanal. Chem., 59:47 (1975).

16. V. R. Ott and F. A. Schultz, Electrochemical reduction of a binuclear dioxo-bridged molybdenum(V) complex with cysteine, J. Electroanal. Chem., 61:81 (1975).

17. F. A. Schultz, D. A. Ledwith, and L. O. Leazenbee, Electro-chemically catalyzed reduction of nitrogenase substrates by binuclear molybdenum(V) complexes. ACS Symp. Ser., 38:78 (1977).

18. D. A. Ledwith and F. A. Schultz, Catalytic electrochemical reduction of acetylene in the presence of a molybdenum-cysteine complex, J. Amer. Chem. Soc., 97:6591 (1975).

19. E. I. Stiefel, Proposed molecular mechanism for the action of molybdenum in enzymes: coupled proton and electron transfer, Proc. Nat. Acad. Sci. USA, 70:988 (1973).

20. G. G. Kneale, A. J. Geddes, Y. Sasaki, T. Shibahara, and A. G. Sykes, Preparation and crystal structure of potassium salt of the binuclear molybdenum(III) complex, μ-acetato-di-μ-hydroxo-μ(NN')ethylenediaminetetraacetato-bis[molybdenum(III)], $K[Mo_2(OH)_2(O_2CMe)(C_{10}H_{12}O_8N_2)]$, J. Chem. Soc., Chem. Commun., 356 (1975).

21. T. Ramasami and A. G. Sykes, Further characterization and aquation of the thiolopentaaquochromium(III) complex, $CrSH^{2+}$, and its equilibration with thiocyanate, Inorg. Chem., 15:1010 (1976).

22. L. E. Asher and E. Deutsch, Effect of added ligands on the rate of chromium-sulfur bond fission in thiolatopentaaquochromium complexes. Evidence for a sulfur-induced trans effect in chromium(III) chemistry, Inorg. Chem., 15:1531 (1976).

23. F. A. Armstrong, T. Shibahara, and A. G. Sykes, Effect of
    μ-sulfido ligands on substitution at molybdenum(V). A
    temperature-jump study of the 1:1 equilibration of
    thiocyanate with di-μ-sulfido-bis[aquooxalatooxomolybde-
    num(V)], Inorg. Chem., 17:189 (1978).

24. J. A. McCleverty, Metal 1,2-dithiolene and related complexes,
    Progr. Inorg. Chem., 10:49 (1968).

25. A. Nieuwpoort and J. J. Steggerda, Eight-coordinated 1,1-dithio
    ligand complexes of molybdenum and tungsten, Rec. Trav. Chim.,
    95:250 (1976).

26. F. A. Schultz and W. E. Newton, unpublished work.

27. J. Zelinka, M. Bartusek, and A. Okac, Polarography of
    o-diphenol chelates of molybdenum, Coll. Czech. Chem.
    Commun., 38:2898 (1973).

28. L. M. Charney and F. A. Schultz, manuscript in preparation.

29. L. M. Charney, H. O. Finklea, and F. A. Schultz, manuscript
    in preparation.

30. V. V. Trachevskii and V. V. Lukachina, Interaction of
    molybdenum(V) with pyrogallol, pyrocatechol, and tiron in
    aqueous medium, Zh. Neorg. Khim., 21:117 (1976).

31. M. Ardon and A. Pernick, Molybdenum(IV) in aqueous solution,
    J. Amer. Chem. Soc., 95:6871 (1973).

32. T. Ramasami, R. S. Taylor, and A. G. Sykes, The monocluear
    nature of aquomolybdenum(IV) ion in solution, J. Amer. Chem.
    Soc., 97:5918 (1975).

33. M. Lamache, Existence and properties of Mo(IV) in aqueous
    solution, J. Less-Common Metals, 39:179 (1975).

34. P. Chalilpoyil and F. C. Anson, Electrochemical properties
    of aquomolybdenum ions in noncomplexing acidic electrolytes,
    Inorg. Chem., 17:2418 (1978).

KINETIC STUDY ON THE DISPROPORTIONATION EQUILIBRIUM OF μ-OXO-BIS-[OXOBIS(DITHIO CHELATE)MOLYBDENUM(V)] AND RELATED COMPOUNDS

Toshio Tanaka, Koji Tanaka, Tsuneo Matsuda, and Kazu-aki Hashi

Department of Applied Chemistry, Faculty of Engineering, Osaka University, Suita, Osaka 565, Japan

INTRODUCTION

Dimeric μ-oxomolybdenum(V) complexes of the $Mo_2O_3L_4$ type (L = dialkyldithiocarbamate, diphenyldithiophosphinate, etc.) are known to undergo disproportionation in solution to establish an equilibrium mixture of oxomolybdenum(IV) and dioxomolybdenum(VI) complexes, $MoOL_2$ and MoO_2L_2, as follows.[1,2]

$$Mo^V_2O_3L_4 \quad \overset{K}{\rightleftharpoons} \quad Mo^{IV}OL_2 \quad + \quad Mo^{VI}O_2L_2$$

The equilibrium constant K for $L = S_2CNPr^n_2$ was determined to be 4×10^{-3} mol/l in chlorobenzene at 41°C.[3] This sort of equilibrium has been suggested possibly to play an important role in both molybdenum reductases and molybdenum oxidases.[1] Thus, it is of interest to examine the kinetics for such systems.

This paper reports a kinetic study of the disproportionation equilibria of five μ-oxo-bis[oxomolybdenum(V)] complexes of the $Mo_2O_3L_4$ type ($L = S_2CNEt_2^4(\underset{\sim}{1})$, $SSeCNEt_2^4(\underset{\sim}{2})$, $Se_2CNEt_2^4(\underset{\sim}{3})$, $S_2P(OEt)_2^2(\underset{\sim}{4})$, and $S_2PPh_2^2(\underset{\sim}{5})$) by the concentration-jump relaxation technique.[5]

DISPROPORTIONATION EQUILIBRIA

Table I lists the infrared ν(Mo=O) frequencies of $\underset{\sim}{1}$-$\underset{\sim}{5}$. All the compounds in Nujol mulls exhibit only one ν(Mo=O) band. In dibromomethane, however, they display three additional bands in the ν(Mo=O) region, of which the highest frequency one and the

Table I. $\nu(Mo=O)$ Frequencies of $Mo_2O_3L_4$ (cm^{-1})

L	Medium	$\nu(Mo=O)$			
S_2CNEt_2	Nujol		936		
	CH_2Br_2	967	936	916	879
$SSeCNEt_2$	Nujol		937		
	CH_2Br_2	956	936	913	874
Se_2CNEt_2	Nujol		930		
	CH_2Br_2	961	931	907	882
$S_2P(OEt)_2$	Nujol		963		
	CH_2Br_2	1008[a]	966	932	898
S_2PPh_2	Nujol		960		
	CH_2Br_2	982	959	930	893

[a]This wavenumber is compared with that of the band
(1000 cm^{-1}) observed in $MoO[S_2P(OEt)_2]_2$, which has
been assigned to $\nu(Mo=O)$; R. N. Jowitt and P. C. H.
Mitchell, J. Chem. Soc., (A), 2631 (1969).   A
recent report has suggested that the $\nu(Mo=O)$ band
appears at 972 cm^{-1}; W. E. Newton and J. N. McDonald,
J. Less-Common Met., 54:51 (1977).   A dibromomethane
solution of $Mo_2O_3[S_2P(OEt)_2]_4$, however, exhibits
no absorption in the 970-980 cm^{-1} range.

lowest frequency two are assigned to $\nu(Mo=O)$ of $MoOL_2$ and MoO_2L_2,
respectively.   These assignments are confirmed by comparison with
the spectra of authentic $MoOL_2$ and MoO_2L_2 in dibromomethane, except
for $MoO_2[S_2P(OEt)_2]_2$ which is an unknown compound.   It is, therefore, suggested that 1-5 undergo a partial disproportionation as
follows.

$$Mo^V_2O_3L_4 \underset{k_{-1}}{\overset{k_1}{\rightleftharpoons}} Mo^{IV}OL_2 + Mo^{VI}O_2L_2 \qquad (1)$$

Electronic spectra of 1, 2, 4, and 5 in 1,2-dichloroethane at
room temperature showed absorption maxima at 512, 526, 499, and
508 nm, respectively, assignable to electronic transitions from
the three-center Mo-O-Mo π-bonding orbital to the corresponding
anti-bonding orbital.[6]   The analogous band of 3 was concealed by
an intense absorption in the higher energy region.   At 6°C,
however, this band appeared at 544 nm as a shoulder on the tail of

the ligand absorption.   This observation is possible because the
equilibrium of Eq. 1 shifts to the left with lowering temperature,
resulting in an increasing amount of the Mo(V) complex.

KINETICS

Concentration-jump experiments were carried out with a UNION
GIKEN stopped-flow spectrophotometer RA-401 under an atmospheric
pressure of dry nitrogen.   Rates of the disproportionation reaction
were determined by monitoring the absorbance around 500 nm, which
does not obey the Lambert-Beer law.   When 1,2-dichloroethane
solutions of the Mo(V) complexes (0.2-2.0 mmol/l) were mixed with
the same volume of 1,2-dichloroethane, the intensity of the band
around 500 nm decreases exponentially with half-lives of 0.1-2.6
sec.   The Guggenheim's plots of the absorbance on the concentration-
jump gave a straight line (the correlation coefficient of 0.999 over
a period of 3τ), the slope of which gave the relaxation time τ
($\tau = 1/k_{obsd}$), as shown in Eq. 2, where A_t and $A_{t+\alpha}$ are absorbances

$$\ln(A_{t+\alpha} - A_t) = -t/\tau + \ln(A_\infty - A_o)(1 - e^{-\alpha/\tau}) \qquad (2)$$

at several times, t, and at a further set of times, $t+\alpha$; A_o and A_∞
are the absorbances at the beginning and at the end of reaction,
respectively.   The τ value in the equilibrium reaction (Eq. 1) is
written as Eq. 3,[7] where $[Mo_2O_3L_4]_o$ stands for the concentration

$$(1/\tau)^2 = 4k_1k_{-1}[Mo_2O_3L_4]_o + (k_1)^2 \qquad (3)$$

of $Mo_2O_3L_4$ when the equilibrium is assumed to be shifted to the
left completely.

Figure 1 shows the plots of $(1/\tau)$ vs. $[Mo_2O_3L_4]_o$ at 25°C, from
which the rate constants k_1 and k_{-1}, and therefore the equilibrium
constant K were calculated.   The results are summarized in Table II,
which reveals that the K value of the carbamate complexes increases
in the order of $\underset{\sim}{1} < \underset{\sim}{2} < \underset{\sim}{3}$; the coordination of selenium displaces the
equilibrium (Eq. 1) to the right more than that of sulfur.   The
large K value for $\underset{\sim}{3}$ is compatible with the fact that the characteristic band of the binuclear μ-oxomolybdenum complex was obscured by
the tail of the ligand absorption at room temperature, as described
above.

The Arrhenius plots for the rate constants of the forward and
reverse reactions of Eq. 1, obtained at four different temperatures
in the 5-35°C range, gave a straight line for each compound, from
which activation parameters were calculated.   The results are
collected in Table III.   The activation enthalpies for the disproportionation of $Mo_2O_3L_4$ are considerably larger than those for the

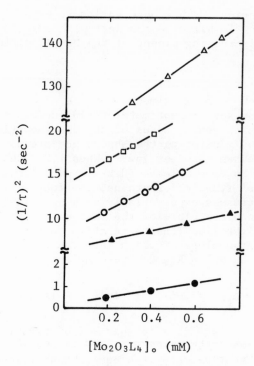

Figure 1.   Plots of $(1/\tau)^2$ against $[Mo_2O_3L_4]_o$ in 1,2-dichloroethane at 25°C; $L = S_2CNEt_2$ (o), SSeCN-Et_2 (□), Se_2CNEt_2 (△), $S_2P(OEt)_2$ (▲), and S_2PPh_2 (●).

Table II.   Rate and Equilibrium Constants for the Disproportionation of $Mo_2O_3L_4$ in 1,2-Dichloroethane at 25°C[a]

L	k_1 (sec^{-1})	$k_{-1} \times 10^{-2}$ (sec^{-1}M^{-1})	$K \times 10^3$ (M^{-1})
S_2CNEt_2	2.93 ±0.29	14.4 ±0.2	2.0 ±0.2
$SSeCNEt_2$	3.78 ±0.11	9.0 ±1.0	4.2 ±0.5
Se_2CNEt_2	10.8 ±0.1	8.3 ±0.1	13.0 ±0.2
$S_2P(OEt)_2$	1.74 ±0.06	4.5 ±0.1	3.9 ±0.2
S_2PPh_2	0.50 ±0.02	8.7 ±0.8	0.58±0.06

[a]Errors were estimated at the 95% confidence level.

Table III.   Activation Parameters for the Disproportionation of $Mo_2O_3L_4$ and the Coupling Reaction of
$MoOL_2$ with MoO_2L_2; the Values for the Latter Reaction
are in Parentheses

	E_a (kJ/mol)	$\Delta H^{\ddagger}_{298}$ (kJ/mol)	$\Delta S^{\ddagger}$ (J/mol K)
S_2CNEt_2	73.8 ±1.7 (23.9 ±1.0)	71.3 (21.4)	0.6 ± 3.7 (−113 ± 2)
$SSeCNEt_2$	81.3 ±1.5 (18.2 ±4.7)	78.8 (15.7)	30.5 ± 0.9 (−135 ±32)
Se_2CNEt_2	56.0 ±4.6 (31.1 ±6.7)	53.5 (28.6)	−45.5 ± 3.5 (−93 ±27)
$S_2P(OEt)_2$	75.3 ±3.1 (18.8 ±1.0)	72.8 (16.3)	8 ±24 (−138 ± 7)
S_2PPh_2	73.6 ±1.1 (46.3 ±0.5)	71.1 (43.8)	−11.2 ± 6.8 (−42.0 ± 3.3)

coupling reaction of $MoOL_2$ with MoO_2L_2.   On the other hand, the
activation entropies for the disproportionation reaction are small
positive or negative values, and those for the coupling reaction
are large negative values.   These results suggest that the Mo-O-Mo
linkage is weakened to a great extent, but the binuclear structure
is still maintained, in the transition state for the disproportionation of the Mo(V) complexes.

THERMODYNAMICS

The molar extinction coefficient of 1 at 512 nm ($\varepsilon_{512 \ nm}$: 24000
$cm^{-1}M^{-1}$) was determined from the equilibrium constant and those of
$Mo^{IV}O(S_2CNEt_2)_2$ and $Mo^{VI}O_2(S_2CNEt_2)_2$ at 512 nm (90 and 75 $cm^{-1}M^{-1}$,
respectively).   This value is in good agreement with that reported
previously (24500 $cm^{-1}M^{-1}$ in benzene[8]).   The values of 2 ($\varepsilon_{526 \ nm}$:
23000 $cm^{-1}M^{-1}$), 3 ($\varepsilon_{544 \ nm}$: 8100 $cm^{-1}M^{-1}$), 4 ($\varepsilon_{499 \ nm}$: 22000
$cm^{-1}M^{-1}$), and 5 ($\varepsilon_{508 \ nm}$: 6400 $cm^{-1}M^{-1}$) were similarly calculated
from the equilibrium constants and the sum of ε values of the
corresponding Mo(IV) and Mo(VI) complexes obtained from the visible
spectra.

Thermodynamic quantities for the disproportionation of the
Mo(V) complexes were obtained from plots of ln k against 1/T.   The
results are listed in Table IV, which shows fairly small ΔH and ΔS

Table IV.   Thermodynamic Quantities for the Disproportionation of $Mo_2O_3L_4$

L	ΔH (kJ/mol)	ΔS (J/mol K)	ΔG_{298} (kJ/mol)
S_2CNEt_2	50.5 ±2.4	116 ±14	15.9 ±4.9
$SSeCNEt_2$	63.1 ±4.8	166 ±29	13.7 ±6.8
Se_2CNEt_2	24.9 ±2.7	47 ±19	10.8 ±4.4
$S_2P(OEt)_2$	56.5 ±3.5	146 ±31	13.0 ±1.6
S_2PPh_2	27.4 ±0.5	31 ±19	18.2 ±0.7

values for 3 and 5 compared with those for 1, 2, and 4.   One
possible explanation for this is that the Mo-O-Mo bonds in 3 and 5
are bent and/or their d_π-p_π-d_π orbitals are twisted with respect
to one another decreasing the contribution of the three-center
π-bond.   This assumption is consistent with the small ε values of
3 and 5.

REFERENCES

1. W. E. Newton, J. L. Corbin, D. C. Bravard, J. E. Searles,
   and J. W. McDonald, Preparation and characterization
   of two series of dimeric molybdenum(V) N,N-dialkyl-
   dithiocarbamates.   Their interrelationship and chemistry as a model for the active site of nitrogenase,
   Inorg. Chem. 13:1100 (1974).
2. G. J.-J. Chen, J. W. McDonald, and W. E. Newton, Synthesis of Mo(IV) and Mo(V) complexes using oxo
   abstraction by phosphines.   Mechanistic implication,
   Inorg. Chem. 15:2612 (1976).
3. R. Barral, C. Bocard, I. S. de Roch, and L. Sajus,
   Activation de l'oxygene moleculaire en phase
   homogene par des complexes oxo du molybdene oxydation
   catalytique selective des phosphines tertiaires,
   Tetrahedron Lett. 1693 (1972).
4. T. Matsuda, K. Tanaka, and T. Tanaka, Kinetic study on
   the disproportionation equilibrium of μ-oxo-bis[oxo-
   bis(N,N-diethyldichalcogenocarbamato)molybdenum(V)],
   Inorg. Chem. 18:454 (1979).
5. C. F. Bernasconi, "Relaxation Kinetics," Academic Press,
   New York (1976), p.240.
6. E. I. Stiefel, The coordination and bioinorganic chemistry of molybdenum in "Progr. Inorg. Chem., Vol. 22,"
   S. J. Lippard, ed., Interscience, New York (1977) 1.

7. Ref. 5, p.3 and p.13.
8. D. B. McDonald and J. I. Shulman, Spectrophotometric
   determination of triphenylphosphine in dilute
   solutions, Anal. Chem. 47:2023 (1975).

OXOMOLYBDENUM COMPLEXES OF CYSTEINE-CONTAINING PEPTIDES:

SOLUTION STRUCTURES AND OXIDATION CATALYSIS

Akira Nakamura and Norikazu Ueyama

Department of Polymer Science, Faculty of Science
Osaka University
Toyonaka, Osaka, Japan 560

INTRODUCTION

Cis-dioxomolybdenum(VI) and di-μoxodimolybdenum(V) species coordinated with cysteine-thiolate groups are important as inorganic models of active sites of molybdenum enzymes.[1] Molybdenum(VI) and (V) complexes of cysteine dianion (cys) or cysteine methyl ester anion (cysOMe) have already been prepared and X-ray structures are reported. Among cysteine-containing peptides, only the glutathione trianion ($^-O_2CCH(NH_2)CH_2CH_2CO$ NHCH(CH_2S^-)CONHCH$_2CO_2^-$) complex of the binuclear dioxo-μ-dioxo-dimolybdenum(V) dication has been characterized.[4] We have initiated a systematic study on a variety of cysteine-containing peptides as chelating ligands for cis-dioxomolybdenum(VI) and binuclear dioxodimolybdenum(V) species[5] The objective is to simulate the coordination environment of molybdenum in the natural molybdoenzymes.

SOLUTION STRUCTURES

The oligopeptides containing cysteine-residue used in this study include:
a) dipeptides with Cys at the N-terminal e.g. Cys-Phe-OMe;
b) dipeptides with Cys at the C-terminal e.g. Phe-Cys-OMe;
c) tripeptides with Cys at the center of the peptide sequence e.g. glutathione (γ-Glu-Cys-Gly-OH);
d) tri- and tetrapeptides with a Cys residue at both terminals e.g. Z-Cys-Ala-Cys-OMe, or Z-Cys-Ala-Ala-Cys-OMe; and
e) di- or tripeptides with His and Cys residues e.g. Z-Cys-His-OMe.
    Preparation of oxomolybdenum complexes of these peptides has been performed by reactions (1) - (3) in aqueous solution, where HS⌒L represents Cys-containing peptides functioning as chelate

ligands.

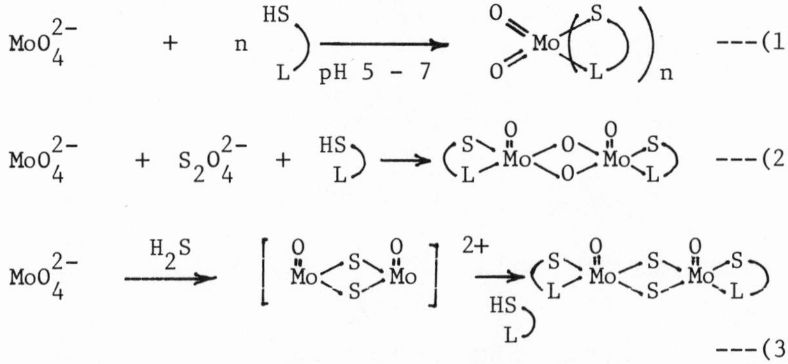

$$MoO_4^{2-} \quad + \quad n \underset{L}{\overset{HS}{\Big)}} \xrightarrow{pH\ 5-7} \underset{O}{\overset{O}{\searrow}}Mo\left(\underset{L}{\overset{S}{\Big)}}\right)_n \quad ---(1)$$

$$MoO_4^{2-} \quad + \quad S_2O_4^{2-} \quad + \quad \underset{L}{\overset{HS}{\Big)}} \longrightarrow \left(\underset{L}{\overset{S}{\searrow}}\overset{O}{\underset{}{Mo}}\overset{O}{\underset{O}{\searrow}}\overset{O}{\underset{}{Mo}}\underset{L}{\overset{S}{\Big)}}\right) \quad ---(2)$$

$$MoO_4^{2-} \xrightarrow{H_2S} \left[\overset{O}{\underset{}{Mo}}\overset{S}{\underset{S}{\searrow}}\overset{O}{\underset{}{Mo}}\right]^{2+} \xrightarrow[\underset{L}{\overset{HS}{\Big)}}]{} \left(\underset{L}{\overset{S}{\searrow}}\overset{O}{\underset{}{Mo}}\overset{S}{\underset{S}{\searrow}}\overset{O}{\underset{}{Mo}}\underset{L}{\overset{S}{\Big)}}\right)$$
$$---(3)$$

The modes of coordination of these peptides to binuclear Mo(V) and mononuclear Mo(VI) species have been investigated by a combination of spectroscopic techniques, in particular ^{1}H- and ^{13}C-NMR and circular dichroism (CD) spectra. Mo=O and Mo-S bonding was detected by IR and Raman spectra in the solid state. Sulfur-bridged species were also investigated in dilute aqueous solutions by the resonance Raman technique. An excitation profile has been obtained for the Mo=O and Mo-S bridge vibrations. An unambiguous assignment of the Mo-S bridge (420 cm^{-1}) and Mo-S-CH$_2$- (350 - 400 cm^{-1}) vibrations was made. Various aspects of these vibrational spectroscopic results will be reported elsewhere.

In the solution NMR study, the proton and carbon signals due to cysteine-thiol group (-CH$_2$-SH) exhibit a conspicuous shift to lower magnetic field on coordination to a Mo(VI) or a Mo(V) center. These NMR shifts can be utilized to diagnose coordination through the cysteine thiolate anion. Thus, the cysteine thiolate in Cys-Phe-OMe, N-acetylcysteine, or Z-Cys-His-OMe is found to ligate to [MoO$_2$]$^{2+}$ or [Mo$_2$O$_2$(μ-S)$_2$]$^{2+}$ ions in d$_7$-DMF solutions. Oligopeptides with Cys at the N-terminal are generally thought to form stable 5-membered N,S-chelate rings for [MoO$_2$]$^{2+}$ or [Mo$_2$O$_2$(μ-S)$_2$]$^{2+}$ the presence of which is readily shown by the observation of two non-equivalent NH proton signals in the NMR spectra (see Fig. 1 for the solution structure.

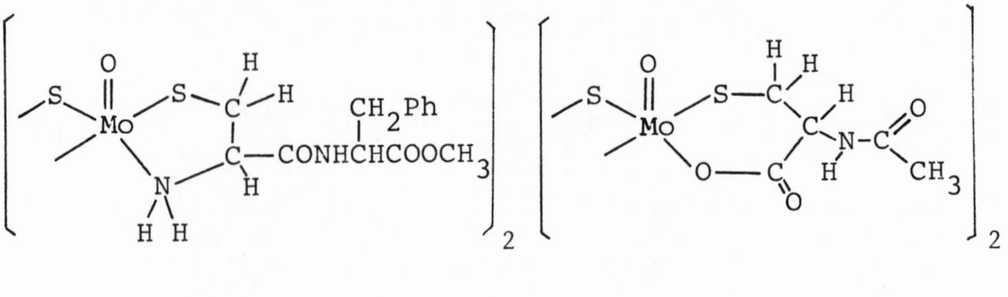

Fig. 1. Structures of Mo$_2$O$_2$(μ-S)$_2$(Cys-Phe-OMe)$_2$ and Mo$_2$O$_2$(μ-S)$_2$(Ac-Cys)$_2$

A somewhat labile 6-membered O,S-chelate ring is formed with N-ace-
tyl cysteine. Here, the NMR spectrum of the $[Mo_2O_2(\mu-S)_2]^{2+}/(N-$
acetylcysteine) complex was analyzed to derive three vicinal cou-
pling constants $(^3J_{N\alpha}, {}^3J_{H\alpha H\beta}, {}^3J_{H\alpha H\beta'})$ which are then utilized to
infer three dihedral angles among three C-H and one NH bonds.
From these results, the structure in Fig. 1 is proposed. When the
anionic coordination of a cysteine C-terminal is blocked by esteri-
fication, the peptide is forced to form an unstable, large chelate
ring. The NMR study of such cases, e.g. Phe-Cys-OMe/$[Mo_2O_2(\mu-S)_2]^{2+}$,
has not yet given any definitive conclusion as to the prevailing
state solution structure. Instead, the coordination environment
of the metal seems labile as judged by the broadness of the ^1H NMR
peaks.

For the glutathione Mo(V) complex, a binuclear O,O-bridged
structure was proposed where the glutathione functions as a triply
anionic, tridentate ligand[4] We have reacted $[Mo_2O_2(\mu-S)_2]^{2+}$ with
glutathione at pH 6. The ^1H and ^{13}C NMR spectra of the resulting
complex reveal an O,S-chelate ring employing the Cys thiolate and
the carboxylate group of the Gly residue. Hence, the structure
shown in Fig. 2 is proposed.

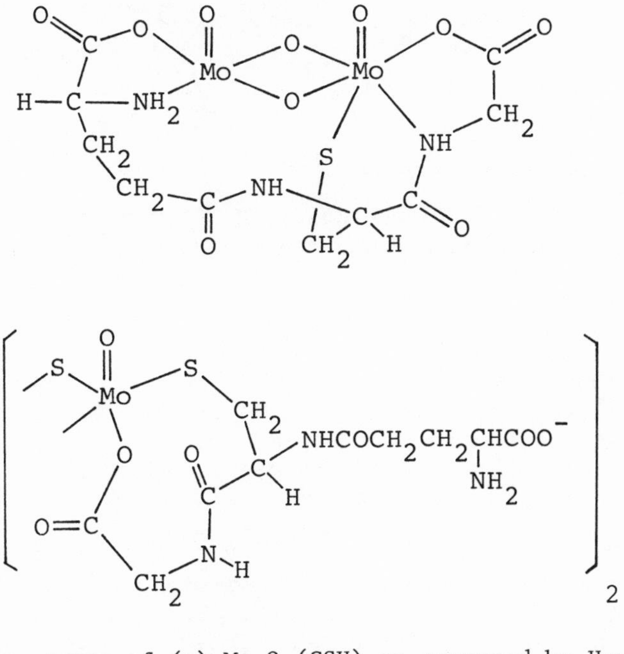

Fig. 2. Structures of (a) $Mo_2O_4(GSH)$ as proposed by Huang and
Haight[4] and (b) $Mo_2O_2(\mu-S)_2(GSH)_2$ from our work.

Peptides with two Cys residues separated by two or three other amino acid residues are important as models of molybdenum containing and iron-containing proteins. We have examined the NMR and CD spectra of Hg(II), Pd(II), and Mo(IV, V) complexes of Z-Cys-(Ala)$_n$-CysOMe (n=1, 2) in d$_6$-DMSO. In the ^{1}H NMR spectra of the Z-Cys-Ala-Ala-CysOMe/(PdCl$_4$)$^{2-}$ system, remarkable shifts of NH and α-CH proton signals were observed.[7] These are considered as evidence for S,S-chelation because no such shifts were detected in the corresponding dimercury(II) complex prepared from HgCl$_2$. The analysis of coupling constants ($J_{N\alpha}$) for the Pd complex also supports the S,S-chelated structure shown in Fig.3. The ^{1}H NMR and CD spectra

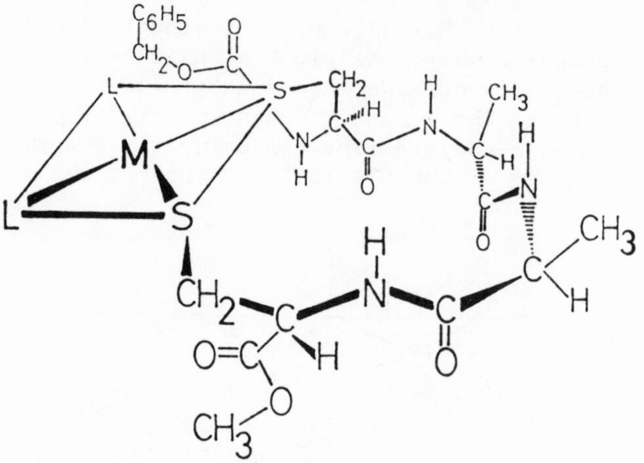

Fig. 3. Proposed structure of ML$_2$(Z-Cys-Ala-Ala-Cys-OMe) (M = Pd, L = Cl; M = MoO, L = μ-S).

of a reaction mixture of Z-Cys-Ala-Ala-CysOMe and [Mo$_2$O$_2$(μ-S)$_2$]$^{2+}$ in d$_6$-DMSO were also measured and interpreted to indicate the S,S-chelated structure shown in Fig.3. In some metalloenzymes, Cys and His residues work together for coordination and for catalysis. The coordinating behavior of Z-Cys-His-OMe towards [MoO$_2$]$^{2+}$ and [Mo$_2$O$_2$(μ-S)$_2$]$^{2+}$ was, therefore, examined. Although strong ligation of the thiolate to the metal was readily shown by NMR, coordination through one of the imidazole nitrogens has not been unambiguously demonstrated as yet.

The CD spectra of the above mentioned peptide complexes were also measured in DMF and/or H$_2$O. Although spectral assignments have not been attempted, the shapes of the CD spectra are useful for deducing the solution structures. For example, the formation of a stable 5-membered N,S-chelate ring with [Mo$_2$O$_2$S$_2$]$^{2+}$/CysPheOMe causes two strong CD extrema at 292 and 335 nm which are also exhibited in the CD spectrum of the known Mo$_2$O$_2$S$_2$(CysOMe)$_2$.[8] The CD curves for labile peptide complexes, such as [Mo$_2$O$_2$S$_2$]$^{2+}$/Phe-

CysOMe, are quite different and involve only low intensities. Absence of stable N,S-chelate rings is, thus, readily shown.

Recently, we have found that the ESR spectra ($g_{average}=1.97$) of DMF solutions prepared by reduction of $MoO_2(CysOEt)_2$ with PPh_3 may be important in the investigation of the solution structures of monomeric oxomolybdenum(V) species coordinated with cysteine thiolate ligands. However, correlation of g-values with structure is not secure enough to warrant reasonable assignment of the structures of solution species at present.

OXIDATION CATALYSIS

It is well known that cis-dioxomolybdenum(VI) species coordinated with two dialkyldithiocarbamate ligands, $MoO_2(R_2dtc)_2$, are excellent homogeneous catalysts for the air oxidation of triphenylphosphine (equation 4).[9] Our systematic study of the oxidation

$$Ph_3P + 1/2\ O_2 \xrightarrow{MoO_2(R_2dtc)_2} Ph_3P=O \text{ --- (4)}$$

tendencies of various dioxomolybdenum chelates, $MoO_2(L-L)$ (L-L = chelating monoanion), revealed that the oxidation of tertiary-phosphines and the dehydrogenation of hydrazobenzene are promoted by the sulfur co-ligands.[10] The relative order is shown in the Table. The relative order may be correlated with the Mo=O

Table: Relative Ease of Reaction of $MoO_2(L-L)_2$ with Ph_3P or $PhNHNHPh$

Reaction	Ligand L-L		
	(S_2C-NEt_2)	thio-oxinato	CysOMe
Ph_3P oxidation	:Rapid at -10°	10 min at 25°	300 min at 30°
$(PhNH)_2$ dehydrogenation	:60 min at 25°	6 h at 25°	

stretching frequencies. Thus, the values, 905 and 878 cm^{-1}, for $MoO_2(Et_2dtc)_2$ are lower than those for both $MoO_2(tox)_2$ (920 and 890 cm^{-1}) and $MoO_2(Cys-OMe)_2$ (912 and 884 cm^{-1}).

Primary and secondary alcohols slowly reduce $MoO_2(S_2CNEt_2)_2$ to give $Mo_2O_3(S_2CNEt_2)_4$. The organic product of this reaction was not readily identifiable for the reaction of MeOH, EtOH, or i-PrOH. PhCHO was identified as a product of the reaction using $PhCH_2OH$, based upon glc examination of the reaction mixture by direct injection. However, PhCHO is not present free in the solution, but it may exist as an unidentified molybdenum adduct. Adduct formation between organic carbonyl compounds and thiols, including dithiocarbamates, are well known. When benzoin, PhCH(OH)COPh, was used in the same reaction with $MoO_2(S_2CNEt_2)_2$, benzil(PhCOCOPh) was identified as a product by its visible spectrum after chromato-graphic separation on Al_2O_3. No adduct formation was detectable between benzil and $Na(S_2CNEt_2)$ or $MoO_2(S_2CNEt_2)_2$.

The stoichiometric oxidation of Ph_3P with $MoO_2(CysOMe)_2$ was investigated by ^{1}H NMR in DMF-CDCl$_3$ (1 : 1) at 30°. A rate equa-tion (5) was found which involves initial complex formation between the reactants followed by the rate-determining oxygen atom transfer step. With initial concentrations of Ph_3P of 3.5×10^{-1}

$$- \frac{d[Ph_3P]}{dt} = \frac{kK[MoO_2(CysOEt)_2][Ph_3P]}{1 + K[MoO_2(CysOEt)_2]} \qquad \text{--- (5)}$$

mole/l and of $MoO_2(CysOEt)_2$ of $(0.36\sim1.1) \times 10^{-1}$ mole/l, we determine $k = 1.3 \times 10^{-3}$ sec^{-1} and $K = 3.3$ mol/l. The following mechanism (equation 6) was proposed.

$$MoO_2(CysOEt)_2 + Ph_3P \underset{K}{\overset{}{\rightleftharpoons}} [MoO_2(CysOEt)_2 \cdot Ph_3P]$$

$$Mo_2O_3(CysOEt)_4 \xleftarrow{+ MoO_2(CysOEt)_2} \Big\downarrow k \qquad \text{--- (6)}$$

$$Ph_3PO$$

The oxidation of Ph_3P proceeds catalytically under an oxygen atmosphere in the presence of a catalytic amount of $MoO_2(CysOR)_2$ or similar complexes. Among others, $MoO_2(CysOEt)_2$ was found to be the most effective and the rate of the catalytic reaction was investigated by measuring the amounts of Ph_3P and Ph_3PO by liquid chromatography. Initial rates were utilized to derive a Michaelis-Menten type rate equation (7). At 30° in DMF/H_2O, we found

$$- \frac{d[Ph_3P]}{dt} = \frac{k'[catalyst][Ph_3P]}{Km + [Ph_3P]} \qquad \text{--- (7)}$$

$Km = 5.7 \times 10^{-2}$ mol/l and $k' = 1.5 \times 10^{-3}$ sec^{-1}. The time-conversion curves showed no deterioration of the catalyst during the course of the reaction for $MoO_2(CysOEt)_2$. But other complexes containing labile chelates, e.g. $MoO_2(AlaCysOMe)_2$, deteriorated rapidly. The effects of substitution in cysteine esters of Cys-containing peptides on the rate were investigated qualitatively and the following relative order was found,

Cys-OEt > CysOMe > AcCys-OH > AlaCysOMe ≃ ZCysHisOMe.

The catalysts prepared in situ by mixing equivalent amounts of Na_2MoO_4 with the chelate ligands in a solvent containing excess Ph_3P were also examined. The ratio, Mo/ligand, was found to be very important. The catalytic activity of a given catalyst system with a 1 : 2 ratio was greatly increased over that with a 1 : 1 ratio but then remained almost the same at 1 : 3. This result

indicates that two N,S-chelating ligands are necessary for the
catalysis. An S,S-chelate, $\underline{o}$-C$_6$H$_4$(CH$_2$SH)$_2$, was also examined and
again the system MoO$_4^{2-}$/dithiol= 0.5, was found to be weakly
effective.

The course of the catalysis with MoO$_2$(CysOEt)$_2$ was examined
by NMR and ESR spectroscopy. Formation of paramagnetic molybdenum
species was shown by the disappearance of the ^{13}C NMR peaks of the
CysOEt ligand when the catalysis was run under oxygen-deficient
conditions. At the same time, well-resolved ESR signals were found
at room temperature with g$_{average}$= 1.97 together with weak satel-
lites caused by [95]Mo and [97]Mo isotopes. The observed g-value is
very close to that reported for reduced xanthine oxidase and
aldehyde oxidase.[11] Huang and Haight postulate only slight dissociation of μ-dioxo-bridged binuclear Mo(V) species based on their
ESR study of Mo(V)/cysteine systems.[12] Our ^{13}C NMR result indicates
that almost all of the Mo(V) species becomes paramagnetic.
Therefore, the Mo(V) species formed by Ph$_3$P is different from those
present in conventional Mo(V)/cysteine systems in water. The
mechanism proposed for the tertiary phosphine oxidation catalysis
by MoO$_2$(CystOEt)$_2$ is shown in Scheme 1 and compared with that of
MoO$_2$(S$_2$CNR$_2$)$_2$.[9]

Scheme 1.   Proposed mechanisms for the catalytic oxidation of Ph$_3$P.

(a) Using MoO$_2$(S$_2$CNR$_2$)$_2$ (ref. 9).

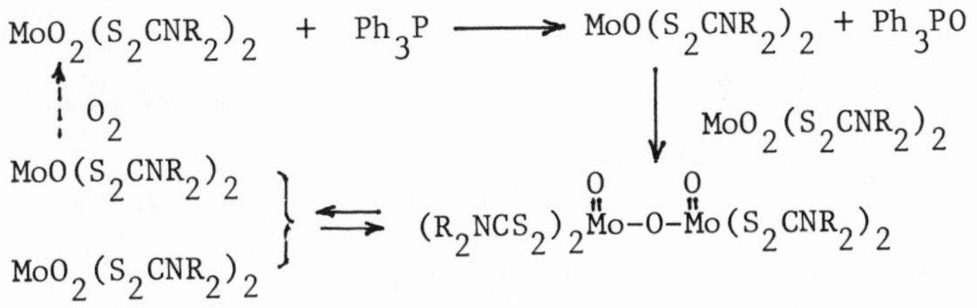

$$MoO_2(S_2CNR_2)_2 \;+\; Ph_3P \longrightarrow MoO(S_2CNR_2)_2 + Ph_3PO$$

$$\Big\uparrow\; O_2$$

$$\downarrow\; MoO_2(S_2CNR_2)_2$$

$$\left.\begin{array}{l} MoO(S_2CNR_2)_2 \\ MoO_2(S_2CNR_2)_2 \end{array}\right\} \rightleftharpoons (R_2NCS_2)_2\overset{O}{Mo}-O-\overset{O}{Mo}(S_2CNR_2)_2$$

(b) With MoO$_2$(CysOEt)$_2$.

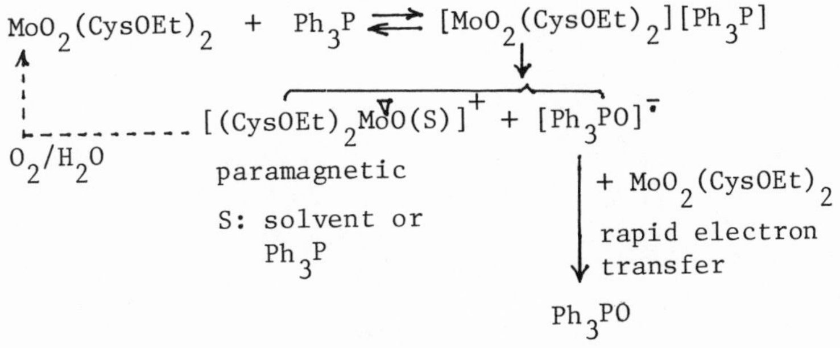

$$MoO_2(CysOEt)_2 \;+\; Ph_3P \rightleftharpoons [MoO_2(CysOEt)_2][Ph_3P]$$

$$\Big\uparrow\; O_2/H_2O \qquad [(CysOEt)_2M\overset{\vee}{o}O(S)]^+ + [Ph_3PO]^{\cdot -}$$

paramagnetic

S: solvent or
      Ph$_3$P

$+$ MoO$_2$(CysOEt)$_2$

rapid electron
transfer

Ph$_3$PO

REFERENCES

1.   (a) S. P. Cramer, H. B. Gray, and K. V. Rajagopalan, The
     molybdenum site of sulfite oxidase. Structural information
     from X-ray absorption spectroscopy, J. Am. Chem. Soc. 101 :
     2772 (1979).
     (b) J. M. Berg, K. O. Hodgson, S. P. Cramer, J. L. Corbin, A.
     Elsberry, N. Pariyadath, and E. I. Stiefel, Structural results
     relevant to the Molybdenum sites in xanthine oxidase and sulfite
     oxidase. Crystal structures of MoO_2L, $L=(SCH_2CH_2)_2NCH_2CH_2X$
     with X = SCH_3, $N(CH_3)_2$, J. Am. Chem. Soc. 101 : 2774 (1979).
     (c) T. D. Tullius, D. M. Kurtz, Jr., S. D. Conradson, and
     K. O. Hodgson, The molybdenum site of xanthine oxidase.
     Structural evidence from X-ray absorption spectroscopy, J. Am.
     Chem. Soc. 101 : 2776 (1979).
2.   J. R. Knox and C. K. Prout, The structure of a cysteine complex
     of molybdenum(V) : $Na_2Mo_2O_4[SCH_2CH(NH_2)CO_2]_2 \cdot 5H_2O$, Acta Cryst.
     B25 : 1857 (1969).
3.   M. G. B. Drew and A. Key, Crystal and molecular structure of
     di-μ-oxo-bis[oxo(L-cysteinato ethyl ester-N,S)-molybdenum(V)],
     J. Chem. Soc. (A) 1846 (1971).
4.   T. J. Huang and G. P. Haight, Jr., Characterization and
     triplet-state electron paramagnetic resonance spectra of
     binuclear [Molybdenum(V)]$_2$-glutathione complex. An example of
     molybdenum-polypeptide complexes, J. Am. Chem. Soc. 93 : 611
     (1971).
5.   A. Nakamura, N. Oguni, N. Ueyama, M. Kamada, and S. Otsuka,
     Electron transfer reductions catalyzed by molybdoferredoxin
     model complexes. Paper presented at 27th Symp. Coord. Chem.,
     Matsumoto, Japan, Sept., 1977, Abst. ID19.
6.   N. Ueyama, M. Nakata, T. Araki, A. Nakamura, S. Yamashita,
     and T. Yamashita, Raman and resonance Raman spectra of sulfur-
     bridged binuclear molybdenum(V) complexes of cysteine-
     containing chelate anions, Chem. Lett. : 421 (1979).
7.   N. Ueyama, M. Nakata, and A. Nakamura, Metal chelate complexes
     of oligopeptides containing two cysteine residues on both
     ends, Paper presented at ACS/CSJ Chemical Congress, Honolulu,
     April, 1979, Abst. INOR34.
8.   K. Saito and Y. Sasaki. Private communication.
9.   R. Barral, C. Bocard, I. Sareé de Roch, and L. Sajus,
     Activation de l'oxygene moleculaire en liquide homogene par
     des complexes oxo du molybdene oxidation catalytique selective
     des phosphines tertiaires, Tetrahedron Lett. 1693 (1972) :
     W. E. Newton, J. L. Corbin, D. C. Bravard, J. E. Searles, and
     J. W. McDonald, Preparation and characterization of two
     series of dimeric molybdenum(V) N,N-dialkyldithiocarbamates.
     Their interrelationship and chemistry as a model for the
     active site of nitrogenase, Inorg. Chem. 13 : 1100 (1974) ;
     D. B. McDonald and J. I. Shulman, Spectrophotometric determi-

nation of triphenylphosphine in dilute solutions, Anal. Chem. 47 : 2023 (1975) ; G. J.-J. Chen, J. W. McDonald, and W. E. Newton, Synthesis of Mo(IV) and Mo(V) complexes using oxo abstraction by phosphines. Mechanistic implications, Inorg. Chem. 15 : 2612 (1976).

10. A. Nakamura, M. Nakayama, K. Sugihashi, and S. Otsuka, Reactivity of oxomolybdenum(VI), -(V), and -(IV) compounds as controlled by sulfur chelate ligands, Inorg. Chem. 18 : 394 (1979).

11. For an extensive review, see E. I. Stiefel, The coordination and bioinorganic chemistry of molybdenum, Progr. Inorg. Chem. 22 : 1 (1977).

12. T. J. Huang and G. P. Haight, A paramagnetic monomeric molybdenum(V)-cysteine complex as a model for molybdenum-enzyme interaction, J. Am. Chem. Soc. 92 : 2336 (1970).

STRUCTURE AND CIRCULAR DICHROISM OF BINUCLEAR

MOLYBDENUM(V) COMPLEXES WITH OPTICALLY ACTIVE LIGANDS

Kazuo Saito, Yoichi Sasaki, Shun'ichiro Ooi[*],
and Kimiko Z. Suzuki

Chemistry Department, Faculty of Science, Tohoku
University, Sendai, 980, Japan
* Chemistry Department, Faculty of Science, Osaka City
University, Sumiyoshi-ku, Osaka, 558, Japan

INTRODUCTION

Molybdenum ions in various oxidation states are involved in [1] active sites of a variety of enzymes catalyzing redox reactions. Many kinds of model complexes have been prepared and studied with reference to their function as catalysts for the reduction of various simple molecules, including acetylene, azobenzene, and nitrogen. [2-4] Most of them are dimeric molybdenum(V) complexes with one or two oxo or sulfido bridges and L-cysteinate and its derivatives as ligands. Some structures have been revealed by X-ray crystallography,[5] but not discussed with reference to their catalytic activity. Since most of the model complexes have optically active ligands, circular dichroism (CD) should be a useful tool for studying their stereo-chemistry. The CD spectra of some molybdenum complexes have been recorded, [6-9] but the relationship to structure was not discussed.

We have measured the CD spectra of three types (Fig.1) of complexes containing natural aminocarboxylates and found that the distorted structure around the binuclear core should be responsible for determining the CD pattern.

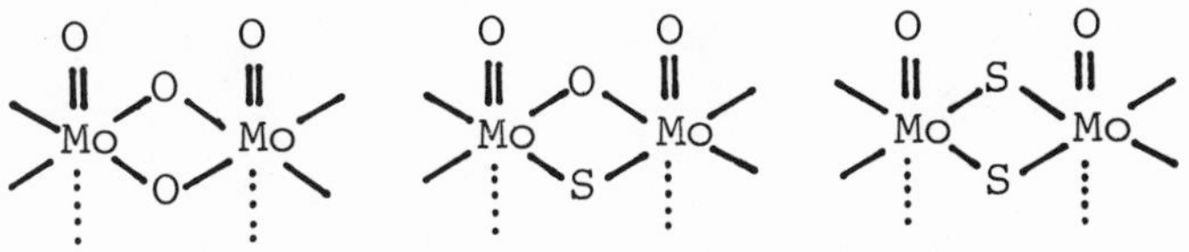

Fig. 1. The three types of complexes considered in this study.

We have, thus, synthesized three new complexes containing R-propyl-
enediaminetetraacetate (R-pdta^{4-}) with the three bridging moieties
in Fig. 1 and have determined their crystal structures by X-ray
diffraction methods. Our assumption was verified by the comparison
of the structure and the CD pattern of these new complexes.

RESULTS

Preparation of the Complexes

The following complexes were prepared by known methods [2] with
necessary modifications; $Na_2[Mo_2O_4(S\text{-cys})_2]\cdot5H_2O$, $[Mo_2O_4(S\text{-Mecys})_2]$,
$[Mo_2O_4(S\text{-Etcys})_2]$, $[Mo_2O_4(S\text{-hist})_2]\cdot2H_2O$, $Na_2[Mo_2O_2S_2(S\text{-cys})_2]\cdot4H_2O$,
$[Mo_2O_2S_2(S\text{-hist})_2]\cdot2H_2O$, $[Mo_2O_2S_2(S\text{-Etcys})_2]$, $Na_2[Mo_2O_3S(S\text{-cys})_2]\cdot$
$4H_2O$, $[Mo_2O_3S(S\text{-Etcys})_2][10]$ and $Na_2[Mo_2O_4(R\text{-pdta})]\cdot3H_2O$. ($S$-cys = S-
cysteinate dianion; Mecys = methylester of cysteine; Etcys = ethyl-
ester of cysteine; S-hist = S-histidinate). The following new
complexes were prepared by modifying the methods for similar comp-
lexes; $Na_2[Mo_2O_4(R\text{-pen})_2]\cdot2H_2O$, $Na_2[Mo_2O_3S(R\text{-pdta})]\cdot4H_2O$, $[Mo_2O_3S-$
$(S\text{-hist})_2]\cdot3H_2O$ and $Na_2[Mo_2O_2S_2(R\text{-pdta})]\cdot4H_2O$. ($R$-pen = R-penicill-
laminate dianion). They were identified by elemental analysis and
IR spectra. [10]

Structure of the Three R-Propylenediaminetetraacetate Complexes

The crystal structures are shown in Fig. 2. The sexadentate
ligands coordinates similarly in all three complexes. Important
bond lengths and angles are listed in Table 1. The crystal data of
the binuclear moiety show no significant differences from those of
the other doubly brdiged molybdenum(V) complexes.   The detailed

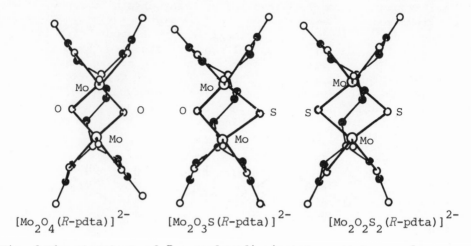

$[Mo_2O_4(R\text{-pdta})]^{2-}$ $[Mo_2O_3S(R\text{-pdta})]^{2-}$ $[Mo_2O_2S_2(R\text{-pdta})]^{2-}$

Fig. 2 The structure of R-propylenediaminetetraacetato complexes

Table 1. Important Bond Distances (Å) and Angles (°) of the Complexes $[Mo_2O_2XY(R\text{-}pdta)]^{2-}$

$$O(1) \quad\quad O(2)$$
$$\underset{\displaystyle\overset{\displaystyle Mo(1)}{}}{}\overset{Y}{\underset{X}{>}}\underset{\displaystyle Mo(2)}{}$$

X and Y	X = Y = O	X = S , Y = O		X = Y = S
Mo(1) - Mo(2)	2.535(2)	2.656(1)		2.807 (2)
Mo(1) - O(1)	1.68(2)	1.691(6)		1.69(1)
Mo(2) - O(2)	1.68(2)	1.676(6)		1.70(1)
Mo(1) - X	1.94(1)	2.318(2)	(S)	2.318(4)
Mo(1) - Y	1.93(2)	1.946(5)	(O)	2.298 (4)
Mo(1)-X-Mo(2)	82.0(5)	69.7(1)		74.3(1)
Mo(1)-Y--Mo(2)	82.4(6)	86.1(2)		75.2(1)
X-Mo(1)-Y	92.8(6)	98.8(2)		103.0(1)
O(1)-Mo(1)-X	106.7(7)	105.9(2)		104.5(4)
O(1)-Mo(1)-Y	112.6(7)	107.1(3)		104.9(4)

structure will be further discussed with reference to the CD spectra.

CD Spectra in Various Media

CD Spectra of all the complexes were measured in various solvents and in KBr disks. The spectra of the four μ-dioxo complexes are shown in Figs. 3 to 6. Although the location and the strength of the CD peaks change slightly depending on the medium, their overall patterns remain almost unchanged. The CD peaks in the 26000 to 29000 cm^{-1} region are always at slightly longer λ region in KBr disk than in solution. Their basic skeletal structure in the crystals should be retained in the solutions, *i.e.*, isomerization and ligand dissociation do not occur upon dissolution, and the relative location of the atoms does not seem to change much.

Nature of the Absorption Bands of Di-μ-oxo Complexes

The absorption spectrum of $[Mo_2O_4(R\text{-}pdta)]^{2-}$ has two bands at 25600 and 33000 cm^{-1} and a shoulder at *ca.* 20500 cm^{-1}. Similar absorption patterns are observed for the complexes, $[Mo_2O_4(H_2O)_6]^{2+}$, $[Mo_2O_4(C_2O_4)_2(H_2O)_2]^{2-}$ [2], $[Mo_2O_4(edta)]^{2-}$ and $[Mo_2O_4(hedta)(H_2O)]^-$ [12], where edta^{4-} and hedta$^-$ stand for the ethylenediamine-tetraacetate and N-2-hydroxyethyl-N,N',N'-triacetate, respectively. All the other complexes also have absorption peaks at *ca.* 32000 cm^{-1}, but none in the visible region. However, they give CD bands in visible region, suggesting the presence of electronic transition of corresponding energies. The electronic transition energies of all the complexes were determined from either the absorption or the CD spectra and are summerized in Table 2.

At least four trnasitions are observed and their energies change only slightly from one complex to another. Therefore, all these transitions must be mainly due to the Mo$_2$O$_4$ core and to a much lesser extent to variation of the other ligands. Such a

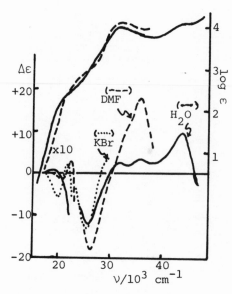

Fig. 3 Absorption and CD
        spectra of $[Mo_2O_4(S\text{-cys})_2]^{2-}$.

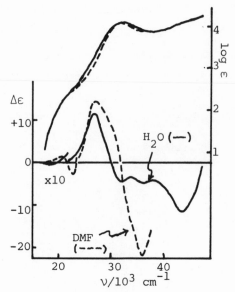

Fig. 4 Absorption and CD
        spectra of $[Mo_2O_4(R\text{-pen})_2]^{2-}$.

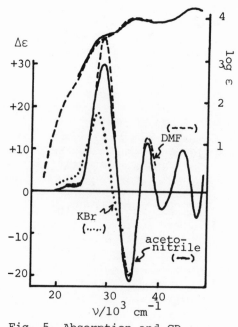

Fig. 5 Absorption and CD
        spectra of $[Mo_2O_4(S\text{-Etcys})_2]$.

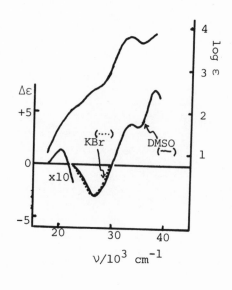

Fig. 6 Absorption and CD
        spectra of $[Mo_2O_4(S\text{-hist})_2]$.

Table 2. The Energy of the Electronic Transitions for the $Mo_2O_4^{2+}$ Complexes Obtained from the Absorption and CD Spectra

Complexes		Energy ($10^{-3} cm^{-1}$ d)				
$[Mo_2O_4(S\text{-cys})_2]^{2-}$	a	20.8*	22.6*	26.4	32.0	36.0*
$[Mo_2O_4(R\text{-pen})_2]^{2-}$	a	20.5*	22.4*	26.8*	31.7	36.0*
$[Mo_2O_4(S\text{-hist})_2]$	b	20.4*		26.9*	33.0	38.0*
$[Mo_2O_4(S\text{-Mecys})_2]$	a	22.4*		29.0*	34.5	37.4*
$[Mo_2O_4(S\text{-Etcys})_2]$	a	22.4*		29.0*	34.4	37.6*
$[Mo_2O_4(R\text{-pdta})]^{2-}$	a	20.6*		25.6	33.0	37.3*
$[Mo_2O_4(edta)]^{2-}$	c			25.6	32.7	
$[Mo_2O_4(hedta)(H_2O)]^-$	c			26.1	33.9	
$[Mo_2O_4(C_2O_4)_2(H_2O)_2]^{2-}$	c	20.4**		26.0	32.8	
$[Mo_2O_4(H_2O)_6]^{2+}$	c	21.9**		26.0	33.8	

a, In dimethylformamide   *b*, In dimethylsulfoxide   *c*, In water
d, * from CD peaks,   ** shoulder.

conclusion is consistent with the result of SCFMO calculation for $[Mo_2O_4(S\text{-cys})_2]^{2-}$ made by Brown *et al.* [8] They predicted five transitions at 13250, 20320, 27390, 30870 and 35700 cm^{-1}, involving the orbitals mainly located on the two molybdenum atoms. There were no appreciable transitions at 13250 cm^{-1} in aqueous solutions less concentrated than 0.1 mol dm^{-1}; however, the absorption or CD peak at this wave length is expected to be very weak, if it exists at all.

Characteristics of the CD Spectra of Di-μ-oxo Complexes

The CD strength at *ca.* 20500 cm^{-1} is small and the pattern depends on the medium. The CD sign of the bands at *ca.* 27000 cm^{-1} is positive for $[Mo_2O_4(S\text{-Mecys})_2]$, $[Mo_2O_4(S\text{-Etcys})_2]$ and $[Mo_2O_4(R\text{-pen})_2]^{2-}$, and negative for the others. For all the complexes sign of the peak at *ca.* 33000 cm^{-1} is always opposite to that at *ca.* 27000 cm^{-1}. Hence, the asymmetric source for these two transitions should be common. However, the absolute configuration of the ligands themselves cannot be responsible for these transitions. The CD peak at *ca.* 36000 cm^{-1} seems to be related to the absolute configuration of the ligands, because the *S*-cys, *S*-Mecys, *S*-Etcys, and *S*-hist complexes give positive peaks while the *R*-pen and *R*-pdta complexes give negative peaks.

Absorption and CD Spectra of the μ-Sulfido Complexes

The absorption and CD spectra of the Mo_2O_4, $Mo_2O_2S_2$ and Mo_2O_3S complexes containing *S*-cys, *S*-Etcys, *S*-hist and *R*-pdta are compared with one another in Figs. 7 to 10. The absorption intensity of the complexes with the given ligands increases remarkably in the region around 27000 cm^{-1} with an increase in number of bridging sulfides. Changes in the CD spectra are much more modest as compared with the big changes in the absorption. In the UV region, they give one to three peaks, but no characteristic trend is seen with increase of S.

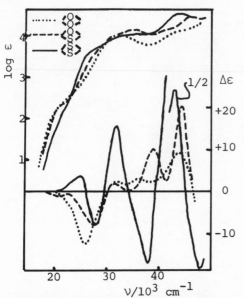

Fig. 7 Absorption and CD spectra of $[Mo_2O_2XY(S-cys)_2]^{2-}$
in water.

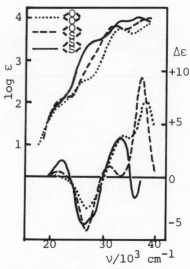

Fig. 8 Absorption and CD spectra of $[Mo_2O_2XY(S-hist)_2]$
in dimethylsulfoxide.

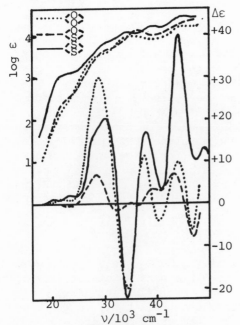

Fig. 9 Absorption and CD spectra of $[Mo_2O_2XY(S-Etcys)_2]$
in acetonitrile.

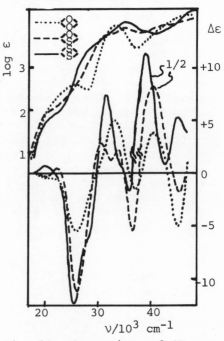

Fig. 10 Absorption and CD spectra of $[Mo_2O_2XY(R-pdta)]^{2-}$
in water.

DISCUSSION

Importance of the Bands at ca. *27000 and 33000 cm⁻¹*

In the following discussion, the CD spectra in the regions around 27000 and 33000 cm^{-1} are mainly considered. The CD pattern in the 26000 cm^{-1} region of the di-μ-oxo complexes is least affected by the medium and the asymmetric anisotropic factor $\Delta\varepsilon/\varepsilon$ is at least five-to-ten times as large as those of the other three transitions. The CD sign at 33000 cm^{-1} is opposite to that at 27000 cm^{-1} for all the di-μ-oxo complexes, and they should be discussed together. The complexes with μ-oxo-μ-sulfido and di-μ-sulfido bridges have apparently complicated CD patterns in these regions. However, the CD peaks in the 27000 cm^{-1} can be understood by considering the appearance of new CD bands, the sign of which is positive for the *S*-cys and its ester complexes and negative for the *S*-hist and *R*-pdta complexes. The bands in common with the di-μ-oxo complexes remain at almost the same position with the same sign, regardless of the change in bridging species. The appearance of the new CD band is consistent with the marked increase in the absorption intensity. However, the $\Delta\varepsilon/\varepsilon$ values at the peak or the shoulder of these new bands are not so large as those of the common band. Thus, we consider the CD sign of the common band to be important for discussing the stereochemistry of all the complexes and we compare them with the structure of crystals.

Di-μ-oxo Complexes

Since the nature of the absorption bands is mainly due to the transitions in the binuclear moiety, local asymmetries, such as asymmetric carbons and the conformation of the chelate ring, are not responsible for transitions in the region of interest. The asymmetry caused by the arrangement of the atoms of the $[Mo_2O_4]^{2+}$

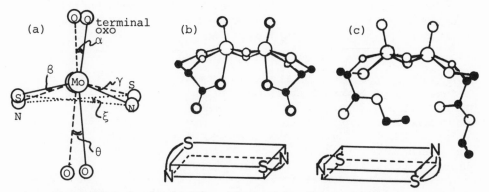

Fig. 11 Asymmetric distortion of the binuclear complexes.
(a) A view along Mo-Mo axis of $[Mo_2O_4(S\text{-cys})_2]^{2-}$. (b) A view down the axis connecting the two μ-oxides of $[Mo_2O_4(S\text{-cys})_2]^{2-}$ and its tetrahedral distortion.   (c) Similar views for $[Mo_2O_4(S\text{-Etcys})_2]$

Table 3   Torsion Angles, α *and* ξ *, and the CD Sign of the Peak at* ca. *27000 cm*$^{-1}$ ^{a}

Complexes	α ^{b}	ξ ^{b}	Sign
$Na_2[Mo_2O_4(S\text{-cys})_2] \cdot 5H_2O$	4.0 (Δ)	0.06 (Δ)	−
$[Mo_2O_4(S\text{-Etcys})_2]$	6.7 (Λ)	28.4 (Λ)	+
$Na_2[Mo_2O_4(R\text{-pdta})] \cdot 3H_2O$	5.4 (Δ)	8.6 (Δ)	−
$Na_2[Mo_2O_2S_2(S\text{-cys})_2] \cdot 2H_2O$	4.7 (Δ)	∼0	−
$[Mo_2O_2S_2(S\text{-Mecys})_2]$	9.2 (Λ)	35.9 (Λ)	+
$[Mo_2O_2S_2(S\text{-hist})_2] \cdot 1.5H_2O$	2.5 (Δ)	4.5 (Δ)	−
$Na_2[Mo_2O_2S_2(R\text{-pdta})] \cdot 4H_2O$	∼0	10.4 (Δ)	−
$Na_2[Mo_2O_3S(R\text{-pdta})] \cdot 4H_2O$	∼0	11.5 (Δ)	−

. See text for definition of α and ξ.
. Absolute configuration of the distorted core moiety in ( ).

core should be very important as should that caused by the arrangement of coordinated ligand atoms. The view down the Mo-Mo axis of $[Mo_2O_4(S\text{-cys})_2]^{2-}$ (Fig. 11) indicates that the two Mo=O bonds are twisted in the Δ direction by an angle α and those bonds cis to Mo=O are twisted similarly by angles β and γ. On the other hand, the bonds trans to Mo=O are twisted in the Λ direction by an angle θ. The S-ethylcysteinato complex exhibits distortion angles α, β and γ twisted oppositely (in the Λ direction) to the S-cysteinato complex but has no θ. However, the bonds trans to Mo=O are long and should contribute less to the overall asymmetry than the other three bonds. The asymmetry of the basal bond (cis to Mo=O) may be more easily expressed by the torsion angle ξ (Fig. 11), which represents the tetrahedral distortion of the four basal atoms. The Δ and Λ distortions expressed by the angles α, β, γ and ξ should be responsible for determining the CD sign in 27000 cm^{-1} region.

In order to confirm such a consideration, we have synthesized and determined the structure of $[Mo_2O_4(R\text{-pdta})]^{2-}$, which has the asymmetric carbon at a very different location from those of the complexes in Fig. 11. Table 3 clearly indicates that the same relationship applies to the R-pdta complex.

μ-*Oxo*-μ-*sulfido and Di*-μ-*sulfido Complexes*

Table 3 includes the absolute configuration and the α and ξ values cauased by the distortions of the sulfido bridged complexes. It also shows the sign of their CD peaks in the 27000 cm^{-1} region, which are in common with the di-μ-oxo complexes. The same relationship is also seen between the CD sign and the configuration. If our assumption concerning the analysis of the CD pattern of sulfido bridged complexes in this region is appropriate, the CD sign in the 27000 cm^{-1} region should reflect the absolute configuration around the doubly bridged binuclear core of the molybdenum(V) complexes.

Not many examples are known in which this type of distortion exerts such a large influence upon the CD pattern. We have seen that CD can provide useful information concerning the twisted structure of metal complexes. If the distortions around a central metal ion markedly influence the electronic state of a complex, they should then be related to the stereo-selective reactions involving this asymmetric complex.[14]

REFERENCES

1. R. A. D. Wentworth, Mechanism for the reactions of molybdenum in enzymes, Coord. Chem. Rev. 18:1 (1976).

2. E. I. Stiefel, The coordination and bioinorganic chemistry of molybdenum, Prog. Inorg. Chem. 22:1 (1976) and references cited therein.

3. J. L. Corbin, N. Pariyadath, and E. I. Stiefel, Ligand effects and product distributions in molybdothiol catalyst systems, J. Am. Chem. Soc. 98:7862 (1977).

4. A. Nakamura, K. Sugihashi, and S. Otsuka, Reductive cleavage of an azo linkage: stoichiometric and catalytic reactions of azobenzene with low valent molybdenum compounds, J. Less-Common Metals 54:495 (1977).

5. B. Spivack and Z. Dori, Structural aspects of molybdenum(IV), molybdenum(V) and molybdenum(VI) complexes, Coord. Chem. Rev. 17:99 (1975) and references cited therein.

6. R. M. Wing and K. P. Callahan, Stereospecific synthesis and absolute configuration of (-)-dioxodi-μ-oxo-D(-)-1,2-propylenediaminetetraacetatodimolybdate(V), Inorg. Chem. 8:2302 (1969).

7. D. H. Brown and J. MacPherson, Complexes of molybdenum(V) with some aldoses and ketoses, J. Inorg. Nucl. Chem. 32:3309 (1970); D. H. Brown and J. MacPherson, Some molybdenum(V) complexes with α-hydroxycarboxylic acid, J. Inorg. Nucl. Chem. 33:4203 (1971); D. H. Brown and J. MacPherson, Molybdenum(V) and (VI) complexes with some naturally occurring ligands, J. Inorg. Nucl. Chem. 34:1705 (1972).

8. D. H. Brown, P. G. Perkins, and J. J. Stewart, The electronic states of the dioxo-di-μ-oxo-dimolybdate(V) group, J. Chem. Soc. Dalton Trans. 1105 (1972).

9. D. H. Brown and P. G. Perkins, The electronic structure and spectra of the di-μ-sulphido-bis[oxo-(L-cysteinato)molybdate (V)]$^{2-}$ ion, Rev. Roum. Chim. 20:515 (1975).

10. V. R. Ott, D. S. Swieter, and F. A. Schultz, Di-μ-oxo, μ-oxo-μ-sulfido, and di-μ-sulfido complexes of molybdenum(V) with EDTA, cysteine, and cysteine ester ligands. Preparation and electrochemical and spectral properties, Inorg. Chem. 16:2538 (1977).

11. J. Dirand-Colin, M. Schappacher, L. Ricard, and R. Weiss Formation and molecular structure of some molybdenum sulfur complexes, J. Less-Common Metals 54:91 (1977).

12. Y. Sasaki and T. S. Morita, Preparation and the acid hydrolysis of N-(2-hydroxyethyl)ethylenediamine-N,N',N'-triacetate complex of di-μ-oxo-bis(oxomolybdenum(V)). Bull. Chem. Soc. Japan 50:1637 (1977).

13. M. G. B. Drew and A. Kay, Crystal and molecular structure of di-μ-sulphido-bis[oxo-(L-cysteinato methyl ester-*N,S*)-molybdenum(V)]. J. Chem. Soc. (A) 1851 (1971).

14. S. Kondo, Y. Sasaki, and K. Saito, Stereoselectivity in outer-sphere redox reaction between dimeric molybdenum(V) complex of *R,S*-propylenediaminetetraacetate and (μ-hyperoxo-μ-amido) tetrakis(ethylenediamine)dicobalt(III) in aqueous solution. manuscript in preparation.

MOLYBDENUM CATALYZED REDUCTIONS OF SMALL MOLECULES:

REQUIREMENTS OF THE ACTIVATED STATE

G. P. Haight, Jr., P. R. Robinson, and T. Imamura

School of Chemical Sciences
University of Illinois
Urbana, Illinois 61801

INTRODUCTION

It is commonly assumed that the active site in nitrogenase contains molybdenum in one of its higher oxidation states. Biological reduction of nitrogen to ammonia occurs without formation of detectable intermediates and is accompanied by the hydrolysis of several molecules of ATP per mole of nitrogen reduced. This paper will examine the role of molybdenum and similar elements in known catalytic processes involving the reduction of oxidized nonmetals, and also in the hydrolysis of polyphosphates to determine the most likely kinetic properties to be found in molybdenum-containing active sites in enzymes. Kinetic requirements will be coupled to recently acquired structural information on molybdenum environments in enzymes in the pursuit of knowledge of the active site.

ACTIVATED STATES FOR MOLYBDENUM-MEDIATED ELECTRON TRANSFER REACTIONS

A general phenomenon has been observed in which early transition metals in higher oxidation states act as templates in which they bind both reducing agents and inert oxidizing agents and also act as electron transfer pathways between the two. Catalytic reduction of ClO_4^- and NO_3^- in the presence of Mo(VI) at dropping mercury cathodes[1,2] is shown in Figure 1. Catalysis occurs in coincidence with the irreversible reduction of Mo(V) to Mo(III). In $HCl/HClO_4$ solution, catalysis is coincident with a reversible reduction of Mo(VI) to Mo(IV).[3] Mo(III) gives no catalytic waves. By a process of elimination, Mo(IV) is seen to be the active catalyst. Analysis of catalytic currents leads to a mechanism,[2]

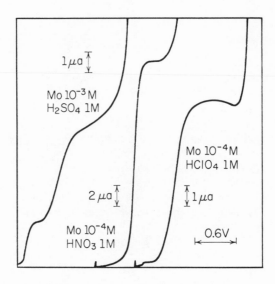

Fig. 1.   Polarographic waves for Mo(VI) in various acids using a
          dropping mercury electrode and mercury pool anode.   Each
          curve begins at 0.0 volts.

$$Mo(VI) \text{ diffuses to Hg electrode} \qquad k_D \qquad (1)$$

$$Mo(VI) \xrightarrow{e^-} Mo(V) \xrightarrow{e^-} Mo(IV) \qquad k_{2e} \qquad (2)$$

$$Mo(IV) + ClO_4^- \rightleftarrows Complex \qquad K_3 \qquad (3)$$

$$Complex \xrightarrow{8e^-} Mo(IV) + Cl^- \qquad k_{4e} \qquad (4)$$

$$2Mo(IV) \rightarrow Mo(V) + Mo(III) \qquad k_5 \qquad (5)$$

$$Mo(IV) \text{ diffuses away from Hg} \qquad k_D \qquad (6)$$

Positively charged Mo(IV) binds to ClO_4^- and to the negative mercury
cathode in the activated state $\left[Hg \vdots Mo - OClO_3 \right]^{\ddagger}$.

   Nb(V), W(VI), Re(VII), and Mo(IV) each catalyze the reduction
of ClO_4^- and NO_3^- by $SnCl_3^-$.[4,5]   Kinetic studies reveal a common
rate law $\dfrac{d[Sn(II)]}{dt} = \dfrac{k_1 k_2 [SnCl_3^-][M][ClO_4^-]}{k_{-1} + k_2 [SnCl_3^-]}$, where M is the oxidized
metal.   In each case, the oxidized nonmetal in the center of the
oxo-anion is protected from direct nucleophilic attack by $SnCl_3^-$ by
oxygen atoms and negative π-electron clouds between the oxygen
atoms.   The mechanism for catalysis involves formation of a complex
between the oxo-anion and the positive transition metal atom (equation 7),

$$M + ClO_4^- \underset{k_{-1}}{\overset{k_1}{\rightleftharpoons}} MOClO_3 \tag{7}$$

followed by nucleophilic attack by $SnCl_3^-$ on the positive transition metal and an innersphere two-electron transfer to oxidized nonmetal (equation 8),

$$MOClO_3 + :SnCl_3^- \xrightarrow{k_2} [Cl_3Sn-M-O-\overset{\overbrace{2e^-}}{ClO_3}]^{\ddagger} \rightarrow products \tag{8}$$

Such polyelectron innersphere transfers have also been observed in Cr(VI) oxidations[6] (equation 9) and are postulated to occur in oxygen atom transfer reactions[6] (equation 10). Table 1 summarizes reactions studied to date which involve transition metal catalyses of these types.

$$HCrO_4^- + 2HSO_3^- + H^+ \rightarrow \left[O_2\overset{2e^-}{\overset{\overbrace{}}{S}}-O-\overset{O}{\underset{O}{Cr}}-O-\overset{1e^-}{\overset{\overbrace{}}{S}}O_2\right]^{\ddagger} \rightarrow Cr(III) + SO_4^{2-}$$
$$+ 2H_2O \qquad\qquad + \tfrac{1}{2}S_2O_6^{2-} \tag{9}$$

$$HSO_3^- + H^+ + ClO_3^- \rightarrow O-\underset{\underset{OH_2}{\uparrow}}{\overset{\overset{O}{\searrow}}{S}}-O-\underset{O}{\overset{2e^-}{\overset{\overbrace{}}{Cl}}}-O^{\ddagger} \rightarrow SO_4^{2-} + ClO_2^- + 2H^+ \tag{10}$$

SPECIAL FEATURES OF MOLYBDENUM CATALYSES

Although homogeneous catalyses are initiated by addition of Mo(VI) or Mo(V), the active catalyst has been deduced to be an unstable form of Mo(IV) which is susceptible to disproportionation. Molybdenum is not catalytic under conditions in which it can be reduced to Mo(III).[3] However, zinc metal which reduces Mo(V) to Mo(III) in H_2SO_4 and HCl can reduce Mo(VI) only partially to Mo(V) when ClO_4^- is present. Sn(II) is known to reduce Mo(VI) to Mo(IV).[4] Cr(II), V(II), and Ti(III) do reduce Mo(VI) to Mo(III) and do not reduce ClO_4^- catalytically. Catalyses initiated by Mo(VI) exhibit induction periods. Eventual steady state kinetics are the same after the Mo(VI)-induced surge and the Mo(V)-induced induction period (Figure 2).[9]

The kinetics of perchlorate reductions require that a weak Mo(IV) complex with ClO_4^- be reduced by Sn(II). In other cases, kinetic studies do not allow one to distinguish between direct substrate reduction by Mo(IV) or Sn(II) attack on a complex of Mo(IV) and substrate as in reaction (2) in the general mechanism above. Reaction (2) does occur in several cases where the catalyst is not oxidizable (W(VI), Nb(V), and Re(VII)) and, by analogy,

Table I

Transition Metal Catalyst	Inert Substrate	Reducing Agents	Products	Ref.
Mo(IV)	$ClO_4^-, ClO_3^-, ClO_2^-$	Sn(II), Zn, Hg(electrode)	Cl^-	1,2,4,9
	NO_3^-		N_2O, NH_3OH^+	1,2,4
	NH_3OH^+ *	Sn(II)	NH_4^+	4
	$NH_2NH_3^+$ *	Sn(II)	NH_4^+	7
	N_2	Sn(II) Zn Cathode	NH_4^+	10
Mo(VI)	Methylene Blue	Hydrazine	Leuco-MB, N_2	8
W(VI)	ClO_4^-	Sn(II) Hg(electrode)	Cl^-	4
	N_2	Sn(II)	NH_4^+	10
Nb(V)	ClO_4^-	Sn(II)	Cl^-	5
Re(VII)	NO_3^-	Sn(II)	?	5

*The study of reduction of hydrazine is in progress. It is catalytically reduced by Sn(II) in the presence of Mo(VI) or Mo(V) with kinetics similar to those of hydroxylamine. Thus, both possible mono- and di-nitrogen intermediates in the reduction of N_2 are susceptible to reduction in this system.

becomes the reaction of choice for Mo(IV) catalyses in the absence of distinguishing evidence to the contrary.

These effects may be significant for active sites in enzymes. Molybdenum in the (IV) state and perhaps in the (V) and (VI) states can serve as an electron transfer medium to oxidized nonmetal substrates. In the presence of two-electron reducing agents, the reduction pathway from Mo(VI) to Mo(III) involves

$$2Mo(VI) \longrightarrow 2Mo(IV) \begin{smallmatrix} Mo(V) \\ Mo(III) \end{smallmatrix} \longrightarrow Mo(III).$$

This process can be arrested at the Mo(IV) stage if Mo(IV), bound in monomeric fashion in the enzyme, is unable to disproportionate.

FACTORS LIMITING EFFECTIVENESS OF Mo(IV)-MEDIATED ELECTRON TRANSFER

The first important factors are the properties of the substrates. Catalytic rates decrease in the order $ClO_3^- \gg NO_3^- > ClO_4^- > NH_3OH^+ \approx N_2H_5^+ \gg N_2$. ClO_3^- (which is slowly reduced by $SnCl_3^-$ without catalysis) can be determined by rapid titration with $SnCl_3^-$ in the presence of $[Mo(VI)]_0 > 10^{-5}M$. The catalytic

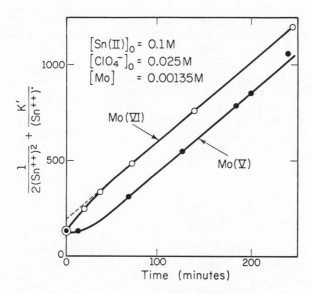

Fig. 2. Plot of the integrated form of the rate law for Sn(II)
reduction of ClO_4^-. Intercept of dashed line measures fast
initial reaction with Mo(VI) present (open dots). Induction period with Mo(V) present is evident (solid dots).
Both runs give the same slope after steady state kinetics
are achieved.

process is several orders of magnitude faster for ClO_3^- than for
ClO_4^-. Chemical intuition would arrange the substrates in the same
order for their relative binding ability to a positive metal center,
although the better coordination of ClO_3^- vs. ClO_4^- would probably
not be so great as to account for the difference observed in their
rates of catalytic reduction.

This tendency for coordination to a positive metal center may
be significant for enzymic active sites. If molybdenum in an active
site is bound to an electron reservoir, it must only present a
labile binding site for anionic substrates, such as nitrate, sulfite,
etc., since these substrates bind competitively even in the open
systems here described. Since nitrite is reducible without molybdenum in the open systems, molybdenum, to be necessary in nitrite
reductase, must provide the point of contact for NO_2^- with the
electron transfer system. In nitrogenase, preference for N_2 and CO
as substrate must involve creation of a nonpolar pocket with geometry
making it inaccessible to competing anionic substrates. The active
site probe cyclopropene described by McKenna[11] supports this conclusion.

The lability of the substrate as an electron acceptor is a dif-
ficult-to-measure property, but its importance can be easily inferred
as a factor determining (a) whether electron transfer can take place
during the lifetime of the ternary activated complex [Sn-Mo-S]‡ and
(b) if oxidants in the system are capable of reoxidizing Mo(III) and
Mo(V) to Mo(VI). While NH_3OH^+ and $N_2H_5^+$ just oxidize $(Mo(V))_2$[10] to
Mo(VI) or induce disproportionation[8] of Mo(V) in the rate limiting
step, aqueous Mo(V) systems have been observed to undergo oxidation
with mechanisms characteristic of the existence of pre-equilibria
(equations 11 and 12).[8,12] For example, N_2 reduction ceases soon
after initiation in the open system because nothing in the system
can oxidize the products, $(Mo(V))_2$ or Mo(III), which are inert, non-
catalytic states of molybdenum.

$$(Mo(V))_2 \rightleftharpoons 2Mo(V) \qquad \text{(epr active monomers)} \qquad (11)$$

$$(Mo(V))_2 \rightleftharpoons Mo(VI) + Mo(IV) \qquad (12)$$

The second major factor is the instability of Mo(IV) in catalytic systems. Sample rate laws for homogeneous molybdenum catalyses
are as follows.

$$\text{Rate} = \frac{k[Mo]^{1/2}[Sn(II)]^{3/2}[C\ell O_4^-]^{3/2}}{1 + K[C\ell O_4^-]} \qquad \text{(from ref. 4)}$$

$$\text{Rate} = \frac{k[Mo]^{1/2}[Sn(II)]^{1/2}[NH\,OH^+]^{3/2}[H^+]^2}{1 + K[NH_3OH^+]} \qquad \text{(from ref. 5)}$$

$$\text{Rate} = k[Mo]^{1/2}[Sn(II)]^{1/2}[N_2H_5^+]^{3/2} \qquad \text{(from ref. 7)}$$

The observed half powers in various reagent concentrations are consistent with disproportionation as the most important process for
removing Mo(IV) from these reacting systems. Monomeric immobile
molybdenum complexes in enzymes could prevent such disproportionation from occurring.

OTHER d^2 SYSTEMS IN N_2 FIXATION

The implication of Mo(IV) as a mediator for electron transfer
is encouraged by the appearance of d^2 species in other systems involving reduction of N_2. Ti(II) is involved in reactions studied
by van Tamelen.[14] Bercaw's stable complex of Zr(II) with N_2 can be
activated for electron transfer by protonation.[15] The latter complex contains two Zr(II) atoms bridged by N_2 and two terminal N_2
ligands. [15]N tracer studies indicate the following mechanism for
H$^+$-induced production of hydrazine (equations 13 and 14). Half the
hydrazine formed comes from bridging N_2 and half from terminal N_2
in the original complex. The second Zr atom is clearly a template

for the formation of an activated state for the electron transfer disproportionation of diimide.

The production of hydrazine in Bercaw's system coupled to the absence of intermediates in the enzymic reduction of N_2 to NH_3 suggests that an electron reservoir of, at least, six electrons is attached to the active site molybdenum in nitrogenase. The existence of such a reservoir has been clearly implied in Burris' paper[16] in this volume.

THE QUESTION OF HYDRIDE FORMATION

The common occurrence of hydrides, even of higher oxidation states of early transition metals (e.g., ReH_9^{2-}), has led to plausible speculations of metal hydride formation as a source of reduction of, and hydrogen addition to, coordinated substrates of nitrogenase (see equation, for example).

$$Mo - N \equiv \overset{+}{N}H \rightarrow Mo = N = NH_2 \qquad (15)$$
$$\overset{|}{H}$$

Chatt's mechanism for reduction of N_2 bound to molybdenum and tungsten suggests protonation of reduced metal as a step in the reduction of N_2. Because N_2 must find a nonpolar binding site and be protonated as it is reduced, it is attractive to postulate that molybdenum acts as an electron mediator and as a proton mediator between substrate and both the electron and proton reservoirs. Mo(IV) would have to be ligated in such a way as to form a pocket accessible to N_2 on one side while being bound to the electron and proton donors on the other side (reaction 16).

$$6e^- \searrow \qquad \qquad$$
$$Mo(IV) - N \equiv N: \longrightarrow 2NH_3 \qquad (16)$$
$$6H^+ \nearrow$$

A recent study shows that insertion of butyne-2 into a $Zr - H$ bond[18] forms a cis product,

$$\underset{H_3C}{\overset{Zr}{>}}C = C\underset{CH_3}{\overset{H}{<}}.$$

The reduction of acetylene by nitrogenase model systems in D_2O which yields cis-dideuteroethylene[14] could, by analogy, be considered evidence for insertion of acetylene into a molybdenum-hydride bond. Conversely, if hydride is used to reduce platinum acetylides, trans-ethylenes are formed (reaction 17).[20]

$$
\underset{\substack{C \\ \diagdown R}}{\overset{\substack{R \\ \diagup \\ C}}{Pt\Big\langle\!\!\!\parallel\!\!\!}} \; + \; HC\ell \;\rightarrow\; Pt \underset{C\ell}{-}C\underset{R}{=}C\overset{R}{\underset{H}{\diagdown}} \tag{17}
$$

Molybdenum(IV) hydride is an attractive postulate for an active
site since the enzyme is a powerful catalyst for hydrogen production
even when nitrogen fixation is also taking place. If the active site
in nitrogenase does involve Mo(IV) to mediate proton and/or electron
transfer, the following properties and characteristics are postulated:

a. Mo(IV) is accessible to N_2 for coordination;

b. Mo(IV) is protected from disproportionation; and

c. Mo(IV) is attached to multielectron and multiproton
    sources to provide continuous reduction of
    $N_2 \rightarrow N_2H_2 \rightarrow N_2H_4 \rightarrow 2NH_3$, such that intermediates
    would be reduced before they could dissociate
    from the activated complex.

EXAFS experiments suggest that molybdenum in the nitrogenase cofactor
may occupy one corner of a molybdenum–iron–sulfur cluster cube.[21]
Such a cluster could be part of the electron transport system in
nitrogenase providing electrons for continuous reduction of bound N_2
to $2NH_3$ without dissociation of intermediates, such as N_2H_2, N_2H_4,
or NH_3OH^+ from the active site. The Fe_3Mo cluster isolated by Holm,
et al.[23] exhibits six-coordinate Mo(III,IV) bound only to sulfur in
the first coordination sphere. Six sulfur atoms bound octahedrally
would effectively shield molybdenum from attack by substrate, but
trigonal prismatic coordination, as observed in MoS_2 and $Mo(mnt)_3$,
would allow approach of substrate through the square faces.

ACTIVATED STATES FOR THE HYDROLYSIS OF POLYPHOSPHATES INDUCED BY THE OXIDATION OF TRANSITION METAL COMPLEXES OF POLYPHOSPHATES

Dramatic enhancement of rates of hydrolysis of linear tripoly-
phosphate ($P_3O_{10}{}^{5-}$ or PPP_i), adenosine triphosphate (ATP) and other
molecules containing P–O–P bonds occurs following oxidation of complexes of VO^{2+} and Mn^{2+} by $MnO_4{}^-$, H_2O_2, and Br_2. Very complicated
systems are produced which hydrolyze to give orthophosphates, oxo-
metallates, etc. With other oxidizable metal ions, processes competitive with polyphosphate hydrolyses dominate. So far, only oxidation of VO^{2+} and Mn^{2+} has produced a significant effect, the
kinetics of which can be studied. Oxidation of Mo(III) has given
a third example of the effect, but we have not succeeded, as yet,
in obtaining reproducible results. Studies of the effect of oxidation of $((VO^{2+})_n(PPP_i)_m)$ complexes on the hydrolyses of PPP_i indicate rather specific requirements for metal ion activation of polyphosphates for hydrolysis.

For the reaction (18), it is observed that:

$$P_3O_{10}{}^{5-} + H_2O \xrightarrow[MnO_4]{VO^{2+}} HP_2O_7{}^{3-} + HPO_4{}^{2-} \tag{18}$$

(1) of the oxidized complexes, $(V(V)PPP_i)$ hydrolyzes at half
the rate of $((V(V))_2PPP_i)$ and that $(V(V)(PPP_i)_2)$ is as inert to hydrolysis as is free $PPP_i)$ at 25° to 40°C; (2) all potential ligands
inhibit hydrolysis of PPP_i; and (3) oxidation of $((VO^{2+})(PPP_i))$ at
0°C gives a colorless complex with a UV band characteristic of tetrahedral $VO_4{}^{3-}$. This last complex does not hydrolyze PPP_i but undergoes hydrolysis to a yellow complex, which then does hydrolyze
PPP_i.[21] A plausible structure of the inert initial tetrahedral V(V)
complex is (I). The inertia of (I) and of $(V(V)(PPP_i)_2)$ to polyphosphate hydrolysis and the general inhibition of such hydrolysis

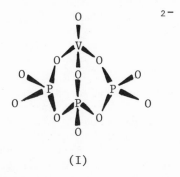

(I)

by other ligands suggests that V(V)-induced hydrolyses occur only
when H_2O (or OH^-) ligated to vanadium is the effective nucleophile
hydrolyzing the P(V) center. An analogous conclusion is reached in
studies of Mn(III) catalyzed PPP_i hydrolyses.

Other observations that pentacoordinate P(V)-containing axial-
equatorial rings are labile and of the apparent equivalence of V(V)
atoms in $((V(V))_2P_3O_{10})$ suggest the activated state (II) for hydrolysis. At room temperature, (I) hydrolyzes to give active

(II)

$(V(V)P_3O_{10})$ with V(V) bound to the end $PO_3{}^-$ group. This complex
hydrolyzes half as fast as (II). $P_2O_7{}^{4-}$ resulting from the hydrolysis of (II) hydrolyzes much faster than $(V(V)P_2O_7{}^{4-})$ formed by oxidation of $((VO^{2+})(P_2O_7{}^{4-}))$. Thus, an end bonded fragment

$$V \overset{O}{\underset{O}{\diamondsuit}} P - O - P - O \quad \text{formed from hydrolysis of (II) is rapidly}$$

hydrolyzed, while chelated (III) does not have ligated OH$^-$ positioned for attack on a P(V) center.

$$
\begin{array}{c}
\text{(structure III)}
\end{array}
$$

(III)

This description of an activated state for phosphoryl transfer indicates that molybdenum in nitrogenase **may** be an active center for hydrolysis of ATP, if it can provide a center of high positive charge (likely) coordinated to H_2O or OH$^-$ cis to a coordination site for the end PO_3^- group in ATP. However, Burris[16] associates ATP hydrolysis with electron transfer from the Fe protein to the Mo-Fe protein of nitrogenase. It is possible that oxidation of iron-sulfur clusters bound to ATP labilizes P-O-P bonds in the same way as oxidation of vanadium(IV) described above. It seems highly improbable that Mo(IV) bound at one corner of a (MoFe$_3$S$_4$) cube could be held tightly enough to avoid self-destruction by either disproportionation or formation of binuclear complexes, **and** permit labile ligands to exchange with ATP, H_2O, and N_2, forming an all-purpose activated state for N_2 (or H_2) reduction and ATP hydrolysis. However, the ability of Mo(IV), a transition metal atom of high positive charge, to expand its co-ordination number to 8, to coordinate CN$^-$ isoelectronic to N_2, and to catalyze reduction of oxidized nonmetals, makes it a strong candidate for an effective center for either electron transfer to N_2 or ATP hydrolysis -- or both. Molybdenum in Holm's compound[23] may

illustrate a protected form of the active site. Dimeric dioxo bridged Mo(V) dimers ($Mo_2O_4^{2+}$) are easily dissociated to give very active Mo(V) monomers[8] or they disproportionate to give Mo(VI) and active Mo(IV) under attack by nucleophiles.

ACKNOWLEDGEMENT

This work has been supported by grants from the National Institutes of Health and from the Research Corporation.

REFERENCES

1. G. P. Haight, Jr., A rapid polarographic determination of molybdate, Anal. Chem. 23:1505 (1951).
2. G. P. Haight, Jr., Mechanism of the molybdate catalyzed reduction of perchlorate and nitrate ions at the dropping mercury electrode, Acta Chem. Scand. 15:2012 (1961).
3. G. P. Haight, Jr., Polarography of hexavalent molybdenum in hydrochloric acid, J. Inorg. Nucl. Chem. 24:673 (1962).
4. G. P. Haight, Jr., Use of oxo-mettallates in the reduction of inert compounds of oxidized non-metals, CHEMSA 4:5 (1978).
5. G. P. Haight, Jr., A. C. Swift, and R. A. Scott, Mechanisms of niobate and perrhenate catalyzed reduction of inert oxo-anions, Acta Chem. Scand. A33:47 (1979).
6. R. A. Scott, G. P. Haight, and J. N. Cooper, Proton assisted two-electron transfer from hydroxylammonium ion to chromium(VI) through oxygen bridges, J. Am. Chem. Soc. 96:4136 (1974).
7. C. Landis. Private Communication.
8. T. J. Huang, The oxidation of hydrazine by molybdenum(VI) and methylene blue, Ph.D. Thesis, Utah State University (1968).
9. G. P. Haight and W. F. Sager, Evidence for preferential one-step divalent changes in the molybdate catalyzed reduction of perchlorate by stannous ion in sulfuric acid solution, J. Am. Chem. Soc. 74:6056 (1952).
10. G. P. Haight and A. C. Swift, Kinetics of oxidation of Mo(V) with hydroxylamine, J. Phys. Chem. 65:1921 (1961).
11. C. E. McKenna, T. Nakajima, J. Jones, C. Huang, M.-C. McKenna, H. Evan, and A. Osumi. This volume.
12. P. R. Robinson and C. R. Landis, Kinetics of the disproportionation of complexes of di-μ-oxo-bis[oxomolybdate(V)] in the presence of cyanide ion, Inorg. Chim. Acta. 33:63 (1979).
13. G. P. Haight and R. Scott, Molybdate- and tungstate-catalyzed fixation of nitrogen, J. Am. Chem. Soc. 86:743 (1964).
14. E. E. Van Tamelen, R. B. Fechter, S. W. Schneller, G. Boche, R. H. Grisley, and B. Akermark, Titanium(II) in the fixation-reduction of molecular nitrogen under mild conditions, J. Am. Chem. Soc. 91:1551 (1969).
15. J. Manriquez, R. D. Sanner, R. E. Marsh, and J. E. Bercaw, Reduction of molecular nitrogen to hydrazine. Structure of a dinitrogen complex of bis(pentamethylcyclopentadienyl)zirconium(II) and an ^{15}N labeling study of its reaction with hydrogen chloride, J. Am. Chem. Soc. 98: 3042 (1976).
16. R. Burris and R. V. Hageman. This volume.
17. J. Chatt. This volume.

18. J. Schwartz and J. Labinger, Hydrozinconation: a new transition
     metal reagent for organic synthesis, Angew. Chem. Int. Ed.
     15:333 (1976).
19. G. N. Schrauzer, Nonenzymatic simulation of nitrogenase reac-
     tions and the mechanism of biological nitrogen fixation,
     Angew. Chem. Int. Ed. 14:514 (1975).
20. P. Tripathy and D. Roundhill, Mechanistic aspects of reaction
     of acids with some platinum-acetylene complexes, J. Am. Chem.
     Soc. 92:3825 (1970).
21. S. P. Cramer, K. O. Hodgson, W. O. Gillum, and L. E. Mortenson,
     The molybdenum site of nitrogenase. Preliminary structural
     evidence from X-ray absorption spectroscopy, J. Am. Chem. Soc.
     100:3398 (1978).
22. S. P. Cramer, W. O. Gillum, K. O. Hodgson, L. E. Mortenson,
     E. I. Stiefel, J. R. Chisnell, W. J. Brill, and V. K. Shah,
     The molybdenum site of nitrogenase. 2. A comparative study of
     Mo-Fe proteins and the iron-molybdenum cofactor of X-ray
     absorption spectroscopy, J. Am. Chem. Soc. 100:3814 (1978).
23. R. B. Frankel, T. E. Wolff, J. M. Berg, C. Warrick, K. O.
     Hodgson, and R. H. Holm, The molybdenum-iron-sulfur cluster
     complex $[Mo_2Fe_6S_9(SC_2H_5)_8]^{3-}$. A synthetic approach to the
     molybdenum site in nitrogenase, J. Am. Chem. Soc. 100: 4630
     (1978).

AQUEOUS SOLUTION PROPERTIES OF SULFIDO COMPLEXES OF MOLYBDENUM(VI)

M. Harmer and A.G. Sykes

Department of Inorganic and Structural Chemistry
The University, Leeds LS2 9JT, England

INTRODUCTION

Tetrahedral sulfido (i.e. thiolato) complexes, MS_4^{n-}, are known for a number of d^0 transition metal ions.[1] These may be listed as follows:-

V(V)		
Nb(V)	Mo(VI)	Tc(VII)
Ta(V)	W(VI)	Re(VII)

Preparations involve the reaction of solutions of the oxy-anions with H_2S. With Cr(VI)[2] and Mn(VII), oxidation of S^{2-} occurs and no stable complexes are formed. With Mo(VI), the mixed species $MoOS_3^{2-}$, $MoO_2S_2^{2-}$, MoO_3S^{2-} are well characterized,[3] although the latter is difficult to obtain pure in the crystalline state.[4] Müller and co-workers have considered various properties of these species including infrared[5] and electronic spectra.[6] X-ray crystal studies of $(NH_4)_2MoS_4$ (Mo-S = 2.17 Å)[7] and Cs_2MoOS_3 (Mo-S = 2.18 Å; Mo-O = 1.79 Å)[8] have been reported. The Mo-O distance in the latter is somewhat longer than in the Mo(VI) complexes K_2MoO_4 (1.76 Å),[9] MoO_3(dien) (1.74 Å)[10] and $Mo_2O_5-(C_2O_4)_2(H_2O)_2^{2-}$ (terminal Mo-O 1.69 Å).[11] Present evidence is that Mo(VI)-sulfido complexes are all mononuclear and bridged μ-sulfido complexes do not appear to be readily formed. Thus, there are no counterparts to the polymeric (and heteropoly) species formed (by corner and edge sharing) from MoO_4^{2-} at pH <7. The red crystalline solid obtained on reacting K_2MoS_4 with H_2K_2edta in aqueous solution at pH 6, and by bubbling H_2S through an aqueous solution containing Mo_2O_6(edta)$^{4-}$ at pH 7-8, previously thought to be a μ-sulfido Mo(VI) complex,[12] is now believed to be a Mo(V) μ-oxo-μ-

sulfido complex. With molybdenum(V), di-μ-sulfido complexes are
well characterized,[13] and it is more difficult to obtain complexes
with terminal sulfido ligands.[14]

Kinetic studies on the interconversions of $MoO_xS_{4-x}^{2-}$ species
are described here. Very few quantitative aqueous solution
studies with H_2S as a reactant are to be found in the literature,
and we experienced difficulty initially in defining conditions at
the higher pH's required for this study. Under acid conditions,
the reactions of $Ru(NH_3)_5H_2O^{2+}$ (substitution),[15] and CrO_4^{2-}
(redox),[2] with H_2S have recently been reported. Aquation and
anation (NCS^-) studies on $Cr(H_2O)_5SH^{2+}$ have been described.[16]

EXPERIMENTAL

Materials: Sodium molybdate, Na_2MoO_4, and sodium chloride
were used without further purification. Buffers were made up
from NH_4Cl and NH_3, which was added until the desired pH was
reached (pH 8.0-10.5), and sodium acetate/acetic acid (pH 3.5-6.0).
All these chemicals were BDH, Analar. Cylinder hydrogen sulfide
(Air Products) was used without further purification.

Measurement of pH: A Radiometer (PHM 4) meter fitted with a
combined electrode (GK 2322C) was calibrated at pH 9.2 and 7.0
using appropriate buffer tablets.

Preparation of Complexes: The complexes $(NH_4)_2MoS_4$, $(NH_4)_2$-
MoO_2S_2 and Cs_2MoOS_3 were prepared by literature procedures.[17]
Characterization was by means of known u.v.-visible absorption
spectra (in H_2O),[1] and elemental micro-analyses for N, H and S.
Samples were stored under N_2.

Procedure: All solutions were deoxygenated using N_2, serum
caps, and 2 mm gauge teflon tubing (use of steel needles resulted
in brown precipitates on introducing H_2S). Without this
precaution, solutions of H_2S and buffer gave slight absorbance
changes at ca. 300 nm over a few minutes. Saturated solutions of
H_2S in water were prepared, and the concentration (0.095 M at $25°C$)
determined by the iodometric method. The latter solution was
diluted (micro-syringes) into air-free solutions as required.
Tris and cacodylate buffers were found to react with H_2S and were
not used further. Using 0.25 M NH_3/NH_4^+ buffer, pH's in the
range 9.1-10.1 were found to be constant to ±0.02 throughout runs.
The complex MoS_4^{2-} gave precipitates (probably MoS_2) on treatment
with 0.1 M $HCl/HClO_4$. The liberated H_2S functions as a mild
reducing agent under these conditions of acidity. The $MoO_xS_{4-x}^{2-}$
species exhibit distinctive spectra,[1] and interconversions could
be monitored by conventional spectrophotometry on a Unicam SP500,
using (for the faster reactions) a direct readout chart recorder
with scale amplification. The absorptions of H_2S, MoO_4^{2-} and

NH_3/NH_4^+ are small at wavelengths >275 nm. First and second acid
dissociation constants of H_2S give pK_a's of 7 and 14 respectively.[18] The first protonation constant of MoO_4^{2-} is 3 x 10^3 M^{-1}
at 25°C and I = 1.0 M (NaCl).[19] The ionic strength of reaction
solutions in this study was adjusted to I = 0.50 M (NaCl). When
one of reactants was in at least 10-fold excess, plots of
absorbance (A) changes, $\log(A_\infty-A_t)$, against time were generally
linear to >80% completion. First-order rate constants were
obtained from the slopes (x 2.303).

RESULTS

Determination of the Equilibrium Constant for the Formation of
MoO_3S^{2-}

   Pure samples of MoO_3S^{2-} have not been isolated previously.
The complex is known to have an absorption maximum at ca. 289 nm,
but the precise absorption coefficient (ε) is not known, a
complicating feature being that $MoO_2S_2^{2-}$ has an absorption maximum
at the same wavelength. Our experiments indicated an absorption
maximum at 292 nm for MoO_3S^{2-}. Conditions were, temperature 25°C,
pH 9.1-10.1 (0.25 M NH_3/NH_4^+ buffer), I = 0.5 M (NaCl), $[H_2S]$ = 1.9
x 10^{-4} M and $[MoO_4^{2-}]$ = (0.25 - 2.0) x 10^{-2} M. Absorbance changes
were monitored during the equilibration process and the maximum
absorbance (yielding ε_{obs}, see below) recorded at 292 nm. A
subsequent decrease in absorbance corresponded to formation of
amounts of $MoO_2S_2^{2-}$. For the equilibration (1),

$$MoO_4^{2-} + H^+ + HS^- \underset{\longleftarrow}{\overset{K_{01}}{\longrightarrow}} MoO_3S^{2-} + H_2O \qquad (1)$$

the expression (2)

$$\frac{1}{(\varepsilon_{obs}-\varepsilon_o)} = \frac{1}{K_{01}[H^+](\varepsilon_1-\varepsilon_o)} \frac{1}{[MoO_4^{2-}]} + \frac{1}{(\varepsilon_1-\varepsilon_o)} \qquad (2)$$

can be derived, where ε_{obs}, ε_1 and ε_o are absorption coefficients
(M^{-1} cm^{-1}) for equilibrated solutions, MoO_3S^{2-}, and $MoO_4^{2-}/NH_3/H_2S$
respectively. The value of ε_o was taken as zero. From linear
plots of ε_{obs}^{-1} against $[MoO_4^{2-}]^{-1}$ at pH's 9.17, 9.35, 9.51 and
10.1 the ratio of intercept/slope gave $K_{01}[H^+]$ values. The slope
of a plot of $K_{01}[H^+]$ against $[H^+]$ gave K_{01} = 5.8 x 10^{11} M^{-2}.
This equilibration can be expressed alternatively as (3),

$$MoO_4H^- + HS^- \rightleftharpoons MoO_3S^{2-} + H_2O \qquad (3)$$

in which case K_{01}' = 1.9 x 10^8 M^{-1} (using 3 x 10^3 M^{-1} for the first
protonation constant of MoO_4^{2-}), or as (4),

$$MoO_4^{2-} + H_2S \rightleftharpoons MoO_3S^{2-} + H_2O \tag{4}$$

when $K_{01}'' = 6 \times 10^4$ M^{-1} (using 10^7 M^{-1} for the protonation constants for HS^-). The latter gives a value 3.3×10^6 with $[H_2O] = 55$ M. At the 292 nm peak, MoO_3S^{2-} gives $\varepsilon_1 = 7900$ M^{-1} cm^{-1}, which compares with a value for $MoO_2S_2^{2-}$ at 292 nm of 4600 M^{-1} cm^{-1}.

Kinetics of the 1:1 Equilibration of H_2S and MoO_4^2

The reaction can be expressed as in (5),

$$MoO_4^{2-} + H^+ + HS^- \underset{k_{10}}{\overset{k_{01}}{\longrightleftharpoons}} MoO_3S^{2-} + H_2O \tag{5}$$

with rate constants k_{01} and k_{10} as defined. Conditions were temperature $25°C$, pH 9.2-10.2 (0.25 M NH_3/NH_4^+ buffer), I = 0.50 M (NaCl). The increase in absorbance was monitored at 292 nm ($t_{\frac{1}{2}}$ in range 15-70s), and plots of $\log(A_\infty - A_t)$ against time were linear to 95% completion. Slopes (x 2.303) gave first-order equilibration rate constants k_{eq}. The rate constant obtained is the same (± 5%) whether MoO_4^{2-} or H_2S is in excess, pH 9.67. Observed rate constants were also found to be independent of the concentration of buffer (0.02-0.25 M) at pH 9.2 (± 8%). On varying the $[MoO_4^{2-}]$ in the range 0.003-0.0115 M with $[H_2S] = 1.9 \times 10^{-4}$ M, linear plots of k_{eq} against $[MoO_4^{2-}]$ were obtained at each pH. These give a common intercept (k_{10}), with slopes dependent on $[H^+]$. Thus the equilibration process may be summarized by (6)

$$k_{eq} = k_{01} [H^+] + k_{10} \tag{6}$$

with $k_{01} = 4.0 \times 10^9$ M^{-2} s^{-1} and $k_{10} = 6.5 \times 10^{-3}$ s^{-1}. The ratio $k_{01}/k_{10} = 6.2 \times 10^{11}$ M^{-2} in good agreement with the equilibrium constant $K_{01} = 5.8 \times 10^{11}$ M^{-2} from the previous determination. For a reaction between MoO_4H^- and HS^- (the first protonation constant for MoO_4^{2-} is 3×10^3 M^{-1}), k_{01} converts to $k_{01}' = 1.3 \times 10^6$ M^{-1} s^{-1}. An alternative interpretation in terms of MoO_4^{2-} reacting with H_2S gives $k_{01}'' = 4 \times 10^2$ M^{-1} s^{-1}. This seems unlikely since, from ^{18}O-exchange studies, the rate constant for the reaction of MoO_4^{2-} with H_2O has been determined and is much slower (0.33 s^{-1}). Allowing for the molarity of H_2O, the difference in second-order rate constants is ca. 10^5. An interpretation in terms of k_{01}' and k_{10} is therefore favored.

Kinetics of the Aquation of $MoOS_3^{2-}$

This reaction was studied over two ranges of pH. At the higher pH's in the range 8.27-9.70 (0.25 M NH_3/NH_4^+ buffer), an equilibration process (7), is observed. Absorbance changes were

$$MoOS_3^{2-} + H_2O \underset{k_{23}}{\overset{k_{32}}{\rightleftharpoons}} MoO_2S_2^{2-} + HS^- + H^+ \tag{7}$$

followed at the $MoOS_3^{2-}$ 393 nm peak (ε 8400 M^{-1} cm^{-1}). The
absorbance of $MoO_2S_2^{2-}$ (ε 3200 M^{-1} cm^{-1}), MoO_3S^{2-} (ε 700 M^{-1} cm^{-1})
and MoO_4^{2-} ($\rightarrow$ 0) are much less. Since the aquation reactions of
$MoO_2S_2^{2-}$ and MoO_3S^{2-} are faster, the absorbance A_∞ was assumed to
be zero. Plots of log A_t against time were linear for 30-40
minutes but then became curved. Because H_2S was lost from the
solution and could not readily be retained at a known value, an
alternative approach was adopted in which H_2S was removed by
bubbling a steady stream of N_2 gas through the reaction solutions.
Under these conditions, first-order plots were linear for longer
periods, and k_{32} could be determined. No dependence on pH 8.27-
9.40 was observed, and k_{32} = $\underline{ca}$. 6 x 10^{-5} s^{-1}.

Aquation studies over a pH range 3.6-5.6 were also carried
out using acetate buffer (0.02 M). The same ($\pm$ 5%) rate constant
was obtained on decreasing the concentration of acetate ten-fold.
Within this pH range the aquation process proceeded to near
completion, and an $[H^+]$ dependence (8) was observed. The small

$$k_{obs} = k'_{32}[H^+] + k_{32} \tag{8}$$

intercept $k_{32} \sim 3$ x 10^{-5} s^{-1} corresponds to and is in satisfactory
agreement with the high pH value. From the slope the $[H^+]$-
dependent path gives k'_{32} = 3.0 M^{-1} s^{-1}.

Kinetics of the Aquation of MoS_4^{2-}

Absorbance changes were monitored at the MoS_4^{2-} peak position
at 470 nm (ε 12400 M^{-1} cm^{-1}). Exploratory experiments demonstrated
that the first stage of aquation (9),

$$MoS_4^{2-} + H_2O \xrightarrow{k_{43}} MoOS_3^{2-} + H_2S \tag{9}$$

is an order of magnitude slower than the aquation step k_{32}. Only
$MoOS_3^{2-}$ of the other sulfido complexes absorbs at this wavelength
(ε 2700 M^{-1} cm^{-1}), and it was therefore assumed that the final
absorbance appropriate to this investigation is zero. Conditions
investigated were temperature 25°C, pH 8.9-9.8 (0.25 M NH_3/NH_4^+
buffer), I = 0.50 M (NaCl). Plots gave satisfactory linearity
over the time monitored (generally one $t_\frac{1}{2}$ of $\underline{ca}$. 50 h). No
attempt was made to purge the solution of H_2S formed. The rate
constant from initial slopes was 1.6 x 10^{-6} s^{-1}.

The Reaction of H_2S with $MoOS_3^{2-}$

The conditions used were $25°C$, pH 9.2 (0.25 M NH_3/NH_4^+), I = 0.50 M (NaCl), with $[MoOS_3^{2-}] = 5 \times 10^{-5}$ M, $[H_2S] = 10^{-2}$ M. No change in absorbance at 470 nm was observed over 1 hour, and it is concluded that k_{34} is $<1 \times 10^{-3}$ M^{-1} s^{-1} (reactants $MoOS_3^{2-}$ and HS^-). Alternatively, this limit may be written $<1.6 \times 10^6$ M^{-2} s^{-1} (reactants $MoOS_3^{2-}$, HS^- and H^+), or $<5.3 \times 10^2$ M^{-1} s^{-1} (reactants $MoOS_3H^+$ and HS^-, where the protonation constant is assumed to be 3×10^3 M^{-1}).

Further Studies on the Reactivity of MoS_4^{2-}

A feature of the solution chemistry of MoO_4^{2-} is its tendency to increase its coordination number from 4 to 6.[20] This tendency was tested for in the case of MoS_4^{2-} (4×10^{-5} M) by addition of excess 8-hydroxyquinoline (1×10^{-4} M) at pH 9.3 (0.25 M NH_3/NH_4^+), I = 0.50 M (NaCl). No spectral change was observed. Also under similar conditions with addition of an excess of H_2S (10^{-2} M), no change in spectrum was observed. There is no evidence in these experiments for MoS_4^{2-} having the ability to increase its coordination number.

DISCUSSION

Rate constants determined in this study may be summarized as

Table I

Formation (M^{-1} s^{-1})		Aquation (s^{-1})	
k_{01}	1.3×10^6 ;	k_{10}	6.5×10^{-3}
	;	k_{32} ca.	4×10^{-5}
k_{34}	$<5.3 \times 10^2$;	k_{43}	1.6×10^{-6}

As far as the formation process k_{01} is concerned an interpretation in terms of the pathway $MoO_4H^- + HS^-$ is preferred rather than the alternative $MoO_4^{2-} + H_2S$. The basis of this preference is that reactions of MoO_4^{2-} appear to be slow unless prior protonation occurs. Protonation has the effect of activitating the complex both to substitution and addition. Although in some cases proton ambiguities exist, and the interpretation is not clearcut, in at least one case, that of the $[H^+]$-dependent path in the reaction of $Co(NH_3)_5H_2O^{3+}$ with MoO_4^{2-}, it is clear that MoO_4H^- is involved.[21] With additional data provided by the $H_2^{18}O$ exchange studies on MoO_4^{2-}, the summary ($25°C$) in Table II exists.[22]

Table II

$$MoO_4^{2-} + H_2O \longrightarrow (k = 0.33 \ s^{-1})$$

$$MoO_4^{2-} + OH^- \longrightarrow (k = 2.2 \ M^{-1} \ s^{-1})$$

$$MoO_4H^- + Co(NH_3)_5H_2O^{3+} \longrightarrow (k = 3.2 \times 10^5 \ M^{-1} \ s^{-1})$$

The rate constant k_{01} of $1.3 \times 10^6 \ M^{-1} \ s^{-1}$ for reaction of MoO_4H^- with HS^- is consistent with the latter value, whereas a rate constant $(4 \times 10^2 \ M^{-1} \ s^{-1})$ for the reaction of MoO_4^{2-} with H_2S is more difficult to accept. The influence of H^+ on the reactivity of MoO_4^{2-} is quite remarkable since protonation is by no means extensive, and in this particular instance, protonation of HS^- while small will be more prevalent.

The rate constants for the aquation steps also indicate a slowing down in the substitution process as more sulfides are introduced into the Mo(VI) coordination sphere. In view of the inability of MoS_4^{2-} (as compared to MoO_4^{2-}) to add an additional ligand, e.g. 8-hydroxyquinoline, it is concluded that the size effect of S^{2-} as compared to O^{2-} is crucial. It has been noted above that the Mo(VI)-O bond distance is longer in $MoOS_3^{2-}$ than in any other Mo(VI) complex. While MoO_4^{2-} (certainly when protonated) most probably reacts by an associative mechanism, this becomes more difficult as the number of sulfido groups increases. However, the whole process of substitution at oxy-anions may be much more complicated (simple dissociative and associative processes may not apply), since it is clear that hydrogen bonding plays an important role in the activated complex.[23-25]

The results for the aquation of $MoOS_3^{2-}$ in the lower pH range (3.6-5.6) is of interest in that an $[H^+]$-dependent term is observed. From the acid dissociation (pK_a) value listed in Table III, it is

Table III

	H_2O	H_2S	Ref.
uncoordinated	15.7	7	18
$Ru(NH_3)_5H_2X^{2+}$	13.1	4	15
$Cr(H_2O)_5H_2X^{3+}$	4	$\ll 1$	16

X = O or S

estimated (on the basis of a pK_a of 3.47 for MoO_4H^-) that the pK_a of a sulfide ligand should be ca. -5. Protonation of a sulfide of $MoOS_3^{2-}$ seems a likely mechanistic requirement for the $[H^+]$-dependent aquation.

In the context of these results, it is interesting to recall
that the aquation rate constant $(40°C)$ for $Cr(H_2O)_6^{3+}$ $(3.6 \times 10^{-6}$
$s^{-1})$, $Cr(H_2O)_5SH_2^{3+}$ $(>>3.8 \times 10^{-5}~s^{-1})$ and $Cr(H_2O)_5SH^{2+}$ $(3.0 \times 10^{-5}$
$s^{-1})$ indicate that the sulfido ligand is more labile than H_2O.
Also from studies on the 1:1 NCS^- anation of $Cr(H_2O)_5SH^{2+}$ (replacement of H_2O by NCS^-), it can be concluded that the ratio of rate
constants for the anation of $Cr(H_2O)_5SH_2^{3+}$ to $Cr(H_2O)_6^{3+}$ is $>>10^3$,
and for $Cr(H_2O)_5SH^{2+}$ to $Cr(H_2O)_5OH^{2+}$ is 17. A possible extra-
polation is that O^{2-} and S^{2-} might exhibit similar labilizing
effects. However, in the case of Mo(V), bridging sulfide ligands
in the di-μ-sulfido complex $Mo_2O_2S_2(C_2O_4)_2(H_2O)_2^{2-}$ have a <u>ca.</u> 40-
fold labilizing effect as compared to the corresponding di-μ-oxo
complex.[26]

Finally, attention should be drawn to the conclusion of
Müller et al.[27] that MoO_3S^{2-} and $MoO_2S_2^{2-}$ are formed more quickly
than WO_3S^{2-} and $WO_2S_2^{2-}$. However, no rate constants were reported
in this and other similar previous investigations relating to the
interconversions of $MoO_xS_{4-x}^{2-}$ (and $WO_xS_{4-x}^{2-}$) species.

REFERENCES

1. E. Diemann and A. Müller, Thio and Seleno Compounds of
    Transition Metals with d⁰ Configurations, <u>Coord. Chem. Rev.</u>,
    10: 79 (1973).
2. For a recent kinetic study, see the report B. Banas, Kinetics
    and Mechanism of Cr(III) Coordination to Reactants and Products
    in the Reduction Reaction of Cr(VI) by Hydrogen Sulfide,
    <u>Proceedings XIX International Coordination Chemistry
    Conference</u>, Prague, 1978, p.9.
3. (a) J.C. Bernard and G. Tridot, Contribution to the Study of
         Alkali-metal Thiotungstates and Thiomolybdates:
         Identification and Characterization of Their Ions in
         Aqueous Solution, <u>Bull. Soc. chim. France</u>, 810 (1961) and
    (b) A. Müller and E. Diemann, Preparation of Some Di-, Tri-
         and Tetrathiomolybdates and Tungstates, <u>Z. Naturforschg</u>,
         23B: 1607 (1968).
4. P.J. Aymonino, A.C. Ranade and A. Müller, Evidence for the
    Existence of MoO_3S^{2-} and WO_3S^{2-} Ions in Aqueous Solution,
    <u>Zeit. anorg. allgem. Chemie</u>, 371: 295 (1969).
5. A. Müller, W. Rittner and G. Nagarajan, The Electronic
    Absorption Spectrum of Thiomolybdate and Thiotungstate Ions,
    <u>Z. physik Chem. (Frankfurt)</u>, 54: 229 (1967).
6. A. Müller, E. Diemann, A.C. Ranade and P.J. Aymonino, Electronic
    Spectra of MoO_3S^{2-} and WO_3S^{2-} and Their Comparison with the
    Spectra of the Ions $MoO_xS_{4-x}^{2-}$ and $WO_xS_{4-x}^{2-}$, <u>Z. Naturforschg</u>,
    24B: 1247 (1969).
7. H. Schäfer, G. Schäfer and A. Weiss, The Crystal Structure
    of Ammonium Thiomolybdate, <u>Z. Naturforschg</u>, 23B: 76 (1964).

8. B. Kreks, A. Müller and E. Kindler, Crystal Structure of
    Cs_2MoOS_3, Z. Naturforsch, 25B: 222 (1970).
9. B.M. Gatehouse and P. Leverett, Crystal Structure of Potassium
    Molybdate, K_2MoO_4, J. Chem. Soc.(A), 849 (1969).
10. F.A. Cotton and R.C. Elder, The Crystal and Molecular
    Structure of Trioxo(diethylenetriamine)molybdenum(VI), Inorg.
    Chem., 3: 397 (1964).
11. F.A. Cotton, S.M. Morehouse and J.S. Wood, The Identification
    and Characterization by X-ray Diffraction of a New Binuclear
    Molybdenum(VI) Oxalate Complex, Inorg. Chem., 3: 1603 (1964).
12. B. Spivack and Z. Dori, Sulphur-bridged Complexes of Mo(V)
    and Mo(VI) Containing the Ligands Ethylenediaminetra-acetic
    Acid, Cysteine, and Histidine, J. Chem. Soc. Chem. Comm.,
    1716 (1970).
13. e.g. E.I. Stiefel, The Coordination and Bioinorganic Chemistry
    of Molybdenum, Prog. Inorg. Chem., ed. S.J. Lippard, 22: 55
    (1977).
14. V.R. Ott, D.S. Swieter and F.A. Schultz, Di-μ-oxo, μ-oxo-μ-
    sulfido, and Di-μ-sulfido Complexes of Molybdenu(V) with EDTA,
    Cysteine, and Cysteine Ester Ligands. Preparation and Electro-
    chemical and Spectral Properties, Inorg. Chem., 16, 4538, 1977.
15. C.G. Kuehn and H. Taube, Ammineruthenium Complexes of Hydrogen
    Sulfide and Related Sulfur Ligands, J. Amer. Chem. Soc.,
    98: 689 (1976).
16. T. Ramasami and A.G. Sykes, Further Characterization and
    Aquation of the Thiolopentaaquochromium(III) Complex, $CrSH^{2+}$,
    and Its Equilibration with Thiocyanate, Inorg. Chem., 15:
    1010 (1976).
17. G. Kruss, Concerning the Sulfur Compounds of Molybdenum,
    Ann. Chem., 225: 29 (1884).
18. M. Widmer and G. Schwarzenbach, The Acidity of the Hydrogen
    Sulfide Ion HS^-, Helv. Chim. Act, 47: 266 (1964).
19. J.J. Cruywagen and E.F.C.H. Rohwer, Coordination Number of
    Molybdenum(VI) in Monomeric Molybdic Acid, Inorg. Chem.,
    14: 3136 (1975).
20. P.F. Knowles and H. Diebler, Kinetics of Complex Formation
    of Molybdate with 8-Hydroxyquinoline, Trans. Faraday Soc.,
    64: 977 (1968).
21. R.S. Taylor, Investigation of the Rapid Complexation of
    Aquopentaamminecobalt(III) with Molybdate in Weakly Basic
    Solution, Inorg. Chem., 16: 116 (1977).
22. H. von Felten, B. Wernli, H. Gamsjäger and P. Baertschi,
    Oxygen Exchange Between Oxo-anions and Water in Basic Media:
    Molybdate and Tungstate, J.C.S. Dalton Trans., 496 (1978).
23. R.K. Wharton, R.S. Taylor and A.G. Sykes, Temperature-jump
    Study of the Rapid Complexing of Iodate with Aquometal Ions,
    Inorg. Chem., 14: 33 (1975).
24. A.D. Fowles and D.R. Stranks, Selenitometal Complexes. 2.
    Kinetics and Mechanism of the Reaction of Aquocobalt(III)
    Cations with Hydrogen-selenite Anions, Inorg. Chem., 16:
    1276 (1977).

25. A.D. Fowles and D.R. Stranks, Selenito Complexes. 3. Kinetics
     and Mechanism of the Reaction of Hydroxocobalt(III) and
     Hydroxorhodium(III) Cations with Monomeric and Dimeric
     Selenite Anions, Inorg. Chem., 16: 1282 (1977).
26. F.A. Armstrong, T. Shibahara and A.G. Sykes, Effect of
     μ-Sulfido Ligands on Substitution at Molybdenum(V). A
     Temperature-jump Study of the 1:1 Equilibration of Thio-
     cyanate with Di-μ-sulfido-bis[aquooxalatooxomolybdenum(V)],
     Inorg. Chem., 17: 189 (1978).
27. P.J. Aymonino, A.C. Ranade, E. Diemann and A. Müller,
     Study of Formation and Relative Reaction Rates of Different
     Thioanions of Molybdenum and Tungsten, Zeit. anorg. allgem.
     Chemie, 371: 300 (1969).

LIST OF CONTRIBUTORS

S.R. Acott, Department of Chemistry, Manchester University, Manchester M13 9PL, England.

B.A. Averill, Department of Chemistry, Michigan State University, East Lansing, MI. 48824, USA.

M. Barber, Department of Biochemistry, Duke University, Durham, N.C. 27710, USA.

J.M. Berg, Department of Chemistry, Stanford University, Stanford, CA. 94305, USA.

P. Bosserman, Department of Chemistry, University of California, Riverside, CA. 92521, USA.

R.C. Bray, School of Molecular Sciences, University of Sussex, Falmer, Brighton, BN1 9QJ, England.

W.J. Brill, Department of Bacteriology, University of Wisconsin, Madison, 53706, USA.

J.A. Broomhead, Department of Chemistry, Faculty of Science, Australian National University, Canberra, A.C.T. 2600, Australia.

A.E. Bruce, C.F. Kettering Research Laboratory, Yellow Springs, OH. 45387, USA.

J. Budge, Department of Chemistry, Faculty of Science, Australian National University, Canberra, A.C.T. 2600, Australia.

B.K. Burgess, C.F. Kettering Research Laboratory, Yellow Springs, OH. 45387, USA.

R.H. Burris, Department of Biochemistry, University of Wisconsin-Madison, WI. 53706, USA.

N.C. Chasteen, Department of Chemistry, University of New Hampshire, Durham, NH. 03824, USA.

411

J. Chatt, A.R.C. Unit of Nitrogen Fixation, University of Sussex, Falmer, Brighton, Sussex, BN1 9RQ, England.

G. Christou, Department of Chemistry, Manchester University, Manchester M13 9PL, England.

S.D. Conradson, Department of Chemistry, Stanford University, Stanford, CA. 94305, USA.

J.L. Corbin, C.F. Kettering Research Laboratory, Yellow Springs, OH. 45387, USA.

S.P. Cramer, Exxon Research Corporation, Linden, NJ. 07036, USA.

M.G.B. Drew, Chemistry Department, The University, Whiteknights, Reading, RG6 2AD, England.

M. Emptage, Department of Biochemistry, University of Wisconsin-Madison, Madison, WI. 53706, USA.

J.H. Enemark, Department of Chemistry, University of Arizona, Tucson, AZ. 85721, USA.

H. Eran, Department of Chemistry, University of Southern California, Los Angeles, CA. 90007, USA.

C.D. Garner, Department of Chemistry, Manchester University, Manchester M13 9PL, England.

H.B. Gray, Department of Chemistry, California Institute of Technology, Pasadena, CA. 91109, USA.

R.V. Hageman, Department of Biochemistry, University of Wisconsin-Madison, Madison, WI. 53706, USA.

G.P. Haight, School of Chemical Sciences, University of Illinois, Urbana, IL. 61801, USA.

M. Harmer, Department of Inorganic and Structural Chemistry, The University of Leeds, Leeds LS2 9JT, England.

T. Hase, Department of Biology, Faculty of Science, Osaka University, Osaka, 560 Japan.

K. Hashi, Department of Applied Chemistry, Faculty of Engineering, Osaka University, Osaka, 560 Japan.

A. Hattori, Ocean Research Institute, University of Tokyo, Nakano, Tokyo, 164 Japan.

M. Henzl, Department of Biochemistry, University of Wisconsin-
    Madison, Madison, WI. 53706, USA.

M. Hidai, Department of Industrial Chemistry, University of Tokyo,
    Hongo, Tokyo, Japan.

K.O. Hodgson, Department of Chemistry, Stanford University, Stanford,
    CA. 94305, USA.

C. Huang, Department of Chemistry, University of Southern
    California, Los Angeles, CA. 90007, USA.

J₀T₀ Huneke, Department of Chemistry, University of Arizona,
    Tucson, AZ. 85721, USA.

B.H. Huynh, Freshwater Biological Research Institute, University
    of Minnesota, Navarre, MN. 55392, USA.

I. Imamura, School of Chemical Sciences, University of Illinois,
    Urbana, 61801, USA.

G.S. Jacob, I.B.M. Watson Laboratory, Yorktown, NY. 10598, USA.

K. Jacobson, Department of Biochemistry, University of Wisconsin,
    Madison, WI. 53706, USA.

C.E. Johnson, Oliver Lodge Laboratory, University of Liverpool,
    Liverpool L69 3BX, England.

J. Jones, Department of Chemistry, University of Southern California
    Los Angeles, 90007, USA.

M. Kamata, Department of Chemistry, Faculty of Engineering Science,
    Osaka University, Osaka, 560 Japan.

G.W₀ Kenner, Robert Robinson Laboratories, University of Liverpool,
    Liverpool L69 3BX, England.

T.J. King, Chemistry Department, Nottingham University, Nottingham,
    NG7 2RD, England.

C. Knobler, Chemistry Department, University of Canterbury,
    Christchurch, New Zealand.

L. Leon, Department of Chemistry, University of California, River-
    side, CA. 92521, USA.

L.G. Ljungdahl, Department of Biochemistry, University of Georgia,
    Athens, GA. 30602, USA.

F.E. Mabbs, Department of Chemistry, Manchester University,
    Manchester, M13 9PL, England.

A.J. Matheson, Chemistry Department, University of Canterbury,
    Christchurch, New Zealand.

H. Matsubara, Department of Biology, Faculty of Science, Osaka
    University, Osaka, 560 Japan.

T. Matsuda, Department of Applied Chemistry, Faculty of Engineering,
    Osaka University, Osaka, 560 Japan.

C.E. McKenna, Department of Chemistry, University of Southern
    California, Los Angeles, CA. 90007, USA.

M-C. McKenna, Department of Chemistry, University of Southern
    California, Los Angeles, CA. 90007, USA.

K.F. Miller, C.F. Kettering Research Laboratory, Yellow Springs, OH.
    45387, USA.

R.M. Miller, Department of Chemistry, Manchester University,
    Manchester M13 9PL, England.

W.B. Mims, Bell Laboratories, Murray Hill, NJ. 07974, USA.

M. Minelli, Department of Chemistry and Biochemistry, Utah State
    University, Logan, UT. 84322, USA.

S. Miyake, Department of Applied Chemistry, Faculty of Engineering,
    Osaka University, Osaka, 560 Japan.

Y. Mizobe, Department of Industrial Chemistry, University of Tokyo,
    Hongo, Tokyo, Japan.

E. Münck, Freshwater Biological Research Institute, University of
    Minnesota, Navarre, MN. 55392, USA.

T. Nakajma, Department of Chemistry, University of Southern
    California, Los Angeles, CA. 90007, USA.

A. Nakamura, Department of Polymer Science, Faculty of Science,
    Osaka University, Osaka, 560 Japan.

W.E. Newton, C.F. Kettering Research Laboratory, Yellow Springs, OH.
    45387, USA.

D.J.D. Nicholas, Department of Agricultural Biochemistry, Waite
    Agricultural Research Institute, University of Adelaide,
    Glen Osmond, 5064, South Australia.

M. Ohmori, Ocean Research Institute, University of Tokyo, Nakano, Tokyo, 164 Japan.

S. Ooi, Chemistry Department, Faculty of Science, Tohoku University, Sendai, 980 Japan.

N.R. Orme-Johnson, Department of Biochemistry, University of Wisconsin-Madison, Madison, WI. 53706, USA.

W.H. Orme-Johnson, Department of Biochemistry, University of Wisconsin-Madison, Madison, WI. 53706, USA.

A. Osumi, Department of Chemistry, University of Southern California, Los Angeles, CA. 90007, USA.

S. Otsuka, Department of Chemistry, Faculty of Engineering Science, Osaka University, Osaka, 560 Japan.

N. Pariyadath, C.F. Kettering Research Laboratory, Yellow Springs, OH. 45387, USA.

K.V. Rajagopalan, Department of Biochemistry, Duke University, Durham, N.C. 27710, USA.

J. Rawlings, Department of Biochemistry, University of Wisconsin-Madison, Madison, WI. 53706, USA.

C.A. Rice, Department of Chemistry and Biochemistry, Utah State University, Logan, UT. 84322, USA.

P.R. Robinson, School of Chemical Sciences, University of Illinois, Urbana, IL. 61801, USA.

J.D. Rush, Oliver Lodge Laboratory, University of Liverpool, Liverpool L69 3BX, England.

K. Saito, Chemistry Department , Faculty of Science, Tohoku University, Sendai, 980 Japan.

Y. Sasaki, Chemistry Department, Faculty of Science, Tohoku University, Sendai, 980 Japan.

M. Sato, Department of Industrial Chemistry, University of Tokyo, Hongo, Tokyo, Japan.

D.T. Sawyer, Department of Chemistry, University of California, Riverside, CA. 92521, USA.

F.A. Schultz, Department of Chemistry, Florida Atlantic University, Boca Raton, FL. 33431, USA.

N.S. Scott, Department of Chemistry, California Institute of Technology, Pasadena, CA. 91109, USA.

M. Scullane, Department of Chemistry, University of New Hampshire Durham, NH. 03824, USA.

V.K. Shah, Department of Bacteriology, University of Wisconsin-Madison, Madison, WI. 53706, USA.

H.C. Silvis, Department of Chemistry, Michigan State University, East Lansing, MI. 48824, USA.

B.E. Smith, ARC Unit of Nitrogen Fixation, University of Sussex, Falmer, Brighton BN1 9RQ, England.

J.P. Smith, Department of Biochemistry, University of Wisconsin-Madison, Madison, WI. 53706, USA.

J.T. Spence, Department of Chemistry and Biochemistry, Utah State University, Logan, UT. 84322, USA.

E.I. Stiefel, C.F. Kettering Research Laboratory, Yellow Springs, OH. 45387, USA.

K.Z. Suzuki, Chemistry Department, Faculty of Science, Tohoku University, Sendai, 980 Japan.

A.G. Sykes, Department of Inorganic and Structural Chemistry, The University of Leeds, Leeds LS2 9JT, England.

T. Takahashi, Department of Industrial Chemistry, University of Tokyo, Hongo, Tokyo, Japan.

K. Tanaka, Department of Applied Chemistry, Faculty of Engineering, Osaka University, Osaka, 560 Japan.

T. Tanaka, Department of Applied Chemistry, Faculty of Engineering, Osaka University, Osaka, 560 Japan.

R.H. Tieckelmann, Department of Chemistry, Michigan State University, East Lansing, MI. 48824, USA.

C. Touton, Department of Biochemistry, University of Wisconsin-Madison, Madison, WI. 53706, USA.

T.D. Tullius, Department of Chemistry, Stanford University, Stanford, CA. 94305, USA.

Y. Uchida, Department of Industrial Chemistry, University of Tokyo, Hongo, Tokyo, Japan.

N. Ueyama, Department of Polymer Science, Faculty of Science, Osaka University, Osaka, 560 Japan.

G.D. Watt, C.F. Kettering Research Laboratory, Yellow Springs, OH. 45387, USA.

S. Wherland, C.F. Kettering Research Laboratory, Yellow Springs, OH. 45387, USA.

C.J. Wilkins, Chemistry Department, University of Canterbury, Christchurch, New Zealand.

J.P. Wilshire, Department of Chemistry, University of California, Riverside, CA. 92521, USA.

K. Yamanouchi, Department of Chemistry, Univeristy of Arizonia, Tucson, AZ. 85721, USA.

W.G. Zumft, Department of Botany, Faculty of Biology and Chemistry, University Erlangen-Nuremberg, 8520 Erlangen, West Germany.